Modelling in Geography
A Mathematical Approach

Modelling in Geography

A Mathematical Approach

R W Thomas & R J Huggett

Barnes & Noble Books
Totowa, New Jersey

First published in the U.S.A by
Barnes & Noble Books
81 Adams Drive
Totowa, New Jersey 07512

British Library Cataloguing in Publication Data

Huggett, Richard
    Modelling in geography.
    1. Geography — Mathematical models
    I. Title   II. Thomas, Richard
    910'.01'84   G70.23

ISBN   0-389-20049-2   cased
        0-389-20050-6   paperback

Typeset by Preface Ltd, Salisbury, Wilts.
Printed and bound in the United Kingdom
by Butler and Tanner Ltd

To

Jenny, Richard, and Helen
Jane, Jamie, Sarah, and Edward

Contents

PART III   PROBABILITY MODELS

Preface

This book is designed as a text for the first- and second-year geographical techniques courses taught in most university geography departments. However, certain topics are developed a little beyond this level. The book takes its form from our experience in teaching geographical techniques to students at Manchester University up to the Part I level. This course attempts to complement foundation courses in human and physical geography by demonstrating how many modelling structures can be applied to problems in both the human and physical realm. To a lesser extent some material is taken from third-year courses dealing with various topics in geographical systems analysis.

In taking modelling as our theme, we have deliberately avoided the usual 'techniques-book' format wherein either statistical or cartographical methods are emphasized as research techniques in geography. Rather, we examine geographical research methods through their mathematical, and not statistical, logic, an approach expounded by Wilson and Kirkby (1975) in their book *Mathematics for Geographers and Planners*. However, our book differs from Wilson and Kirkby's text in a number of important ways. Our book is about models; therefore the mathematics we cover is determined by the requirements of the model, not by a conscious attempt to teach geographers mathematics for its own sake. For this reason our exposition of model structures progresses by the use of arithmetic examples, the language of which is more familiar to beginning geography students than is formal algebraic reasoning. For instance, when the logic of a model is underpinned by calculus, it is explained using the arithmetic method of finite differencing; reference to the algebra of differential and integral calculus is kept to an absolute minimum. We devote only one chapter (Chapter 2) to mathematics itself; the remainder deal with mainstream themes in geographical modelling such as mass balance and storage (Chapter 3), spatial interaction (Chapter 5), and decision-making frameworks (Chapter 9).

The need to explain model structures in some detail, together with the usual constraints of time and space, meant that a number of important topics, such as input–output analysis, diffusion, and Markov chains, had to be omitted. For this we apologize and can only hope that the background gained from reading this book will help the student to broach such topics with increased confidence.

A number of people have contributed to the writing of this book. Amongst our colleagues at Manchester, Peter Lloyd first encouraged us to put pen to paper, while David Fox's liberal approach to geographical techniques teaching provided us with ample scope for testing the models in the classroom. At a more practical level, Brenda Masson performed miracles typing a difficult manuscript, and Calvin Palmer spent many hours debugging the original manuscript. Lastly, but most of all, we thank our families for their patience and encouragement.

R W Thomas
R J Huggett
Manchester,
April Fools' Day, 1979

PART I

Introducing Geographical Models

CHAPTER 1

Modelling in Geography: The Setting

SYSTEMS AND MODELS

When looking at the world around us, we all impose some kind of order on what we see. This order enables us to organize and to understand our individual experiences. Geographers search for order in the patterns they observe in the landscape. One useful way of establishing order is to pick out geographical structures which appear to function in a meaningful manner and to have a stable form. In composition, these structures may be seen as a set of inter-related components. Cities, for instance, may be thought of as inter-related collections of individuals and organizations (Wilson 1974). The individuals create a demand for residences, jobs, shopping, and recreational facilities, and these needs are supplied by a variety of private enterprises and public institutions. The two components of urban structure are bound together by flows of people, goods, money, and information. In physical geography, the drainage basin is often picked out as a fundamental structure. The components of the drainage basin are storage units such as the soil, vegetation, ground-water, and streams. These units are linked together by flows of energy, minerals, and water which gradually change the composition and form of the landscape. It has become fashionable to refer to these basic structures as systems, and thus we speak of urban systems, social systems, ecosystems, and atmospherical circulation systems.

It is important to stress that individual geographers will view the same system in different ways: they will choose to emphasize the importance of different components and linkages. They will also choose to study very different scales of analysis, or levels of resolution: some may try to unravel the intricacies of a national system of cities, while others might be drawn to the operations of the housing market which itself is a sub-system of the urban system.

Anyone who attempts to isolate the components of a geographical system will soon realize that most of the systems that interest geographers are enormously complex no matter what the scale of analysis. Therefore, the study of a geographical system inevitably requires a degree of abstraction or simplification to be made. Scientists have coined a phrase to describe the process by which we simplify reality to manageable proportions: they term this process *model-building*. For the moment we shall define a model as any simplified representation of reality. The design and construction of a

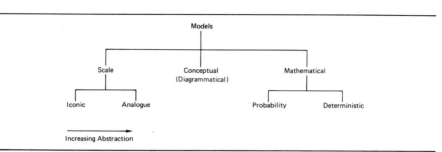

Figure 1.1.   Some types of model.

model can proceed with varying degrees of abstraction (figure 1.1). The simplest level of abstraction occurs when we transform reality only in terms of scale. The *scale model* differs from the real object only in size. The child's toy or the architect's model of a building are examples of scale models. They enable the essential characteristics of an object to be seen and studied. Scale models, which are miniature copies of reality, are sometimes termed *iconic models.* A more abstract refinement of the iconic model is the *analogue model.* Here we not only alter the size of the system but also transform some of its properties. An example of analogue model is a small-scale reconstruction of a valley glacier in which the clay, kaolin, is used instead of ice. It is found that many of the features of valley glaciers, such as crevasses and step faults, gradually develop in the clay. Thus the model allows us to observe in miniature an accelerated represent-ation of glacier behaviour. The most familiar analogue model to geographers is the map. The surface features of the landscape are not only reduced in scale, but are also represented in symbolic form: roads are depicted as red lines, relief is picked out by contours, and buildings and monuments are represented by point symbols.

Essentially, scale models help us to isolate the main components of a system of interest and to organize our initial thoughts and ideas. However, for understanding to progress, a further abstraction is necessary which might well involve the use of a *conceptual model.* We construct a conceptual model by setting down, in a preliminary way, our ideas about how the system functions; in other words, we concentrate on the relationships between the system components. Conceptual models are usually expres-sed as box-and-arrow diagrams in which the boxes represent the system components and the arrows depict the important links and relationships between the components. For example, our previous descriptions of an urban system and a drainage basin are shown as conceptual box-and-arrow diagrams in figure 1.2(*a* and *b*). Since conceptual models help us to clarify our loose thoughts about how a system is made up and how it operates, they are often a first step in the construction of *mathematical models*, which are the subject of this book.

MATHEMATICAL MODELS

In their purest form, mathematical models are constructed by translating the ideas contained in a conceptual model into the formal, symbolic logic of mathematics. To

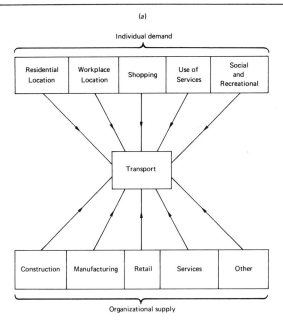

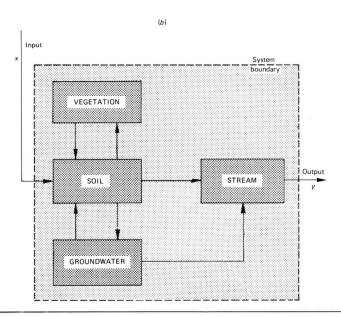

Figure 1.2 Two simple conceptual models. (*a*) Components and links in an urban system (after Wilson 1974). (*b*) Some components and links in a drainage basin.

see how this translation is made, consider the conceptual model of the drainage basin shown in figure 1.2(*b*). The input arrow is denoted by the symbol *x* which represents the amount of rainfall entering the basin. Similarly, the output arrow is denoted by *y* which represents the stream discharge. A very general model of the basin is expressed by the equation

$$y = T(x) \tag{1.1}$$

which states that stream discharge is equal to some transformed, *T*, value of rainfall. In practice, the value of *T* would be obtained from a set of equations which describe how the water moves between each of the storage units. For the sake of simplicity, the equations that make up *T* will not be derived here. The important point to grasp is that this mathematical model enables us to make numerical predictions about the relationship between precipitation and stream discharge. For a given amount of rainfall we may predict the expected amount of discharge.

It is their ability to predict that makes mathematical models preferable to conceptual models. The conceptual model represents a body of ideas about a geographical system which cannot be tested in any formal way. However, once our ideas are translated into mathematical symbolism the validity of the model can be assessed by matching its predictions against the yardstick of observed data. In this sense, the mathematical model plays a role in the development of theory. If a model equation is found to replicate observed data accurately, then the assumptions used to derive the equation provide a theoretical explanation for the behaviour of the system. The development of theories which purport to explain observed phenomena is the essence of scientific endeavour. It is hoped that by the continual process of model-building, model testing, and model redesign we will gradually obtain better explanations for the behaviour of geographical systems.

Constructing and testing a mathematical model

To illustrate our discussion so far we shall explain the various stages of reasoning needed to construct a simple model to predict population change. The various steps in the model-building process shown in figure 1.3 are common to most of the models described in this book.

To begin we must identify some geographical system that seems suited to mathematical analysis. By way of example, let us tackle the problem of predicting how the population of a village has changed from 1900 to 1980. Take as an initial hypothesis the idea that the village population has grown at a constant rate per unit of population. The next step is to translate the main components of this system into mathematical terminology and then make some assumptions about the factors which control the system's behaviour. We shall use the symbol *x* to denote the population size of the village and the label *t* to identify any particular point in time. Therefore x_t will be the predicted population size of the village at a particular point in time. Assume that, at any point in time, the population of the village will change only if there is a difference between the village birth rate, denoted by *b*, and the village death rate, denoted by *d*. In accordance with our initial hypothesis, we shall assume that the values of *b* and *d* will remain constant through time.

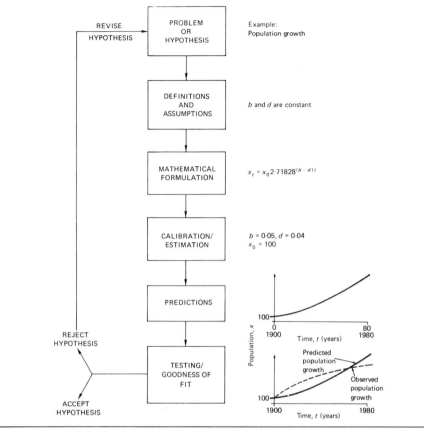

Figure 1.3. The steps in mathematical model building.

The third step in our flow diagram is to deduce a model equation from our assumptions using formal mathematical reasoning. Our assumptions in this case refer to the rate of population growth which we shall denote by dx/dt. This symbol is read as the value of some small increase in population, dx, in some small time interval, dt. We have assumed dx/dt has, at any instant, a value determined by the difference between village births and village deaths at that instant. This assumption may be written mathematically as follows:

$$\frac{dx}{dt} = (b - d)x_t. \tag{1.2}$$

Verbally, this equation states that

$$\begin{array}{l} \text{Rate} \\ \text{of village} \\ \text{population growth} \end{array} = \left(\begin{array}{l} \text{Village} \\ \text{birth} \\ \text{rate} \end{array} - \begin{array}{l} \text{Village} \\ \text{death} \\ \text{rate} \end{array} \right) \times \begin{array}{l} \text{Village} \\ \text{population.} \end{array}$$

Because we wish to predict the village population, x_t, it is necessary to perform some formal mathematical procedures on equation (1.2) so that x appears on the left-hand side of the equation. This rearrangement will then give us the model equation for predicting the village population size at any time t. We shall discuss the mathematical methods needed to rearrange equations such as (1.2) later in the book; here we simply ask the reader to take it on trust that the solution to the problem is

$$x_t = x_0 \times 2 \cdot 71828^{(b - d)t}. \tag{1.3}$$

In this equation x_0 is the village population in 1900, and t refers to time periods expressed in relation to the base date of 1900 when $t = 0$. For example, if the time interval is a year, then $t = 2$ refers to the year 1902.

Having deduced the model equation, the next step is to test the adequacy of the predictions obtained from this equation. The *calibration* of the model refers to the process of giving values to the terms (parameters) which appear on the right-hand side of the model equation; in this case x_0, b, and d. To obtain these parameter values, past data about the village population are used. From a study of these data we find that $x_0 = 100$ people, the value of b is equal to $0 \cdot 05$ births per 100 people per year, and d is equal to $0 \cdot 04$ deaths per 100 people per year. To obtain predicted populations from our model equation we simply substitute these values into formula (1.3) together with the time period t. For example, the predicted population for 1902 ($t = 2$) is obtained as

$$x_2 = 100 \times 2 \cdot 71828^{(0 \cdot 05 - 0 \cdot 04)2}$$

$$= 100 \times 2 \cdot 71828^{0 \cdot 02} = 102.$$

The formal testing of the model is the comparison of the predicted values with observed values; in this instance the predicted population x_t would be compared with actual village populations for a variety of time intervals. If the agreement between observed and predicted system behaviour is good then the hypothesis may be provisionally accepted; conversely, if agreement is poor then the hypothesis is rejected, and the whole modelling procedure has to be repeated with an amended hypothesis. In the present example it is more than likely that the initial hypothesis of a constant proportional rate of population growth would be rejected, because the birth and death rates of the village population are unlikely to remain constant for long periods of time. Therefore, a revised hypothesis should make more detailed assumptions about components of population growth. For instance, we might make assumptions about birth and death rates for different age groups, or introduce migration rates into the model assumptions. (Population models incorporating these and other assumptions are discussed in Chapter 3.) This example typifies the way in which mathematical models are tested and redesigned. The research worker begins by deducing models from very general assumptions about system behaviour and gradually adds more specific assumptions as he learns to identify the key relationships between system components.

Deterministic and probabilistic models

A quick glance at the contents list will reveal that this book is organized in two main parts — deterministic models and probability models — terms which refer to two styles of mathematical modelling open to the geographer. Deterministic models are founded on the notion that the behaviour of geographical systems is controlled by natural, physical laws and therefore, once these laws have been unearthed, the system behaviour can be predicted *exactly*. The deterministic modelling style is most commonly applied in physical geography where laws governing the movement and storage of raw materials are used as a starting point for constructing models which predict the behaviour of physical systems. The structure of such models is explained in Chapters 3 and 4. Chapter 3 outlines the laws which control the movement and storage of raw materials with reference to the storage of carbon in an ecosystem and the problem of predicting population change. This framework is extended in Chapter 4 to include space explicitly as part of the physical system. The problems treated include the modelling of water flow in rock and soil, and the modelling of heat flow in a soil profile.

Systems in human geography are not so well suited to the deterministic style of modelling because exact laws of human behaviour do not exist. Nevertheless, progress can be made if we treat human beings in the aggregate as groups, rather than as free-willed individuals. It is quite permissible to convert observed trends in group behaviour into the law-like statements needed for model construction. Chapter 5 explores this idea in the context of spatial interaction models which use the observed decline in the amount of movement with increasing distance between places as a basis for constructing models of migration, shopping behaviour and the journey-to-work. An alternative approach to studying human systems is to construct *normative* models based on assumptions about how human beings *ought* to behave. For instance, we might assume that the operators of a transport system want to minimize the total cost of transporting a commodity from factory to market place. Given this premise, it is possible to construct spatial allocation models which predict the least-cost assignment of commodity shipments. By comparing the predicted 'optimum' pattern of shipment against the observed pattern, it is possible to assess the rationality of transport operators in accordance with our simplifying assumption. Models of this type are discussed in Chapter 6.

By adopting the deterministic modelling style, the geographer is searching for exact mechanical processes to explain the behaviour of spatial systems. But for many geographical systems we suspect that the components are related in a chance-like way and therefore cannot be explained by physical laws. For instance, the spread of an epidemic through a human population may be thought of as a system of chance-like contacts between members of art infected population and members of a susceptible population. To construct model equations for such systems we express our initial assumptions as a set of probabilities. Thus we might assume there is, say, a 50 per cent chance that an infected person will contact a susceptible person over a stated distance within a defined time interval. To convert such assumptions into probability models we use the mathematical language of probability theory. Because chance is involved, the predictions from a probability model are stated with a known degree of error or tolerance. In this way probability models focus our attention on a range of possible outcomes rather than a single prediction.

In Chapter 7 we examine in some detail the formal mathematical logic of probability theory, together with some of its simpler geographical applications. Chapter 8 extends this framework to encompass the occurrence of chance-like events in space. The topics covered include the identification of the processes which control the evolution of point patterns, and ways of assessing the strength of a relationship between important geographical variables such as the relationship between race and illiteracy over a system of regions. Finally, in Chapter 9, we examine one of the more recent developments in probability theory, namely, decision models. Here the system of interest is a decision-maker who has to choose between different courses of action in the face of uncertain environmental conditions. In particular, we shall examine how probability theory can be used to help the farmer decide which crops to plant in the face of uncertain weather conditions.

It is important to stress that distinction between deterministic and probability models is partly a matter of pedagogic convenience: it is easier to grasp the mathematical essentials when the two topics are studied separately. Yet the objectives of both styles are the same: to isolate and explicate the mathematical structure of a geographical system. Indeed, some of the most convincing geographical models are mixtures of both approaches. For the most part, these mixed models are beyond the scope of this book. In Chapter 5, however, the reader will find an example of such a model: the entropy-maximizing journey-to-work model mixes a deterministic representation of the urban fabric with a probabilistic approach to the citizen's trip-making behaviour.

CHAPTER 2

Some Mathematical Preliminaries

Students embarking on the study of geography at university bring with them varied mathematical backgrounds and attitudes. Some view the x's, y's and i's of mathematical notation as nothing more than an exercise in obfuscation, while others feel at home with the logic and symbolism. Just as it is impossible to appreciate literature written in a foreign tongue without some idea of how the language is assembled, so it is impossible to appreciate a mathematical model without some knowledge of the nuts and bolts of mathematical reasoning. In this chapter we set down the mathematical foundations needed to understand the geographical models which follow later in the book.

We have separated these foundations into six basic ingredients: ways of taking measurements on geographical objects; ways of defining objects as variables; ways of summarizing observations on geographical variables; ways of expressing and manipulating variables in matrix (tabular) form; ways of expressing relationships between variables using simple mathematical functions; and, finally, analysing the rate of change of variables in space and time. The first four topics deal, by and large, with arithmetic operations, while the last two involve algebraic manipulation. The material is therefore graded from the simple to the more complex. However, the reader need not feel obliged to commit this material to memory. Simply flip through and familiarize yourself with the terminology and then refer back if, subsequently, difficulties arise.

MEASUREMENT

At some stage or other in the model-building process we require *data*. The validity of a deterministic model often depends on the degree of correspondence between the model predictions and actual observations. Models involving probability theory often have extensive data requirements. This demand for data involves the process of *measurement*, which in turn begs the question: What is it we do when we measure? Definitions of measurement are numerous. Krumbein and Graybill (1965, p34), a geologist and a statistician, define the process as the assignment of a number to the amount, degree, extent, magnitude, quantity, or quality of an object in accordance with definite rules. Similarly, the psychologist Nunnally (1967, p2) calls measurement rules for assigning objects to numbers to represent quantities of attributes. These, and many other definitions, express the notion that measurement must follow a set of pre-defined rules, although which particular set of rules should be followed is the subject of an endless philosophical debate. At a less abstract level, four types of measurement rules are commonly recognized which define an hierarchy of measurement scales; these are the nominal scale, the ordinal scale, the interval scale, and the ratio scale.

Nominal scaling

The chief scientific usage of the *nominal scale* is to identify objects and events by numbering them. The numbers on the jerseys of football players or the labelling of a set of regions by the numbers 1,2,3, . . . , *n* are examples of identification. Clearly the numbers in a nominal scale cannot be subjected to arithmetical manipulation: the region numbered 5 minus the region numbered 2 does not equal the region numbered 3. For this reason it is doubted by some whether nominal scaling constitutes a measurement system in its own right. Nominal scales simply recognize that the identified objects are equivalent to one another according to one of their attributes; for instance, communist countries have a political system in common.

A derivative of nominal scaling is counting the numbers of objects that fall in some particular class. For example, we might choose a rainfall figure to decide whether a day has been wet or dry. Given this criterion, we can count the number of wet and dry days in some time period. Counting the number of objects which fall into particular classes provides frequencies which can be manipulated arithmetically. Counting the number of objects in classificatory boxes is a different procedure from identifying each box by a number. The frequencies obtained by counting refer to an attribute of the class of objects rather than an attribute of the object itself. Although nominal scaling is a weak form of measurement, virtually any variable we care to think of can be divided with a set of classes. These variables may be the familiar schemes for classifying land uses, soils, vegetation, climates, and landforms, or less familiar subjective schemes for classifying the qualities of an environment as say noisy or quiet, beautiful or ugly.

Ordinal scaling

An *ordinal scale* is one where objects or events can be ordered from most to least, but where there is no numerical information about the intervals separating the measured attribute. Thus ordinal scales involve ranking *n* objects from highest to lowest in terms of the criterion greater than or less than. For example, an ordinal scale would be obtained if we asked an individual to rank *n* cities based on his preference for each city as a place of residence. From this information we should obtain a preference order, but nothing about the degree of preference for each city. Thus we could not say he preferred the place ranked 1 twice as much as the place ranked 2. We may distinguish between completely ordered scales where each object is given a unique rank, and weak ordering when objects are assigned to an ordered hierarchy of categories. For instance, assigning each of *n* settlements to the hierarchy city—town—village—hamlet is an

Table 2.1. Scales of measurement.

Scale	Determinative operation	Examples
Nominal	Equality Equivalence	Land-use types (woodland, arable, etc); presence—absence data; planning regions designated by numbers.
Ordinal	Greater or lesser (rankings) as well as equivalence	Size classes of objects such as city—town—village—hamlet; land-use capability classes; stream order; the rankings of cities based on a person's preference for each city as a place in which they would like to live.
Interval	Equality of intervals or differences	Temperature in $^\circ$C; time in years AD or BP.
Ratio	Equality of ratios	Population density; temperature in degrees absolute (K); plant biomass in grams; spot heights in metres.

example of weak ordering. Quite sophisticated analytical procedures can be applied to ordinal scales.

Interval and ratio scales

In the *interval scale*, as well as equivalence and rank, the dimensions of steps between ranked objects are known and, in consequence, the ratio of any two intervals, say between 10 and 20 $^\circ$C, is known. In the temperature example the ratio would be 10:20 or 1:2. However, the ratio between two points on an interval scale depends on the units of measurement – 0 $^\circ$C to 100 $^\circ$C and 32 $^\circ$F to 212 $^\circ$F are not the same ratio.

In the fourth or *ratio scale*, all the properties of the interval scale are present. Additionally, the ratio of any two scale values is known and the same, irrespective of the units of measurement. So 4 grams of salt is twice the mass of 2 grams of salt: 4 pounds of salt is twice the mass of 2 pounds of salt. All ratio scales have an absolute zero point, whereas zero points on the interval scale are arbitrary; this is why 20 $^\circ$C is not twice the temperature of 10 $^\circ$C, but 20 K (temperature on the absolute temperature scale) is twice the temperature of 10 K. Table 2.1 gives some geographical examples of scales of measurement.

VARIABLES AND NOTATION

Any measured attribute of an object which varies is, appropriately, called a *variable*. We may label a variable with any symbol we care to choose[1], but it is customary to use letters such as x, y, and z. For example, the population of any British city in 1961 may be called variable x.

To quantify the values a variable can take on requires us to make observations, or individual measurements. An individual observation, say the population of Liverpool, may be denoted by the symbol x_i. The subscript i is simply a number which enables a particular observation to be identified amongst a whole set, or collection, of observations on the variable; i is in fact a nominal variable and has no mathematical properties other than identity. For example, the populations of three British cities in 1961 may be written as

Blackburn x_1 = 106 000 people

Blackpool x_2 = 152 000 people

Liverpool x_3 = 747 000 people.

Clearly, the i's enable each of the three cities to be identified: $i = 1$ is Blackburn, $i = 2$ is Blackpool, and $i = 3$ is Liverpool.

Sigma notation

The Greek capital letter Σ (sigma) is widely used as notation for the addition of values. It is in fact a capital 'S' and stands for *sum of*. So

$$\sum_{i}^{n} x_i$$

is an instruction in mathematical shorthand to sum the observations on the variable x, beginning with the ith observation (the i under the Σ tells us this) and ending with the nth observation (the n above the Σ tells us this). In the example of the three British cities, when $i = 2$ and $n = 3$, the instruction is

$$\sum_{2}^{3} x_i = x_2 + x_3$$

$$= 152\ 000 + 747\ 000$$

$$= 899\ 000.$$

When $i = 1$ and $n = 3$, then

$$\sum_{1}^{3} x_i = x_1 + x_2 + x_3$$

$$= 106\ 000 + 152\ 000 + 747\ 000$$

$$= 1\ 005\ 000.$$

If sigma is written as Σ, without the values of i and n stated, the implication is that all n values of observations on x should be summed.

Other operations we shall encounter in this book which use a sigma notation are $(\Sigma x_i)^2$ and Σx_i^2. The first of these, $(\Sigma x_i)^2$, means add up all the values of x then square the total. The second, Σx_i^2, means square each value of x *before* adding them up.

Pi notation

The Greek capital letter Π (pi) is used as an abbreviation for multiplication. It is in fact a capital 'P' and stands for *product of*. So

$$\prod_i^n x_i$$

is an instruction in mathematical shorthand to multiply the observations on variable x, starting with observation i and ending with observation n. For the three British cities

$$\prod_1^3 x_i = x_1 \times x_2 \times x_3$$

$$= 106\,000 \times 152\,000 \times 747\,000$$

$$= 12\,035\,664\,000.$$

Of course, there is not normally any call to multiply city populations, but the pi notation has some important applications in modelling and will be used later in the book.

Superscripts

Normally, a number appearing above a variable indicates that the value of x must be raised to the power denoted by that number, for example, $x_i^2 = x_i \times x_i$. Sometimes, however, the numbers above a variable take the form of superscripts which are used to identify further properties of x. Two examples of superscripts will be found in this book. One is in the form x_i^* where the asterisk denotes an observed value of x^* which can be compared with a predicted value, x_i, from some model. Another instance is where the value of x_i is obtained after a sequence of stages of calculation; in this case we use $x_i^{(1)}$ to denote the value of x after the first stage, $x_i^{(2)}$ for the value after the second stage, and so on.

SUMMARIZING OBSERVATIONS – DESCRIPTIVE STATISTICS

Often in geographical modelling it is necessary to analyse the properties of a large number of observations taken on a particular variable. The map of illiteracy rates in the $n = 11$ counties of the hypothetical country is a typical geographical variable (figure 2.1*a*). As usual, the observed illiteracy rate in each county is denoted by a value x_i, and we denote the collection of $n = 11$ observations by the symbol $\{x_i\}$. Here the curly brackets signify a pre-defined collection, or *set*, of observations. Many of the probability models discussed in Part III of this book require us to work with indices which summarize the major numerical characteristics of observations. These indices are usually termed descriptive statistics.

Essentially, descriptive statistics summarize numerical aspects of the frequency distribution of a set of observations, each denoted by x_i. By the term *frequency distribution* we mean the number of times that x takes on a particular value, or range

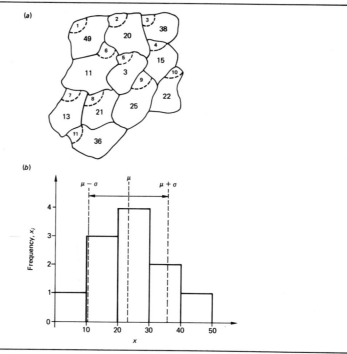

Figure 2.1.   The frequency distribution of a geographical variable. (*a*) Map pattern: the percentage of the adult population x_i who are illiterate in $n = 11$ counties of an imaginary country. (*b*) Histogram.

of values. A visual impression of a frequency distribution is obtained by constructing a *histogram* which is a bar graph recording the number of observations falling within pre-defined class intervals. A histogram of the illiteracy data is shown in figure 2.1(*b*). In this example, a class interval of 10 per cent has been chosen and therefore five classes (0–9, 10–19, ... , 40–49) are needed[2] to describe the range of adult illiteracy from $x_5 = 3$ per cent to $x_1 = 49$ per cent. The frequencies, freq(x_i), are found by counting the number of observations that fall within each class. In this instance we find one observation, $x_5 = 3$, in the 0–9 class, three observations in the 10–19 class, etc.

Descriptive statistics attempt to summarize the main properties of the frequency distribution with just one or two numbers. The two most important properties of a frequency distribution are termed *central tendency* and *dispersion*. Measures of central tendency are designed to locate a central value of x within the frequency distribution, while indices of dispersion measure the spread of the values around the central point.

Central tendency: the mean

The measure of central tendency used in this book is the mean, which is familiar to most people as the average value of a set of observations. The mean of a set of observations may be calculated by adding all their values together and then dividing this sum by the total number of observations. This definition may be written more succinctly by employing sigma notation. If we denote the mean value of any set of observations $\{x_i\}$ by the Greek letter μ (mu), then the general formula for the mean is given by

$$\mu = \left(\sum_{i=1}^{n} x_i \right) \Big/ n. \tag{2.1}$$

The term inside the brackets is simply an instruction to sum all the observations from x_1 to x_n. For our illiteracy variable, where $n = 11$, the mean is calculated from formula (2.1) as

$$\mu = (x_1 + x_2 + x_3 + x_4 + \ldots, + x_{11})/n$$
$$= (49 + 20 + 38 + 15, + \ldots, + 36)/11$$
$$= (253)/11 = 23.$$

This result states that in the 11 counties an average of 23 per cent of the adult population is illiterate. The idea of the mean as a measure of central tendency is clearly seen when we plot its value on the histogram (figure 2.1*b*). Notice the mean value is drawn towards the middling adult illiteracy rates and is located amongst the class with the highest frequency[3].

Dispersion

Descriptive statistics which measure dispersion attempt to give a numerical indication of the width, or spread, of the observed values forming the frequency distribution. An extremely simple numerical expression of dispersion is the *range*, R, of a frequency which is defined verbally as

$$R = \text{largest observation} - \text{smallest observation.} \tag{2.2}$$

Therefore, the range of our illiteracy is

$$R = x_1 - x_5 = 49 - 3 = 46.$$

Clearly, range is a weak measure because it ignores the values of all the other observations forming the frequency distribution. Other measures of dispersion are constructed in terms of deviation from the centre of the frequency distribution and in the ensuing discussion we shall define the centre to be the mean value. If the mean is the central value, the deviation of an observation about the centre is the difference $x_i - \mu$; this quantity is negative if x_i is smaller than the mean and positive if x_i is greater than the mean. Because all n observations have a deviation about the mean, measures of dispersion can be constructed by averaging these deviations. Unfortunately, the average of all the deviations $x_i - \mu$ is always zero because the sum of the negative deviations cancels out the sum of the positive deviations. To circumvent this problem, a measure of dispersion can be constructed using the squares of the individual deviations which eliminates the negative deviation problem. The average of the squared deviation is termed the *variance*, σ^2, and is defined by the formula

$$\sigma^2 = \left[\sum_{i=1}^{n} (x_i - \mu)^2 \right] / n. \tag{2.3}$$

The sigma term inside the square brackets is often referred to as the *sum of squares*, and the variance of our illiteracy data is calculated from formula (2.3) as

$$\sigma^2 = [(x_1 - \mu)^2 + (x_2 - \mu)^2 + (x_3 - \mu)^2 + (x_4 - \mu)^2 +, \ldots, + (x_{11} - \mu)^2]/n$$

$$= [(49 - 23)^2 + (20 - 23)^2 + (38 - 23)^2 + (15 - 23)^2 +, \ldots,$$

$$+ (36 - 23)^2]/11$$

$$= [1796]/11 = 163 \cdot 27.$$

Thus the sum of squares has a value 1796 and the variance a value of 163·27. Because the variance uses squared deviations, its numerical value is difficult to comprehend intuitively. (For our example the variance is 163·27 units of the squared percentage of the adult population that is illiterate.) For this reason it is common practice to work with the square root of the variance which is called the standard deviation, σ, and is defined by

$$\sigma = \sqrt{\left[\sum_{i=1}^{n} (x_i - \mu)^2 \right] \Big/ n} = \sqrt{\sigma^2}. \tag{2.4}$$

The standard deviation is expressed in the same units as the observations x_i. The standard deviation of our illiteracy data is

$$\sigma = \sqrt{163 \cdot 27} = 12 \cdot 78,$$

and measures the spread in units of the percentage of the adult population that is illiterate. A simple visual impression of how the standard deviation measures dispersion around the mean value is gained by superimposing the values $\mu - \sigma$ and $\mu + \sigma$ on the histogram of the frequency distribution (figure 2.1*b*).

Shapes of frequency distributions

A number of statistical terms are used to describe the shapes of the histograms found for many geographical variables. The term *normal* (or symmetrical) is used to describe a frequency distribution where the observations occur most frequently around the mean value and then decline in a bell-shape on either side of the mean (figure 2.2*a*). Observations which measure the quality of some object often follow a symmetrical distribution. For example, wheat yeilds for a set of farms are likely to be characterized by large numbers of average yields and relatively few high and low yields. The age of buildings in a city is likely to follow the same pattern. A *positively skewed* frequency distribution (figure 2.2*b*) is characterized by large numbers of low values and relatively few high values. In such cases the mean will lie to the right of the main bunch of observations. Variables measuring the size or magnitude of an object often follow this form. Thus there are many villages but few cities, many short streams but few long rivers, many fine-grained particles but few stones, etc. In contrast, a *negatively skewed* distribution has many large observations but relatively few low values, and the mean lies to the left of the main bunch (figure 2.2*c*). Negatively skewed frequency distributions are relatively rare in geography. One example is the frequency distribution of daily rainfall in a tropical rain forest. Here there will be many wet days but few dry days.

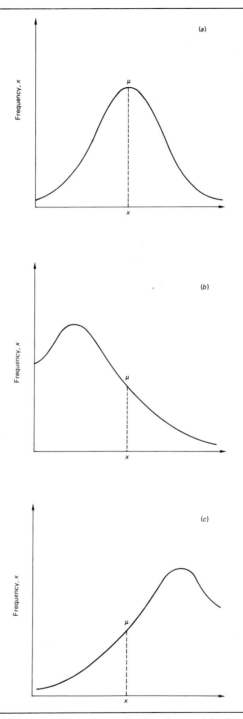

Figure 2.2   Some common frequency distributions. (*a*) Symmetrical (normal).
(*b*) Positively skewed. (*c*) Negatively skewed.

MATRICES

Many geographical models and bodies of data are effectively presented as *matrices* and analysed using *matrix algebra operations.* In this section we shall explore the language of matrix algebra and matrix operations in a geographical setting.

Basic ideas

Five towns — α, β, γ, δ and ϵ — linked by bus and rail routes are shown in figure 2.3. The information contained in this map may be represented as a table. With α to ϵ as column headings and α to ϵ as row headings, we may write

			to		
	α	β	γ	δ	ϵ
α	0	1	1	0	0
β	1	0	0	1	0
from γ	1	0	0	1	1
δ	0	1	1	0	1
ϵ	0	0	1	1	0

where a 1 denotes a direct bus link between two towns and a 0 denotes the absence of such a link. Similarly we may tabulate rail connections between pairs of towns to give

			to		
	α	β	γ	δ	ϵ
α	0	0	1	0	0
β	0	0	1	0	0
from γ	1	1	0	0	1
δ	0	0	0	0	0
ϵ	0	0	1	0	0

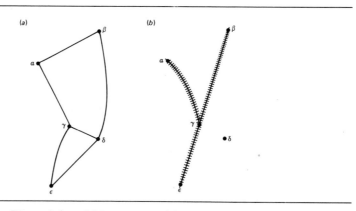

Figure 2.3.   (*a*) Bus routes. (*b*) Rail routes.

The bus and rail networks have been arranged as square *arrays* of five rows and five columns each. Dropping the row and column headings and putting the numbers in brackets, *matrices* are produced, the bus one looking like this

$$\mathbf{B} = \begin{bmatrix} 0 & 1 & 1 & 0 & 0 \\ 1 & 0 & 0 & 1 & 0 \\ 1 & 0 & 0 & 1 & 1 \\ 0 & 1 & 1 & 0 & 1 \\ 0 & 0 & 1 & 1 & 0 \end{bmatrix}$$

and the rail one looking like this

$$\mathbf{R} = \begin{bmatrix} 0 & 0 & 1 & 0 & 0 \\ 0 & 0 & 1 & 0 & 0 \\ 1 & 1 & 0 & 0 & 1 \\ 0 & 0 & 0 & 0 & 0 \\ 0 & 0 & 1 & 0 & 0 \end{bmatrix}.$$

The numbers in each of the matrices are called *elements*. Each matrix may be referred to as a single unit and denoted by a single letter: the bus route matrix can be designated $\mathbf{B}$ and the rail route matrix $\mathbf{R}$. Notice that bold type is used; this is to help distinguish matrices from ordinary variables. The matrix $\mathbf{B}$ contains all the information about bus routes between the five towns; and the matrix $\mathbf{R}$ contains all the information about the rail routes.

If a matrix contains m rows and n columns, it is an $m \times n$ matrix; the quantity $m \times n$ being its *size* or *order*. The bus and rail route matrices are both of size 5×5. In stating the size of a matrix the number of rows is always given first. A matrix with just one row is called a *row vector*; for instance, the links between town α and all other towns in the bus route matrix $\mathbf{B}$ is represented by the row vector $[0\ 1\ 1\ 0\ 0]$. Similarly, a matrix with just one column is a *column vector*; for example, column α in the bus route matrix $\mathbf{B}$, which shows links from all towns to town α, is a column vector

$$\begin{bmatrix} 0 \\ 1 \\ 1 \\ 0 \\ 0 \end{bmatrix}$$

which may be written $\{0\ 1\ 1\ 0\ 0\}'$, the curly brackets and prime indicating that it is a column vector and not a row one. An ordinary number is a 1×1 matrix with only one row and one column, and is known as a *scalar*. A matrix with the same number of rows and columns, like our bus and rail route matrices, is a *square matrix*, four particular types of which are:

$$\mathbf{I} = \begin{bmatrix} 1 & 0 \\ 0 & 1 \end{bmatrix}$$

the *identity* or *unit matrix* which contains 1's in the principal diagonal (the elements in the first row and first column, second row and second column, and so forth), and 0's elsewhere;

$$\mathbf{O} = \begin{bmatrix} 0 & 0 \\ 0 & 0 \end{bmatrix}$$

the *null* or *zero matrix* in which all elements are zero;

$$\mathbf{A} = \begin{bmatrix} 9 & -4 \\ -4 & 9 \end{bmatrix}$$

a *symmetrical matrix* in which the elements of the principal diagonal may assume any value but the element in the ith row and jth column is equal to the element in the jth row and ith column; and

$$\mathbf{B} = \begin{bmatrix} 9 & 2 \\ 1 & -5 \end{bmatrix}$$

an *asymmetrical matrix* in which all elements may assume any value.

It is convenient to use bold capital letters for matrices, bold small letters for row and column vectors, and italic letters for elements. Matrix elements may be identified thus

$$
\begin{array}{c}
\begin{array}{ccccc} & j \rightarrow & & & n \\ 1 & 2 & 3 & 4 & 5 \end{array} \\
\begin{array}{c} i \downarrow \\ \\ \mathbf{B} = \\ \\ m \end{array}
\begin{array}{c} 1 \\ 2 \\ 3 \\ 4 \\ 5 \end{array}
\begin{bmatrix}
b_{11} & b_{12} & b_{13} & b_{14} & b_{15} \\
b_{21} & b_{22} & b_{23} & b_{24} & b_{25} \\
b_{31} & b_{32} & b_{33} & b_{34} & b_{35} \\
b_{41} & b_{42} & b_{43} & b_{44} & b_{45} \\
b_{51} & b_{52} & b_{53} & b_{54} & b_{55}
\end{bmatrix}
\end{array}
$$

where, for instance, b_{11} is the element in the position row one column one; b_{21} is the element in position row two column one; and so on. In the bus route matrix, b_{31} is the element indicating a link between town γ and town α and in the general case b_{ij} is the element showing a bus link (or absence of one) between towns j and i.

We may wish to sum the elements in individual rows or columns. For instance, the sum of row one in the bus route matrix $\mathbf{B}$ is 2; this can be interpreted as town α's having bus links to two other towns. The procedure to find the sum of this row vector is obviously

$$0 + 1 + 1 + 0 + 0 = 2$$

or, in symbols

$$b_{11} + b_{12} + b_{13} + b_{14} + b_{15} = \sum_{j=1}^{5} b_{1j}.$$

Similarly, the sum of the second row vector is

$$b_{21} + b_{22} + b_{23} + b_{24} + b_{25} = \sum_{j=1}^{5} b_{2j}$$

and substituting actual values

$$1 + 0 + 0 + 1 + 0 = 2.$$

The formula for summing the ith row vector is

$$b_{i1} + b_{i2} + b_{i3} + b_{i4} + b_{i5} = \sum_{j=1}^{5} b_{ij}.$$

A row total is sometimes denoted by $b_{i.}$; for example,

$$b_{1.} = 2, \text{ and } b_{2.} = 2.$$

Alternatively, we could sum the column vectors to find the number of bus links to a town from other towns. As the matrix is symmetrical, column vector sums will be the same as their corresponding row vector sums. We have for the first column vector

$$0 + 1 + 1 + 0 + 0 = 2$$

or, in symbols

$$b_{11} + b_{21} + b_{31} + b_{41} + b_{51} = \sum_{i=1}^{5} b_{i1}.$$

The general formula for the jth column vector sum is

$$b_{1j} + b_{2j} + b_{3j} + b_{4j} + b_{5j} = \sum_{i=1}^{5} b_{ij}.$$

A column total is sometimes denoted by $b_{.j}$; for example, $b_{.1} = 2$.

The row or column vector sums can themselves be totalled to give the number of links in the bus network. Adding row totals, we have

$$2 + 2 + 3 + 3 + 2 = 12.$$

(This total counts a link between a pair of towns as 2, one link in each direction). In statistical notation the procedure is written

$$\sum_{j=1}^{5} b_{1j} + \sum_{j=1}^{5} b_{2j} + \sum_{j=1}^{5} b_{3j} + \sum_{j=1}^{5} b_{4j} + \sum_{j=1}^{5} b_{5j} = \sum_{i=1}^{5} \left(\sum_{j=1}^{5} b_{ij} \right),$$

which is often written in abbreviated form as

$$\sum_{i=1}^{5} \sum_{j=1}^{5} b_{ij}.$$

In like manner, the procedure for adding column sums may be expressed as

$$\sum_{i=1}^{5} b_{i1} + \sum_{i=1}^{5} b_{i2} + \sum_{i=1}^{5} b_{i3} + \sum_{i=1}^{5} b_{i4} + \sum_{i=1}^{5} b_{i5} = \sum_{j=1}^{5} \left(\sum_{i=1}^{5} b_{ij} \right)$$

Table 2.2. Matrix summations.

		$j \longrightarrow$				Row totals
		1	2	$\cdots$	n	
i	1	b_{11}	b_{12}	$\cdots$	b_{1n}	$b_{1.} = \sum\limits_{j=1}^{n} b_{1j}$
$\downarrow$	2	b_{21}	b_{22}	$\cdots$	b_{2n}	$b_{2.} = \sum\limits_{j=1}^{n} b_{2j}$
	$\vdots$	$\vdots$	$\vdots$		$\vdots$	$\vdots$
	m	b_{m1}	b_{m2}	$\cdots$	b_{mn}	$b_{m.} = \sum\limits_{j=1}^{n} b_{mj}$
		$b_{.1}$	$b_{.2}$	$\cdots$	$b_{.n}$	Grand total
Column totals		$= \sum\limits_{i=1}^{m} b_{i1}$	$= \sum\limits_{i=1}^{m} b_{i2}$	$\cdots$	$= \sum\limits_{i=1}^{m} b_{in}$	$\sum\limits_{i}^{m} \sum\limits_{j}^{n} b_{ij}$

which is more succinctly put as

$$\sum_{j=1}^{5} \sum_{i=1}^{5} b_{ij}.$$

Table 2.2 summarizes the notation for summing elements in a matrix with $i = 1, 2, \ldots, m$ rows and $j = 1, 2, \ldots, n$ columns.

Matrix addition

Looking at the bus and rail route maps (figure 2.3) we can see that adding the two produces the combined route map shown in figure 2.4. Consider how we might do the same thing using matrices. The combined bus and rail route map may be put in tabular form as

	α	β	γ	δ	ϵ
α	0	1	2	0	0
β	1	0	1	1	0
γ	2	1	0	0	2
δ	0	1	0	0	1
ϵ	0	0	2	1	0

where the elements are the number of links, bus plus rail, between pairs of towns. This information may be stated as a matrix which we shall call **N**

$$\mathbf{N} = \begin{bmatrix} 0 & 1 & 2 & 0 & 0 \\ 1 & 0 & 1 & 1 & 0 \\ 2 & 1 & 0 & 0 & 2 \\ 0 & 1 & 0 & 0 & 1 \\ 0 & 0 & 2 & 1 & 0 \end{bmatrix}$$

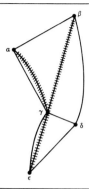

Figure 2.4. Combined bus and rail network.

To derive **N** from **B** and **R** the process is to add corresponding elements from each matrix. Thus

$$b_{12} + r_{12} = n_{12}$$

and substituting actual values

$$1 + 0 = 1.$$

In the general case

$$b_{ij} + r_{ij} = n_{ij} \qquad (i = j = 1, 5).$$

Matrix multiplication

The procedures by which matrices are multiplied are more tricky than the process of matrix addition. Consider again the combined route map (figure 2.4). Assume we wish to take a train trip from one town to another and then take a bus ride to another town. We may, for example, take a train from town α to town γ; and from town γ we may take a bus to town δ, town ϵ, or back to town α — three possibilities. In other words, we may take a train journey followed by a bus journey from town α to any of the towns α, δ, or ϵ. By the same reasoning we can work out the other towns between which it is possible to take a double trip. The information is presented as a matrix **M**:

$$\mathbf{M} = \begin{bmatrix} 1 & 0 & 0 & 1 & 1 \\ 1 & 0 & 0 & 1 & 1 \\ 1 & 0 & 2 & 2 & 0 \\ 0 & 0 & 0 & 0 & 0 \\ 1 & 0 & 0 & 1 & 1 \end{bmatrix}$$

Notice that this matrix is asymmetrical. The element m_{33} has the value 2; this means that there are two different ways of taking the double trip from town γ back to town γ.

M can be derived by multiplying **R** and **B**. Comparing like elements in **B**, **R**, and **M** it is evident that **M** is not derived by multiplying corresponding elements from **B** and **R**. The procedure is in fact to take the elements in the ith row of **R**, multiply each of them by the corresponding elements in the jth column of **B**, and sum the products to give the element m_{ij} in **M**. The first row of **R** is

$$[0 \quad 0 \quad 1 \quad 0 \quad 0]$$

and the first column of **B** is

$$\begin{bmatrix} 0 \\ 1 \\ 1 \\ 0 \\ 0 \end{bmatrix}$$

We multiply them like this

$$[0 \quad 0 \quad 1 \quad 0 \quad 0]\begin{bmatrix} 0 \\ 1 \\ 1 \\ 0 \\ 0 \end{bmatrix} = (0 \times 0) + (0 \times 1) + (1 \times 1) + (0 \times 0) + (0 \times 0)$$

$$= 1$$

to get element m_{11}. All other elements of **M** may be evaluated in the same way using appropriate rows of **R** and columns of **B**. Note that matrices can be multiplied only if the number of columns in the first matrix equals the number of rows in the second matrix. A full example of matrix multiplication is at the end of this chapter[4].

It is important to note that in matrix algebra the order of multiplication is important. The product of the rail and bus matrices, **RB** = **M**, is not the same as the product of the bus and rail matrices, **BR** = **M**, which would tell us between which towns it is possible to take a double trip starting with a bus ride and finishing on a train. In the first case, **RB** = **M**, **B** is said to be *pre-multiplied* by **R**; or **R** is said to be *post-multiplied* by **B**. If a matrix is multiplied by a single number, each element in the matrix is multiplied by the number according to the usual rules of multiplication. For instance,

$$2 \times \begin{bmatrix} 2 & 3 \\ 1 & 4 \end{bmatrix} = \begin{bmatrix} 4 & 6 \\ 2 & 8 \end{bmatrix}.$$

Matrix transposition

In our five-town example we had routes from a town corresponding to rows, and routes to a town corresponding to columns; this is purely arbitrary. The meaning is not changed if rows are used for 'to a town' and columns are used for 'from a town'.

Transposing the double-trip matrix **M** in this way we should have

$$\begin{bmatrix} 1 & 1 & 1 & 0 & 1 \\ 0 & 0 & 0 & 0 & 0 \\ 0 & 0 & 2 & 0 & 0 \\ 1 & 1 & 2 & 0 & 1 \\ 1 & 1 & 0 & 0 & 1 \end{bmatrix}$$

which is known as the *transpose* of the original matrix **M** and is conventionally denoted as **M'** or **M^T**. By transposition, any element in the original matrix in row i and column j will, in the transposed matrix, be in row j and column i:

$$m_{ij} = m_{ji}^T.$$

The principal diagonal is unaffected by transposition. If the transposed matrix is itself transposed, the result is the original matrix; symbolically,

$$(\mathbf{M}^T)^T = \mathbf{M}.$$

Linear equations written and solved as matrices

Many geographical problems involve the specification of sets of *simultaneous, linear equations*. In agricultural geography, a common problem is to guarantee maximum farm production assuming the worst possible weather conditions prevail. Found (1971) gave the example of a plantation of pumpkins whose owner has the resources to plant 100 acres. Under the worst weather conditions, the best yield the plantation owner can expect is 22·4 tons of pumpkins per acre, but this can only be obtained by planting optimum acreages in valley-bottom and low-hill fields. So how can the optimum acreages be calculated? It is known that in wet years the pumpkin yield will be 6 tons per acre in valley-bottom fields and 31 tons per acre in low-hill yields; whereas in dry years, valley-bottom fields will yield 50 tons per acre and low-hill fields 8 tons per acre. From all this information we may write two simultaneous equations

$$\begin{array}{ll} 31x_1 + 6x_2 = 22\!\cdot\!4 & \text{(wet year)} \\ 8x_1 + 50x_2 = 22\!\cdot\!4 & \text{(dry year)} \end{array} \qquad (2.5)$$

low-hill   valley   maximum
fields   bottom   yield
         fields

where x_1 is the proportion of planting in low-hill fields and x_2 is the proportion of planting in valley-bottom fields. Equations (2.5) can be written in matrix form. The parameters on the left-hand side (31, 6, 8, and 50) may be expressed as a 2 × 2 matrix which we shall call **A**; the x's may be expressed as a 2 × 1 column vector **x**, and the right-hand side may also be expressed as a 2 × 1 column vector, **g**

$$\begin{bmatrix} 31 & 6 \\ 8 & 50 \end{bmatrix} \begin{bmatrix} x_1 \\ x_2 \end{bmatrix} = \begin{bmatrix} 22\!\cdot\!4 \\ 22\!\cdot\!4 \end{bmatrix} \qquad (2.6)$$

  A   **x** = **g**
parameters

The reader should verify that the parameter matrix **A**, post-multiplied by the vector **x** yields the left-hand side of equations (2.5), that is

$$\begin{bmatrix} 31 & 6 \\ 8 & 50 \end{bmatrix} \begin{bmatrix} x_1 \\ x_2 \end{bmatrix} = \begin{bmatrix} 31x_1 + 6x_2 \\ 8x_1 + 50x_2 \end{bmatrix}$$

We now need to solve equations (2.6) for x_1 and x_2 since this will tell us what proportion of pumpkins to plant in low-hill and valley-bottom fields.

If we had a single equation of the form

$$ax = b$$

it would be solved by multiplying both sides by $1/a$. We should have

$$\frac{1}{a} \times ax = \frac{1}{a} \times b$$

$$1x = \frac{b}{a}$$

$$\therefore \qquad x = b/a.$$

The essential features of this seemingly trivial solution are that: $1x = x$, the number 1 having the property of, when being multiplied by any value, leaving the value unchanged. In multiplication of numbers, 1 is known as the identity element; $(1/a) \times a = 1$ and $1/a$ is the inverse of a commonly denoted by a^{-1}. These same principles apply in solving simultaneous equations such as equations (2.6). We need to find a matrix which, when multiplied by another matrix, leaves the other matrix unchanged — the identity matrix is such a one. Next we must find a matrix which when multiplied by the parameter matrix **A** yields the identity matrix — the *inverse* of the parameter matrix, written $\mathbf{A}^{-1}$, fits the bill. Before proceeding to solve the equations we must find $\mathbf{A}^{-1}$.

The inverse of **A** is defined so that

$$\mathbf{A}^{-1}\mathbf{A} = \mathbf{I},$$

that is,

$$\mathbf{A}^{-1} \begin{bmatrix} 31 & 6 \\ 8 & 50 \end{bmatrix} = \begin{bmatrix} 1 & 0 \\ 0 & 1 \end{bmatrix}$$

In other words, **A**, pre-multiplied by its inverse, yields the identity matrix. $\mathbf{A}^{-1}$ will be the same size as **A**, in our case a 2 x 2 matrix. To find the inverse of a matrix, if it exists, a series of matrices are used which, step by step, change **A** into the identity matrix. Short-cutting this procedure we may state that the inverse of 2 x 2 matrix **A**,

$$\mathbf{A} = \begin{bmatrix} a_{11} & a_{12} \\ a_{21} & a_{22} \end{bmatrix}$$

is given by

$$\mathbf{A}^{-1} = \frac{1}{a_{11}a_{22} - a_{12}a_{21}} \begin{bmatrix} a_{22} & -a_{12} \\ -a_{21} & a_{11} \end{bmatrix}. \tag{2.7}$$

To derive this three changes are made to **A**. Firstly, the elements on the main diagonal are interchanged. Secondly, the sign of the elements on the other diagonal are reversed. Thirdly, the matrix is pre-multiplied by a factor $1/(a_{11}a_{22} - a_{12}a_{21})$, the bracketed portion of which is known as the determinant of the matrix. In our case

$$A = \begin{bmatrix} 31 & 6 \\ 8 & 50 \end{bmatrix}.$$

Swapping entires on the main diagonal gives

$$\begin{bmatrix} 50 & 6 \\ 8 & 31 \end{bmatrix}.$$

Changing signs on the other diagonal produces

$$\begin{bmatrix} 50 & -6 \\ -8 & 31 \end{bmatrix}.$$

The determinant is $(31 \times 50) - (6 \times 8) = 1502$, and so

$$A^{-1} = \frac{1}{1502} \begin{bmatrix} 50 & -6 \\ -8 & 31 \end{bmatrix}$$

$$= \begin{bmatrix} \dfrac{50}{1502} & \dfrac{-6}{1502} \\ \dfrac{-8}{1502} & \dfrac{31}{1502} \end{bmatrix}.$$

This can be checked using the relation $A^{-1} A = I$; we have

$$\begin{bmatrix} \dfrac{50}{1502} & \dfrac{-6}{1502} \\ \dfrac{-8}{1502} & \dfrac{31}{1502} \end{bmatrix} \begin{bmatrix} 31 & 6 \\ 8 & 50 \end{bmatrix} = \begin{bmatrix} 1 & 0 \\ 0 & 1 \end{bmatrix}.$$

Finding the inverse of larger matrices requires recourse to a computer because the calculations become cumbersome and time-consuming.

Having found A^{-1}, we can now solve the matrix equation $Ax = g$ (equations 2.6). The steps are as follows. Firstly, multiply both sides of the matrix equation by A^{-1}

$$A^{-1} Ax = A^{-1} g.$$

Secondly, since $A^{-1} A = I$, we may write

$$Ix = A^{-1} g.$$

Thirdly, since $Ix = x$, we have

$$x = A^{-1} g.$$

Substituting for $\mathbf{A}^{-1}$ and $\mathbf{g}$ for our case gives

$$\begin{bmatrix} \dfrac{50}{1502} & \dfrac{-6}{1502} \\[2ex] \dfrac{-8}{1502} & \dfrac{31}{1502} \end{bmatrix} \begin{bmatrix} 22{\cdot}4 \\ 22{\cdot}4 \end{bmatrix} = \begin{bmatrix} x_1 \\ x_2 \end{bmatrix}.$$

So by pre-multiplying $\mathbf{g}$ by $\mathbf{A}^{-1}$ we shall find the values of $\mathbf{x}$ which is what we are seeking:

$$\begin{bmatrix} \dfrac{50}{1502} & \dfrac{-6}{1502} \\[2ex] \dfrac{-8}{1502} & \dfrac{31}{1502} \end{bmatrix} \begin{bmatrix} 22{\cdot}4 \\ 22{\cdot}4 \end{bmatrix} = \begin{bmatrix} 0{\cdot}656 \\ 0{\cdot}343 \end{bmatrix}.$$

Therefore $x_1 = 0{\cdot}656$ and $x_2 = 0{\cdot}343$ which means that the plantation owner should plant 65·6 acres of pumpkins in low-hill fields and 34·3 acres in valley-bottom fields.

FUNCTIONS

If X is a set of values, each of which is, say, distance from a city centre; and Y is a set of values, each of which is, say, population density; then a *function* is defined as a rule or relation which associates any element of set X, distance from city centre, with an element of set Y, population density. Using symbols we may write

$$y = f(x), \tag{2.8}$$

where f stands for *some function of*. Instead of the letter f, other letters are sometimes used including g and ϕ. Equation (2.8) is read as y is some function of x; in our example population density is some function of distance from a city centre. Mathematically, the equation tells us that a variable quantity y depends on a variable

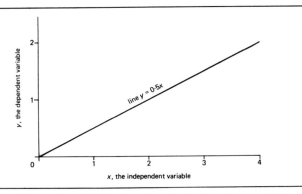

Figure 2.5. A graphical plot of a simple function, $y = 0{\cdot}5x$.

quantity x, such that for any given value of x the value of y is fixed. Let us suppose that the function f were *multiplication by one half*; in this instance we should write

$$y = 0 \cdot 5x.$$

Clearly, if x is 1 then y is $0 \cdot 5 \times 1 = 0 \cdot 5$; if x is 2 then y is $0 \cdot 5 \times 2 = 1$; and so forth. The function may be plotted on sectional paper with x on the horizontal axis and y on the vertical axis. The resulting curve, which is defined by the function, shows how y varies as x varies (figure 2.5). This helps to explain why x is called an *independent variable* and y is called a *dependent variable*. In the example given, the graph is a straight line so the equation $y = 0 \cdot 5x$ is an example of a linear function. Many other types of function are used in geography and we shall now explore them.

Linear functions

In the general case, the equation of a line may be written

$$y = a \pm bx. \tag{2.9}$$

This equation has two terms — a and b — which we have not met before. For any particular line, these terms have just one value each and are therefore called *constant coefficients* or simply *constants*. Looking at figure 2.6 we can see that the value of a tells us where the line crosses or intercepts the y axis; in other words, it is the value of y when x is zero; it can be positive, zero, or negative. The value of b tells us the amount by which y increases or decreases per unit increase in x; in other words, it is the *slope* or *gradient* of the line. If b is thought of as the tangent of the angle θ (figure 2.6), then with $b = 1$, the line will have an upwards slope of 45 degrees to the x axis, so that a unit increase in x will be matched by a unit increase in y: the gradient is 1-in-1 (figure 2.7). With b greater than 1, y will increase more rapidly than x increases, so the line will be at an angle in excess of 45 degrees. With b less than 1, y will increase

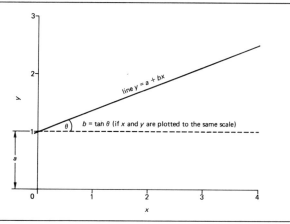

Figure 2.6. The general equation for a line, $y = a + bx$. Note that a is the value of y when x is zero.

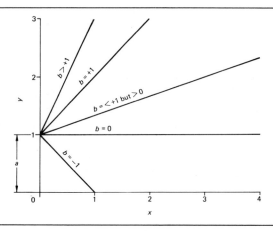

Figure 2.7.   Various lines, all with the same intercept value a, but with different b parameters. Notice that if b is positive the line slopes up; if b is zero the line is horizontal; and if b is negative the line slopes down.

less rapidly than x increases so the line will slope at a lesser angle than 45 degrees. The b coefficient will be negative in those cases where y decreases per unit increase in x and the line slopes down.

Although for any specific linear function a and b are constant, they may vary for specific cases; this being so, they are termed *parameters*. A concrete example will clarify this point. It was shown by Von Thünen that locational rent of agricultural land (defined as gross income less all production and transport costs) in a uniform area surrounding a central market town is a linear function of distance from the market. If we call the locational rent y and distance x, the general relation is described by the function

$$\underset{\substack{\text{locational}\\\text{rent}}}{y} = \underset{\text{parameters}}{a - b} \quad \underset{\text{distance}}{x}.$$

(2.10)

The parameters in this equation can be given a real meaning. The parameter a, being the value of the locational rent at the market, is defined as crop yield, L, times the market price, P, less the crop yield times the production cost, C

$$a = LP - LC$$

$$= L(P - C).$$

In effect, this is the maximum locational rent in the area because a farm situated anywhere but immediately by the market will incur transport costs. The parameter b determines the rate at which locational rent declines from the market owing to transport costs. It is defined as the product of crop yield, L, transport rate, F, and distance, x:

$$b = LFx.$$

Substituting the expressions for a and b into equation (2.10) we get

$$y = L(P - C) - LFx.$$

Table 2.3. Values for vegetables and wheat.

Crop	Crop yield, *L*	Market price *P*†	Production cost, *C*†	Transport cost, *F*‡
Vegetables	1	50	25	2·5
Wheat	1	12	6	0·5

†£ per hectare of production.
‡£ per hectare of production per kilometre.

As crop yield, production costs, market price, and transport costs are different for different crops, the parameters too vary with crop type. Take a hypothetical example for vegetables and wheat. Yields, costs, and market prices are shown in table 2.3. The locational rent for vegetables at the market ($x = 0$) is given by $L(P - C)$ and, as shown in figure 2.8, this comes to £25 per hectare. Similarly, the locational rent for wheat at the market is £6 per hectare. Farms located some distance from the market incur transport costs in taking wheat and vegetables to market. The further from market, the greater the transport costs. For vegetables the transport costs are £2.50 per hectare of production per kilometre; so, as we are considering the yield for one hectare, for every kilometre the farm is from the market, £2.50 is deducted from the locational rent at the market. For farms located 10 km from the market the locational rent for vegetables is zero. Were a farmer to grow vegetables beyond this point he would stand to lose money. For wheat the transport costs are 50p per hectare of production per kilometre, and so wheat may be profitably produced up to 12 km from the market.

Notice that, in terms of the general linear equation (equation 2.9), the locational rent at the market is the *a* parameter – the intercept of the line with the *y* axis. And the slope of the line (the transport costs) is the *b* parameter. The larger the *b* value

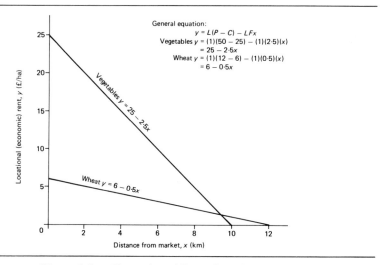

Figure 2.8.   Examples of linear functions.

(transport costs), the more rapid the decline in locational rent with increasing distance from the market and the less widespread the zone of profitable production.

Logarithmic functions

Logarithmic functions are possibly the most widely applied functions in geography. To understand the significance of logarithmic functions it is useful to have a grasp of what a logarithm is. Though it may at first seem unintelligible to the non-mathematician, a formal definition of a logarithm is a helpful starting point: the logarithm of x is the exponent or the power to which a number constant, a, called the base, must be raised in order to produce the natural number or antilogarithm, x. Thus, in the equation

$$a^n = x,$$

a is the base, n is the exponent or logarithm of x, and x is the natural number. Any positive number other than zero or one can be used as the base. Most readers will be familiar with common or Briggsian logarithms which have ten as a base. So we have

$$10^n = x.$$

Letting $n = 2$,

$$10^2 = x$$

and clearly, therefore, $x = 100$. To get the natural number 100, ten has to be raised to the power 2. Therefore, by our definition of a logarithm, log 100 = 2. Likewise, the logarithm of 1 000 000 is the number by which 10 has to be raised to give one million; this is of course 6, since 10^6 is 1 000 000. Not all exponents are whole numbers. The logarithm of 50, for example, will be a decimal number in the range between 1 (that is, log 10) and 2 (that is, log 100). Using log tables we find that log 50 is in fact 1·69897.

It will be noticed that logarithms of numbers increase in a regular way:

log 1   = 0
log 10   = 1
log 100 = 2
log 1000 = 3, and so on.

Logarithms thus give a scale of measurement. In changing from a length of 1 to a length of 4 in logarithmic units, the length is increased by three logarithmic steps; this, rather than being a threefold increase in length, is a 10^3 or thousandfold increase in length. Decimal point values create no problems: an increase from 2·5 log units to 3·2 log units is an increase of 0·7 log units or a $10^{0·7}$-fold increase. Logarithmic scales can be very useful because they have the effect of contracting wide-ranging orders of magnitude to more manageable proportions. Geographers are interested in areas ranging from the surface area of the Earth, through continental areas, national areas and so on, down to a few square metres or, in physical geography, far less. One means of accommodating this vast range of areas is the application of a logarithmic scale of area. The zero point on the scale is the surface area of the earth, G_a, which is $5·1 \times 10^8$ km². The next step down on the scale is $(5·1 \times 10^8/10)$ which is $5·1 \times 10^7$ km²; the next step down is $5·1 \times 10^6$ km²; and so forth. Each of these steps

Table 2.4. Derivation of G-values.

G-value	Subdivision of Earth's surface	Area (10^8 km^2)
0	$G_a \times 10^0$	5·1
1	$G_a \times 10^{-1}$	0·51
2	$G_a \times 10^{-2}$	0·051
3	$G_a \times 10^{-3}$	0·0051
n	$G_a \times 10^{-n}$	5·1 $\times 10^{-n}$

is a reference point on the scale; 1 is $5·1 \times 10^7$, 2 is $5·1 \times 10^6$, 3 is $5·1 \times 10^5$, and so on. The log steps are called G-values (Haggett *et al* 1965). G-values form a logarithmic scale of measurement – the G-scale. Notice that the area decreases tenfold at every step (table 2.4).

To find the G-value of an area, R_a, the following formula may be used

$$G = \log (G_a/R_a),$$

where both the area under investigation, R_a, and the area of the Earth's surface, G_a, are in the same units. Thus to find the G-value of 100 km^2 we have

$$G = \log (5·1 \times 10^8/100)$$

$$= \log (5·1 \times 10^6)$$

$$= \log (5\ 100\ 000)$$

$$= 6·7074.$$

Alternatively, the G-value may be found by subtracting the log of the area under investigation from the log of the Earth's surface area (log $G_a = 8·7074$ when G_a is measured in square kilometres); this yields

$$G = 8·7074 - \log R_a$$

$$= 8·7074 - \log 100$$

$$= 8·7074 - 2·0$$

$$= 6·7074.$$

G-scale values can be converted back to conventional units in the following manner. To convert the G-value 6·7074 to square kilometres

$$6·7074 = 8·8074 - \log R_a.$$

Therefore

$$\log R_a = 8·7074 - 6·7074$$

$$= 2·0.$$

$$R_a = \text{antilog } 2·0$$

$$= 100 \text{ km}^2.$$

Examples of logarithmic scales used by geographers include the pH scale, which measures soil acidity, and the phi-scale, which measures the diameter of sedimentary particles.

Double-log functions. One of the most common logarithmic functions used by geographers is the *double-log* or *log-log* function. This function is an equation in which the logarithm of the dependent variable, $\log y$, is expressed as a function, usually a linear one, of the logarithm of the independent variable, $\log x$. In the general case

$$\log y = \log a \pm b \log x. \tag{2.11}$$

For example, the central area retail floorspace (m^2) in 12 British cities, y, is related to city population, x, by the following double-log function:

$$\log y = \log 0 \cdot 609 + 0 \cdot 671 \log x. \tag{2.12}$$

Double-log functions are also expressible as power functions which are obtained by delogarizing equation (2.11) to give

$$y = ax^b. \tag{2.13}$$

In power notation, equation (2.12) becomes

$$y = 0 \cdot 609 x^{0 \cdot 671} \tag{2.14}$$

If equation (2.12) is plotted on arithmetic sectional paper, the log of x and y being looked up in log tables, a straight line will result which intercepts the y axis at $y = \log 0 \cdot 609$ and has a slope $b = 0 \cdot 671$. Alternatively, equation (2.14) can be used to plot the function. In this case a plot on arithmetic sectional paper would produce a curve (figure 2.9a). To get a straight line, the function must be plotted on log-log graph paper (figure 2.9b) which converts to log values for you. The parameter a in the power function, 0·609, does not represent the intercept value but the value of y when $x = 1$; this follows because with $x = 1$ in equation (2.13),

$$y = a1^b,$$

and since 1 raised to any power equals 1

$$y = a.$$

To plot the function given by equation (2.14), it is necessary to evaluate $x^{0 \cdot 671}$ for various values of x; this can be done using log tables. Thus, for $x = 1\ 000\ 000$ we have

$$\log 1\ 000\ 000^{0 \cdot 671} = 6 \times 0 \cdot 671$$
$$= 4 \cdot 026.$$

Taking the antilog of 4·026 gives the value we seek:

$$\text{antilog } 4 \cdot 026 = 10\ 617.$$

Substituting 10 617 in equation (2.14) yields

$$y = 0 \cdot 609 \times 10\ 617$$
$$= 6465 \cdot 7.$$

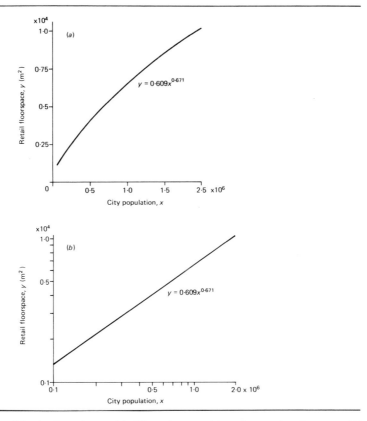

Figure 2.9. Logarithmic functions. (*a*) Plotted on arithmetic sectional paper. (*b*) Plotted on double-log paper.

Other values of y can be found in the same way for a range of x values and the curve built up. The b parameter in the power equation (2.13) determines the form of the curve. If b equals 1, the curve will be a straight line; if b is less than 1 the curve will bend down, a unit increase in x begin met by a less than unit increase in y; if b is greater than 1, the curve will bend up, a unit increase in x being met by a more than unit increase in y.

In some cases it is useful to plot the logarithm of a variable against the natural number of another; such a relation is a *semi-logarithmic* function, an example of which will be given in the next section.

Exponential functions. Another form of logarithmic function is the *exponential function*. The most important exponential function in geography takes the general form

$$y = ae^{bx} \tag{2.15}$$

where y is the dependent variable, x the independent variable, a and b are parameters, and e is the base of the Naperian or hyperbolic logarithms – 2·71828 (more or less).

The point of having a system of logarithms with a base of 2·71828 may elude the non-mathematician. To explain the significance of e we may think of the growth of invested money. If £1 were to grow at simple interest then the increase at any instant is always proportional to the original sum of £1 invested. Assume the rate of interest is 100 per cent per year. Clearly, one year after the investment was first made the £1 will have grown to £2. With compound interest, the increase at any instant is proportional to the total sum of money at that instant. So growing at 100 per cent per year in a compound fashion, after one year the £1 will have grown to £2·71828. Thus the value e relates to compound or proportional growth processes, which are not just found in finance, but in many other growth and spatial processes in geography. It is these processes that may be described by exponential functions.

Take the example of the decline of population density away from a city centre; this may be described by the function

$$y = ae^{-bx},$$

where y is population density and x is distance from the city centre. The parameter a is the value of y when $x = 0$; in other words, it is the population density at the city centre and may be designated y_0; so

$$y = y_0 e^{-bx}.$$

The parameter b determines the rate of decline of population density with distance from the city centre; the larger its value, the steeper will be the population density gradient. If the population density at the city centre, y_0, is 100 persons per hectare and $b = 0·3$, we have

$$y = 100 \, e^{-0·3x}.$$

To evaluate the population density, y, at successive distances from the centre of the city, tables of exponential functions can be used (or the appropriate function button

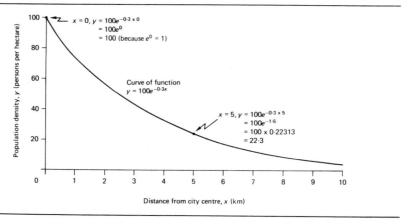

Figure 2.10. Population density against distance from city centre as described by the function $y = 100e^{-0·3x}$. Shown on the graph are the calculations for $x = 0$ and $x = 5$.

on a calculator can be pressed). For instance, to find y when x is 5 km, we have

$$y = 100 \, e^{\,(-0\cdot3)(5\cdot0)}.$$

Multiplying the two exponents,

$$y = 100 \, e^{-1\cdot5}.$$

Looking up $e^{-1\cdot5}$ in exponential tables (Appendix A) we find the value 0·22313. Substituting this value in the previous equation gives

$$y = 100 \times 0\cdot22313$$

$$= 22\cdot313 \text{ persons per hectare at 5 km.}$$

Values of population density at any other distance can be found in like manner and the function plotted (figure 2.10).

If the natural logarithm of population density were plotted against distance a straight line would result; equation (2.15) may also be thought of therefore as a semi-logarithmic function.

Higher-order functions

If we take the function $y = a + bx$ and add to it another term, cx^2, where c is another parameter, we get

$$y = a + bx + cx^2, \tag{2.16}$$

which function will describe a curve. This is clearly so because, providing c is not equal to zero, cx^2 will increase with increasing x in a geometrical rather than an arithmetical manner. Since equation (2.16) does not define a straight line, it is a *nonlinear equation*; and because it contains no powers of x above x^2 it is a *quadratic equation*. Such equations, though not unknown in human geography, are more commonly found in physical geography. The U-shaped section of glaciated valleys can be described by a quadratic function in which the parameters a and b are zero, so equation (2.16) reduces to

$$y = cx^2,$$

where y is height of the land surface and x is distance. The value of x is taken as zero at the lowermost point in the valley cross profile (figure 2.11). The parameter c varies from one locality to another, the one for figure 2.11 being 0·05; the smaller the value of c, the broader and less steep sided the U-shaped valley. And another example, it has been found that the hydraulic conductivity (the rate at which water can move through the soil), y, in the Monona silt loam is defined by the following quadratic function of volumetric soil moisture content, x:

$$y = 46\cdot281 - 265\cdot913x + 379\cdot856x^2 \text{ (cm/h).}$$

Should another term, dx^3, be added to a quadratic equation we will have

$$y = a + bx + cx^2 + dx^3. \tag{2.17}$$

This equation again describes a curve, but a curve with two kinks in it (as opposed to

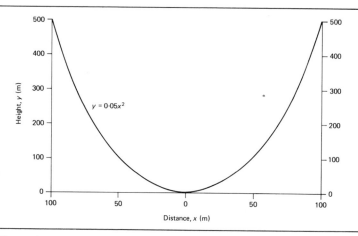

Figure 2.11.   The curve of a glaciated, U-shaped valley: $y = 0.05\, x^2$.

the quadratic equation which has just one kink). As no power of x greater than x^3 appears in equation (2.17) it is a *cubic equation*. By successively adding terms $-ex^4$, fx^5, gx^6, and so on — *quartic, quintic, sextic* equations are formed which describe successively more kinky curves. Geographers are seldom concerned with these save in the case of describing slope profiles. Humphreys (1973) fitted quartic curves to cross-valley profiles in Great Langdale in the Lake District of England; one of his equations is

$$h = 4{\cdot}31601 - 0{\cdot}1162282x + 0{\cdot}00702328x^2 + 0{\cdot}0000001\,1665x^3$$
$$+ 0{\cdot}000000000279x^4,$$

where h is height of the land surface and x is distance. The parameters d and e are small because x^3 and x^4 may be big numbers.

Higher-order exponential functions can be used to give more realistic descriptions of population density decline from a city centre. The function used earlier in the chapter (see figure 2.10) was

$$y = y_0 e^{-bx} \tag{2.18}$$

but this equation makes no allowance for the population density crater characteristic of many cities. A nonlinear form of equation (2.18) gives a better fit to observed variations in population density. For instance, we might define population density decline as proportional to the square of distance from city centre to give the *quadratic exponential equation*

$$y = y_0 e^{-cx^2}, \tag{2.19}$$

the general form of which is shown in figure 2.12(c). Taking natural logs (written $\log_e$ or $\ln$) of both sides of equation (2.19) gives an alternative way of writing it:

$$\ln y = \ln y_0 - cx^2.$$

Even this function does not describe the fall in population density near the city centre.

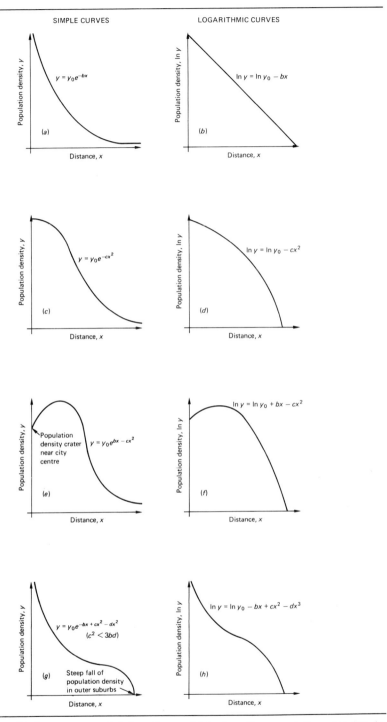

SIMPLE CURVES

LOGARITHMIC CURVES

(a) $y = y_0 e^{-bx}$

(b) $\ln y = \ln y_0 - bx$

(c) $y = y_0 e^{-cx^2}$

(d) $\ln y = \ln y_0 - cx^2$

(e) Population density crater near city centre $y = y_0 e^{bx - cx^2}$

(f) $\ln y = \ln y_0 + bx - cx^2$

(g) $y = y_0 e^{-bx + cx^2 - dx^2}$ $(c^2 < 3bd)$ Steep fall of population density in outer suburbs

(h) $\ln y = \ln y_0 - bx + cx^2 - dx^3$

Population density, y / Distance, x

Figure 2.12. Various exponential functions which describe population density decline away from a city centre. (Based on A M Warnes 1975.)

We might therefore look at a more complex version (figure 2.12(*e*)):

$$y = y_0 e^{bx - cx^2},$$

or what is the same thing (figure 2.12(*f*)):

$$\ln y = \ln y_0 + bx - cx^2,$$

which does seem to describe better observed patterns of population density from a city, providing the parameter *b* is greater than zero, and the parameter *c* is less than zero (that is, negative). Warnes (1975) fitted the following function for population density variations in Manchester in 1966

$$y = 2 \cdot 32 e^{0 \cdot 128x - 0 \cdot 01x^2},$$

where *y* is population density in persons per hectare and *x* is in kilometres. For the case of employment density gradients, which decline increasingly slowly from a city centre but then drop precipitously in the outer suburbs, Warnes (1975) found that

$$y = y_0 e^{-bx + cx^2 - dx^3}.$$

a cubic exponential equation (figure 2.12(*g*)), also written

$$\ln y = \ln y_0 - bx + cx^2 - dx^3,$$

was suitable so long as the parameter *c* was less than the value $\sqrt{(3bd)}$. For employment in 1966 in Manchester, the relation was described by the function

$$y = 5 \cdot 59 e^{-0 \cdot 828 + 0 \cdot 067x^2 - 0 \cdot 002x^3},$$

where *y* is employment density in persons per hectare and *x* is in kilometres.

RATES OF CHANGE IN SPACE AND TIME

Many geographical models of the mathematical variety, especially deterministic ones, deal with the *rates of change* of variables in space and time. An entire branch of mathematics — differential calculus — concerns itself with rates of change. We shall in this section outline those facets of differential calculus which are needed to understand some of the models discussed later in the book. The mathematical principles are demonstrated with slopes, surfaces, and populations because these will be more familiar to students of geography than the abstract quantities dealt with in most mathematical tomes.

Slope profiles

We have met the functional notation $y = f(x)$ and established that this is a shorthand way of saying that a variable quantity *y* depends upon a variable quantity *x* in so far as, for any given value of *x*, the value of *y* is fixed. In the case of a slope profile, let *x* be horizontal distance and *h* the height of the land surface. The curve $h = f(x)$ will describe the way in which the height of the land surface varies with distance; in other words, it will describe a slope profile.

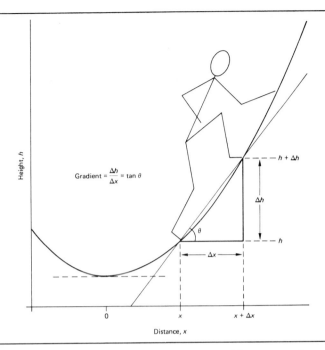

Figure 2.13. The gradient of a valley-side slope profile expressed as finite increments. (NB. The gradient will equal tan θ only if height and distance are plotted at the same scale.)

For the time being, the exact relation between h and x need not concern us. Now, if a hill-walker were to start at the valley bottom and move up the right-hand slope then, as x increased, so would h – he would rise by a certain height. A measure of how fast h increased per unit increase in x is the *slope* or *gradient*. Graphically speaking, the gradient is given by the tangent of the angle θ (see figure 2.13). Clearly, the steeper the gradient, the more sensitive h is to changes in x – climbing the hill-side becomes more strenuous. Note that where the curve reaches its minimum value, that is, at the valley bottom, the gradient is zero (horizontal) and h does not change with distance at this point. Similarly, the gradient is zero where the curve attains its maximum value, that is, on the watershed. The minimum and maximum points on the curve are known as stationary points. The gradient may be measured as the ratio between the height increment for one step taken and the horizontal distance covered by that step and, unless the slope profile is straight, this ratio will change from point to point. To give a more precise definition of the gradient, let us call the horizontal increment of distance taken in one step Δx, and the corresponding increment of height Δh (figure 2.13). The Δ's mean 'a little bit of', so that Δx means a little bit of x, the total horizontal extent of the slope profile, and Δh means a little bit of h, the difference in height between the valley bottom and the watershed. Remembering that the gradient is the height increment divided by the distance increment we can write it as:

$$\text{gradient} = \frac{\text{height increment}}{\text{distance increment}} = \frac{\Delta h}{\Delta x}, \qquad (2.20)$$

which expresses the fact that, to find the gradient for any section of the slope profile a little bit of h is to be divided by a little bit of x.

Finding the gradient. The slope profile is a continuous line. However most survey methods cannot cope with this continuous variation of height with distance. Instead, a series of spot heights are usually measured at regular distance intervals. It is from this series of heights that the approximate gradient may be calculated. A slope profile may have a horizontal extent of say 35 distance units. The height of the land surface along this profile has been surveyed at 8 points, the spacing between points being a small distance Δx. Any spot height along the profile can be identified by a label; if there are n distance increments in all, then any point along the x axis can be designated x_i and the spot height corresponding to this point may be written h_i. The points either side of x_i are x_{i-1} and x_{i+1}, and the heights at these points are h_{i-1} and h_{i+1}, respectively (figures 2.14). So if the point in question is x_4, then the height at that point is h_4 and the heights of adjacent points are h_3 and h_5. By definition, $\Delta x = x_i - x_{i-1}$. To obtain the approximate gradient of the slope profile at point x_i, the height increment over the next distance increment can be used. If a hill-walker should take one step of horizontal distance Δx from x_i, he will arrive at point x_{i+1} and will rise to height h_{i+1}. The difference in height between points x_i and x_{i+1}, Δh, is given by $h_{i+1} - h_i$. The gradient is the height increment divided by the distance increment so we may write

$$\text{gradient} = \frac{\Delta h}{\Delta x} = \frac{h_{i+1} - h_i}{\Delta x}. \tag{2.21}$$

Substituting the example heights shown in figure 2.15, that is $h_i = 2$ dm, $h_{i+1} = 4$ dm

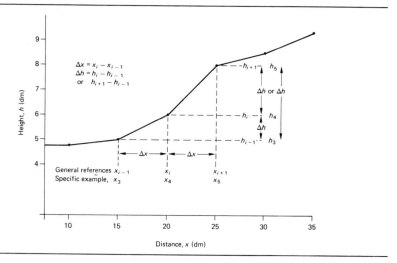

Figure 2.14.   Reference system for points on a valley-side slope.

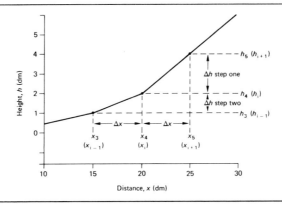

Figure 2.15. Spot heights required to calculate slope curvature at point x_4.

and $\Delta x = 5$ dm,

$$\frac{\Delta h}{\Delta x} = \frac{h_{i+1} - h_i}{\Delta x} = \frac{h_5 - h_4}{\Delta x} = \frac{4 - 2}{5} = + 0{\cdot}4.$$

The gradient is thus $+0{\cdot}4$; this value is a tangent and is approximately 22 degrees. In some applications, the approximate gradient of the slope profile through the point x_i is found by using the height increment between points x_{i-1} and x_{i+1}; this means $\Delta h = h_{i+1} - h_{i-1}$. In this case the height difference has been taken over two distance increments so

$$\frac{\Delta h}{\Delta x} = \frac{h_{i+1} - h_{i-1}}{2\Delta x} = \frac{h_5 - h_3}{2\Delta x} = \frac{4 - 1}{10} = +0{\cdot}3. \qquad (2.22)$$

which is about 17 degrees. Notice that the two methods of obtaining the approximate gradient give slightly different results.

Curvature. As we have seen, $\Delta h/\Delta x$ is the gradient of a curve. Unless the curve happens to be straight, the gradient may increase or decrease with distance and such changes of gradient are defined in terms of *slope curvature*.

To understand what curvature is we may resort to hill-walking again. Take first the case of a hill-walker ascending the slope profile shown in figure 2.15. All in all we shall need to consider the height at three points: h_{i-1}, the height at which we first meet the hill-walker; h_i, the height after one step; and h_{i+1}, the height after two steps. Starting at point x_{i-1}, the hill-walker takes a step uphill which carries him forward a horizontal distance Δx and takes him to point x_i. In doing so he will have risen a height $\Delta h = h_i - h_{i-1}$. The gradient of the small section of slope covered by his step is, as always, the height increment divided by the distance increment, that is

$$\left.\frac{\Delta h}{\Delta x}\right|_{\text{step one}} = \frac{h_i - h_{i-1}}{\Delta x}. \qquad (2.23)$$

Substituting the example values from figure 2.15,

$$\frac{\Delta h}{\Delta x}\bigg|_{\text{step one}} = \frac{h_4 - h_3}{\Delta x} = \frac{2-1}{5} = 0 \cdot 2,$$

or roughly 11 degrees. After taking a second step of horizontal distance Δx, the hill-walker will have between points x_i and x_{i+1} risen a height $\Delta h = h_{i+1} - h_i$. Therefore the gradient of the section of slope covered by his second step is

$$\frac{\Delta h}{\Delta x}\bigg|_{\text{step two}} = \frac{h_{i+1} - h_i}{\Delta x} \tag{2.24}$$

which, using values from figure 2.15, is

$$\frac{\Delta h}{\Delta x}\bigg|_{\text{step two}} = \frac{h_5 - h_4}{\Delta x} = \frac{4-2}{5} = +0 \cdot 4,$$

or approximately 22 degrees. Now in this example the gradient of the ground covered by the first step is less than the gradient of the ground covered by the second step: in moving through point x_4 the gradient changes. Slope curvature measures the amount of change in the gradient. So how shall we define curvature? The method is to evaluate the difference between the two gradients either side of point x_4. Specifically, the second gradient formula (2.24) is subtracted from the first gradient formula (2.23): $\Delta h / \Delta x \big|_{\text{step two}}$ is subtracted from $\Delta h / \Delta x \big|_{\text{step one}}$. The difference is

$$\frac{(h_{i+1} - h_i)}{\Delta x} - \frac{(h_i - h_{i-1})}{\Delta x} = \frac{h_{i+1} - 2h_i + h_{i-1}}{\Delta x}.$$

Because this difference results from moving a distance Δx uphill, to find the rate of change of the gradient the expression must be divided by Δx. Doing so we obtain

$$\frac{h_{i+1} - 2h_i + h_{i-1}}{\Delta x\, \Delta x} = \frac{h_{i+1} - 2h_i + h_{i-1}}{(\Delta x)^2}. \tag{2.25}$$

This expression defines the approximate slope curvature through point x_i and thus tells us how the slope profile is bending. Substituting the example values in figure 2.15, we get

$$\frac{\Delta^2 h}{\Delta x^2} = \frac{h_{i+1} - 2h_i + h_{i-1}}{(\Delta x)^2} = \frac{h_5 - 2h_4 + h_3}{(\Delta x)^2} = \frac{4-4+1}{5^2} = \frac{1}{25} = 0 \cdot 04,$$

where $\Delta^2 h / \Delta x^2$ is a general expression for curvature.

Slope gradient and slope curvature may have positive, negative, or zero values. A slope with a positive gradient runs uphill in the x direction; a slope with a negative gradient runs downhill in the x direction; a slope with zero gradient is horizontal. For both uphill and downhill cases, concave slopes have positive curvature, convex slopes have negative curvature, and straight slopes have zero curvature. Combinations of slope gradient and slope curvature are depicted in figure 2.16.

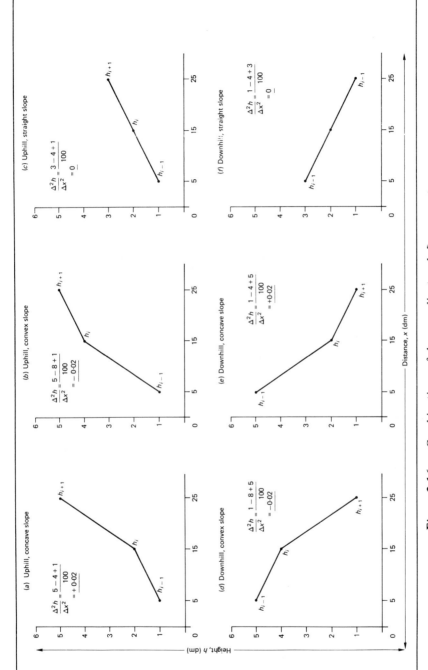

(a) Uphill, concave slope

$$\frac{\Delta^2 h}{\Delta x^2} = \frac{5 - 4 + 1}{100} = +0.02$$

(b) Uphill, convex slope

$$\frac{\Delta^2 h}{\Delta x^2} = \frac{5 - 8 + 1}{100} = -0.02$$

(c) Uphill, straight slope

$$\frac{\Delta^2 h}{\Delta x^2} = \frac{3 - 4 + 1}{100} = 0$$

(d) Downhill, convex slope

$$\frac{\Delta^2 h}{\Delta x^2} = \frac{1 - 8 + 5}{100} = -0.02$$

(e) Downhill, concave slope

$$\frac{\Delta^2 h}{\Delta x^2} = \frac{1 - 4 + 5}{100} = +0.02$$

(f) Downhill, straight slope

$$\frac{\Delta^2 h}{\Delta x^2} = \frac{1 - 4 + 3}{100} = 0$$

Height, h (dm)

Distance, x (dm)

Figure 2.16. Combinations of slope gradient and slope curvature.

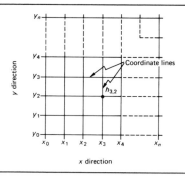

Figure 2.17.   Coordinate grid.

Surfaces

Height varies in directions other than the x direction. In general, the height of the land surface can be represented as a function of two independent variables, x and y, which may be defined as rectangular coordinates like the eastings and northings of a map:

$$h = f(x, y). \tag{2.26}$$

Take x and y to be coordinates of a map and assume that the area covered by the map has been surveyed to obtain spot heights at various points on the land surface. Any spot height will have the general grid reference $h(x, y)$. It is very important for the reader to understand the coordinate notation we shall use to locate and identify spot heights. Figure 2.17 shows how the part of map is divided by a grid of intersecting coordinate lines; just five lines in the x direction and five lines in the y direction are shown. Each coordinate line in the x direction may be referred to by a numerical subscript, x_0 being on the far west of the map, x_1 being parallel to x_0 but distance Δx to the west, x_2 being parallel to x_0 but distance $2\Delta x$ to the west and so on. In general, any coordinate line in the x direction may be designated x_i, where i can be any whole number between 0 and the maximum number of coordinate lines in the x direction; in our example 4, but in the general case n. The coordinate lines in the y direction may be similarly referenced but as $y_0, y_1, y_2, \ldots, y_n$. The spot heights have

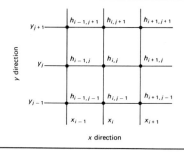

Figure 2.18   General references to nodes in grid about point $h_{i,j}$.

been measured at all the intersections of the x and y direction coordinate lines. The points where the lines intersect, or nodes as they are often called, can be referenced in a similar manner to the way in which map references are given. Thus the point marked h on the diagram lies at the intersection of coordinate line x_3 and coordinate line y_2. We may therefore write $h(x_3, y_2)$ or, as x is always placed first, $h_{3,2}$. In general, the location of any point is $h_{i,j}$ The scheme we shall use to reference the points surrounding any point $h_{i,j}$ is shown in figure 2.18.

The point $h_{i,j}$ lies at the intersection of coordinate lines x_i and y_j. There are eight points which surround $h_{i,j}$. The point due east lies at the intersection of coordinate lines x_{i+1} and y_j; it is therefore labelled $h_{i+1,j}$. The point due west of $h_{i,j}$ lies at the intersection of coordinate lines $x_{i-1,1}$ and y_j; it is therefore labelled $h_{i-1,j}$. The point north-west of $h_{i,j}$ lies at the intersection of coordinate lines x_{i-1} and y_{j+1}; it is therefore labelled $h_{i-1,j+1}$. All other points may be labelled using the same principle. To show how this works out in a specific case the reader should study figure 2.19, which gives the appropriate labels for the points surrounding $h_{3,4}$.

Calculating gradients and curvatures on surfaces. Figure 2.19 shows a small portion of a landscape. The nine spot heights are identified using the $h_{i,j}$ notation. Let us place our tireless hill-walker at point (x_{i-1}, y_j). If he were to take a step in the x direction (due east) which covered a horizontal distance Δx, he would arrive at point (x_i, y_j) and have risen a height $\Delta h = (h_{i,j} - h_{i-1,j})$. The gradient of the land surface between points (x_{i-1}, y_j) and (x_i, y_j) is therefore the height increment in the x direction divided by the distance increment in the x direction, that is,

$$\frac{\Delta h}{\Delta x} = \frac{(h_{i,j} - h_{i-1,j})}{\Delta x}.$$

If the hill-walker were to take another step due east he would arrive at point (x_{i+1}, y_j) and during this step he would have risen a height $\Delta h = (h_{i+1,j} - h_{i,j})$. The gradient in the x direction between points (x_i, y_j) and (x_{i+1}, y_j) is

$$\frac{\Delta h}{\Delta x} = \frac{(h_{i+1,j} - h_{i,j})}{\Delta x}.$$

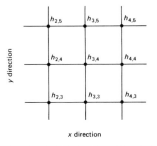

Figure 2.19. Specific references to nodes around point $h_{3,4}$.

The difference between these two gradients is

$$\left.\frac{\Delta h}{\Delta x}\right|_{\text{step two}} - \left.\frac{\Delta h}{\Delta x}\right|_{\text{step one}} = \frac{(h_{i+1,j} - h_{i,j})}{\Delta x} - \frac{(h_{i,j} - h_{i-1,j})}{\Delta x}$$

$$= \frac{h_{i+1,j} - 2h_{i,j} + h_{i-1,j}}{\Delta x}.$$

As we saw earlier (p46), since this difference results from moving a small distance, Δx, uphill, to find the rate of change of gradient, or curvature, in the x direction the expression must be divided by Δx, and this operation gives

$$\frac{\Delta^2 h}{\Delta x^2} = \frac{(h_{i+1,j} - 2h_{i,j} - h_{i-1,j})}{(\Delta x)^2}. \tag{2.27}$$

Using the spot heights given in figure 2.20, the two gradients either side of, and slope curvature through, point (x_i, y_j) are:

$$\left.\frac{\Delta h}{\Delta x}\right|_{\text{step one}} = \frac{(h_{i,j} - h_{i-1,j})}{\Delta x} = \frac{7 - 6}{5} = +0 \cdot 2$$

$$\left.\frac{\Delta h}{\Delta x}\right|_{\text{step two}} = \frac{(h_{i+1,j} - h_{i,j})}{\Delta x} = \frac{9 - 7}{5} = +0 \cdot 4$$

$$\frac{\Delta^2 h}{\Delta x^2} = \frac{(h_{i+1,j} - 2h_{i,j} + h_{i-1,j})}{(\Delta x)^2} = \frac{9 - 14 + 6}{5^2} = +\frac{1}{25} = +0 \cdot 04.$$

If the hill walker were now to start from point (x_i, y_{j-1}) and move to point (x_i, y_j); in other words, take one step of horizontal distance Δy in the y direction (due north), he would rise a height $\Delta h = h_{i,j} - h_{i,j-1}$. Thus the gradient of the land surface between points (x_i, y_{j-1}) and (x_i, y_i) is

$$\left.\frac{\Delta h}{\Delta y}\right|_{\text{step one}} = \frac{h_{i,j} - h_{i,j-1}}{\Delta y} = \frac{7 - 6}{5} = +0 \cdot 2.$$

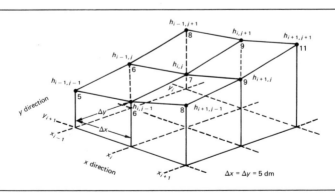

Figure 2.20.   An example of a surface. Heights are in decimetres.

If he were to take a second step due north from point (x_i, y_j) of horizontal distance Δy, he would arrive at point (x_i, y_{j+1}) and would rise a height $\Delta h = h_{i,\,j+1} - h_{i,\,j}$. Thus the gradient of the land surface between points (x_i, y_j) and (x_i, y_{i+1}) is

$$\left.\frac{\Delta h}{\Delta y}\right|_{\text{step two}} = \frac{h_{i,j+1} - h_{i,j}}{\Delta y} = \frac{9-7}{5} = +0{\cdot}4.$$

The slope curvature in the y direction through point (x_i, y_i) is given by the difference in gradients on either side of that point, divided by the distance increment in the y direction, Δy. This gives

$$\frac{\Delta^2 h}{\Delta y^2} = \frac{(h_{i,j+1} - 2h_{i,j} + h_{i,j-1})}{(\Delta y)^2} = \frac{9-14+6}{5^2} = +\frac{1}{25} = +0{\cdot}04.$$

The limiting case: from differences to differentials

The discussion so far has dealt with finite bits of height and distance, and with discrete points along a line or over a map. But in reality height varies continuously along a slope profile and across a surface, and so therefore do gradient and curvature. The method we have used for finding the gradient and curvature does *not* enable us to calculate the gradients and so on at *any point* along a profile or over a surface: we are confined to a finite number of discrete points. How then can we cope with continuous variation of a variable? The easiest way to answer this is to return to the general equation of a slope profile: $h = f(x)$; this equation describes the continuous variation of h with regard to x. The question now is, if we know the functional relation between height and distance, can we derive the gradient and curvature at any point on the curve? The answer, almost

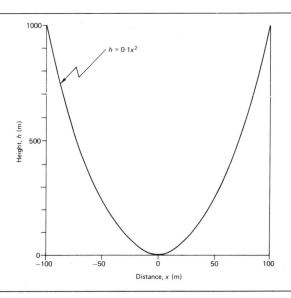

Figure 2.21.   A plot of the curve $h = 0{\cdot}1x^2$.

invariably, is yes. Consider a glaciated, U-shaped valley. The U-shaped slope profile is a parabola and may be described by the equation

$$h = bx^2 .$$

In other words, height is proportional to the square of distance. The coefficient b varies from one locality to another; for instance, it is 0·08 for two valleys in North Wales, 0·04 for Bishopdale in Yorkshire, and 0·1 for Grisedale in the Lake District. The Grisedale curve, $h = 0·1x^2$, is plotted in figure 2.21. Note that $x = 0$ at the lowermost point of the valley floor. Now we have seen that the gradient for particular sections of a slope profile is given by

$$\text{gradient} = \frac{\text{height increment}}{\text{distance increment}} = \frac{\Delta h}{\Delta x} ,$$

where Δh and Δx are finite bits of height and distance. Consider though what happens as Δh and Δx become smaller and smaller. Clearly, the total number of height and distance increments into which the slope profile is divided becomes larger and larger. In the limit, as Δh and Δx approach zero, the number of height and distance increments becomes infinite, and the length of the increments is infinitesimally small. In this situation we have access to any point on the slope profile, which is what we are seeking. Notice we still have height and distance increments, though they are infinitesimally small in size. To distinguish these very small bits of x and h from Δh and Δx, the symbols dh and dx are used. Our definition of gradient still holds but the symbols are now

$$\text{gradient} = \frac{\text{height increment}}{\text{distance increment}} = \frac{dh}{dx} \qquad (2.28)$$

and it means that, to find the gradient at *any point* on the curve, a *tiny* bit of h is to be divided by a *tiny* bit of x. This gradient is a ratio which, in mathematical parlance, goes by the name of a *differential coefficient*, and dh and dx are termed differentials[5]. The reason for its being called a differential coefficient is plain to see; for as this ratio equals the gradient

$$\text{gradient} = \frac{dh}{dx} ,$$

then it follows that

$$dh = \text{gradient } dx$$

and it can be seen that the gradient is the value by which the differential dx has to be multiplied to give the corresponding value dh, that the gradient is therefore a differential coefficient.

To obtain the gradient at any point on our slope profile, we must find the value of dh/dx, and the process by which this is done is called *differentiating*. We have already seen how this is achieved for finite differences; the technique for infinitesimally small differences is not unrelated, but possibly more difficult to visualize. If a hill-walker starts at say point x on the U-shaped valley profile and takes an uphill step there is no problem: we can readily appreciate the geometry of the situation. What we must now try to conceive is that the hill-walker, starting at point x, takes an infinitesimally small

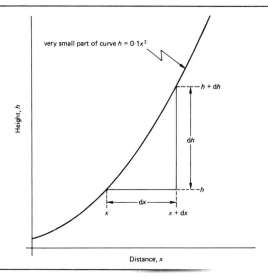

Figure 2.22. An interpretation of differentials, d*h* and d*x*.

uphill step of length d*x*. It might help to think of an ant standing at point *x* and taking an ant-sized, uphill step; this is clearly much smaller than the hill-walker's step, and, though still not infinitely small, will serve to illustrate the method. In taking that very small step, the ant moves forward a very small distance horizontally, d*x*, and rises a very small height, d*h* (figure 2.22). Because we know the relation between *h* and *x*, we can see that the enlarged *h* will be equal to 0·1 times the square of the enlarged *x*:

$$h + dh = 0{\cdot}1\,(x + dx)^2.$$

Doing the squaring we get

$$h + dh = 0{\cdot}1\,\{x^2 + (2x)(dx) + (dx)^2\}$$
$$= 0{\cdot}1x^2 + (0{\cdot}2x)(dx) + 0{\cdot}1(dx)^2.$$

As the term $(dx)^2$ means a very little bit of a very little bit of *x*, it is negligible compared with the other values and can be left out; thus, with $0{\cdot}1\,(dx)^2$ omitted, we have

$$h + dh = 0{\cdot}1x^2 + (0{\cdot}2x)(dx),$$

which can be written as

$$dh = 0{\cdot}1x^2 + (0{\cdot}2x)(dx) - h.$$

Since $h = 0{\cdot}1x^2$, we get

$$dh = (0{\cdot}2x)(dx),$$

or, dividing through by d*x*,

$$\frac{dh}{dx} = 0{\cdot}2x,$$

which is what we set out to find. It means that the ratio of increase in height, dh, to the increase in distance dx, is, in our example, $0.2x$. And this applies to any point on the curve. For example, when $x = -3$,

$$\frac{dh}{dx} = (0.2)(-3)$$

$$= -0.6,$$

which is a downhill slope of 31 degrees. When $x = 2$,

$$\frac{dh}{dx} = (0.2)(2)$$

$$= 0.4,$$

which is an uphill slope of 22 degrees. Notice that when $x = 0$,

$$\frac{dh}{dx} = (0.2)(0)$$

$$= 0,$$

that is, the slope at the lowermost point of the profile is horizontal. In the general case, the differential coefficient of

$$h = a + bx^n,$$

where n is any real number, is

$$\frac{dh}{dx} = nbx^{n-1}. \tag{2.29}$$

The gradient, dh/dx, is known as a first-order derivative. If, in turn, the gradient is differentiated, the second-order derivative is found:

$$\frac{d}{dx}\left(\frac{dh}{dx}\right) = \text{second-order derivative.} \tag{2.30}$$

The term on the left-hand side of equation (2.30) may be written d^2h/dx^2 and this may be recognized as being similar to the finite-difference expression of slope curvature $\Delta^2 h/\Delta x^2$. And that is what the second derivative is — the slope curvature. In the case of the Grisedale valley, the curvature may be found by differentiating the gradient, dh/dx. Using the general formula (equation 2.29), and remembering that $dh/dx = 0.2x$, we have

$$\frac{d}{dx}(0.2x) = 0.2.$$

Therefore

$$\frac{d^2h}{dx^2} = 0.2.$$

This means that the slope curvature, or rate of change of the gradient, is constant at all points along the profile. It also shows that the curvature is at all points positive; in other words, the profile is entirely concave.

Surfaces. The differential notation can be extended to surfaces and it enables us to find the derivatives at any point on the surface. We have to imagine that the grid network forms a mesh, but a mesh that is infinitesimally fine. Now in the case of the slope profile, the distance increment was written dx. However, in the case of a surface, the distance increment in the x direction — if you like, the distance between eastings — is not written as dx but as ∂x, signifying that a minute step has been taken in the x direction while holding the y direction constant. The symbol ∂x should not be confused with δx which is sometimes used synonymously as Δx. Similarly, the distance increment in the y direction — between the northings — is not written as dy, as one might expect, but as ∂y, the significance of which is that the minutest of steps has been taken in the y direction while the x coordinate has remained fixed. In this situation, the gradient of the land surface at any point consists of two components. The first component is the gradient in the x direction; this is not written as dh/dx but as $\partial h/\partial x$, which signifies that the differentiation has been performed with respect to just *one* of the independent variables; that is, it has been performed partially. Because of this, $\partial h/\partial x$ is known as a *partial differential* coefficient and ∂h and ∂x are *partial differentials*. The other component is the gradient in the y direction, which also being a partial derivative, is written as $\partial h/\partial y$. It is sometimes helpful to write a partial differential coefficient as, say, $(\partial h/\partial x)_y$, the subscript y indicating which independent variable(s) has been held constant during the process of differentiation. (See Sumner (1978) and Thompson (1917) for details of how to differentiate, partially or otherwise.)

Times rate of change

Many problems with which geographers deal include time as an independent variable. In population growth, for instance, population may be considered as a quantity which varies when time varies. If we call the size or density[6] of a population x and call time t, then x is a function of t

$$x = f(t).$$

Clearly, if the population grows by 100 individuals in a year, then the growth rate is 100 per year. The curve shown in figure 2.23 shows the change in population size, x, with time, t. The population at time t is x_t and the population after a very small time interval, dt, is x_{t+dt}. The gradient of the curve over the time interval dt is thus

$$\text{gradient} = \frac{(x_{t+dt} - x_t)}{dt}.$$

Because $x_{t+dt} - x_t$ can be written as a differential, dx, which represents a small increment of population, we have

$$\text{gradient} = \frac{dx}{dt}.$$

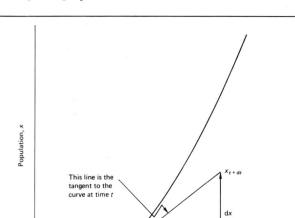

Figure 2.23.   An interpretation of the rate of change of population at instant time t.

Notice that we are speaking of a *gradient in time* and not a spatial gradient. The differential notation for rates of change in time shows that, if the population is growing, then the amount grown in a short time interval, dt, is dx; and the rate of growing is dx/dt. If the population is declining the rate is expressed as $-dx/dt$ because dx is a decrement and not an increment.

Differentiating the rate of population growth, dx/dt, yields

$$\frac{d}{dt}\left(\frac{dx}{dt}\right) = \frac{d^2x}{dt^2} \, ,$$

which is the second-order derivative of population with respect to time. Should d^2x/dt^2 be positive, the population is growing (or declining) at an increasing rate; should it be negative, the population is growing (or declining) at a decreasing rate; should it be zero, the population is growing (or declining) at a constant rate.

Notes

1   Many tests and papers care to choose Greek letters, so here for the non-classicist is the Greek alphabet.

A	α	alpha	H	η	eta	N	ν	nu	T	τ	tau
B	β	beta	Θ	θ	theta	Ξ	ξ	xi	Υ	υ	upsilon
Γ	γ	gamma	I	ι	iota	O	o	omicron	Φ	ϕ	phi
Δ	δ	delta	K	κ	kappa	Π	π	pi	X	χ	chi
E	ϵ	epsilon	Λ	λ	lambda	P	ρ	rho	Ψ	ψ	psi
Z	ζ	zeta	M	μ	mu	Σ	σ	sigma	Ω	ω	omega

2   As a rule of thumb, most frequency distributions can be adequately graphed using no more than 15 classes. Moreover, the class interval should be selected so that the main groupings of observations are clearly shown on the histogram.

3   Two other measures of central tendency are commonly used by statisticians although they are not applied in this book. These are the *mode*, the most frequently occurring observation, and the *median*, which is defined as the middle value of x_i in the ranking of all observations from lowest value to highest value.

4   Defining two matrices

$$C = \begin{bmatrix} 1 & 3 \\ 2 & 2 \\ 3 & 1 \end{bmatrix} \qquad D = \begin{bmatrix} 0 & 2 & 4 & 2 \\ 1 & 3 & 3 & 1 \end{bmatrix}$$

we see that the number of columns in **C** equals the number of rows in **D** so a product matrix, **CD**, can be formed.

$$CD = \begin{bmatrix} 1 & 3 \\ 2 & 2 \\ 3 & 1 \end{bmatrix} \begin{bmatrix} 0 & 2 & 4 & 2 \\ 1 & 3 & 3 & 1 \end{bmatrix}$$

$$= \begin{bmatrix} [1 \;\; 3]\begin{bmatrix}0\\1\end{bmatrix} & [1 \;\; 3]\begin{bmatrix}2\\3\end{bmatrix} & [1 \;\; 3]\begin{bmatrix}4\\3\end{bmatrix} & [1 \;\; 3]\begin{bmatrix}2\\1\end{bmatrix} \\ [2 \;\; 2]\begin{bmatrix}0\\1\end{bmatrix} & [2 \;\; 2]\begin{bmatrix}2\\3\end{bmatrix} & [2 \;\; 2]\begin{bmatrix}4\\3\end{bmatrix} & [2 \;\; 2]\begin{bmatrix}2\\1\end{bmatrix} \\ [3 \;\; 1]\begin{bmatrix}0\\1\end{bmatrix} & [3 \;\; 1]\begin{bmatrix}2\\3\end{bmatrix} & [3 \;\; 1]\begin{bmatrix}4\\3\end{bmatrix} & [3 \;\; 1]\begin{bmatrix}2\\1\end{bmatrix} \end{bmatrix}$$

$$= \begin{bmatrix} (1 \times 0) + (3 \times 1) & (1 \times 2) + (3 \times 3) & (1 \times 4) + (3 \times 3) & (1 \times 2) + (3 \times 1) \\ (2 \times 0) + (2 \times 1) & (2 \times 2) + (2 \times 3) & (2 \times 4) + (2 \times 3) & (2 \times 2) + (2 \times 1) \\ (3 \times 0) + (1 \times 1) & (3 \times 2) + (1 \times 3) & (3 \times 4) + (1 \times 3) & (3 \times 2) + (1 \times 1) \end{bmatrix}$$

$$= \begin{bmatrix} 3 & 11 & 13 & 5 \\ 2 & 10 & 14 & 6 \\ 1 & 9 & 15 & 7 \end{bmatrix}.$$

5   When d's are used, the equations are in their differential form; when Δ's are used, the equations are in their difference form.

6   Population size is given as number of individuals; population density is given in units of number of individuals per unit area.

PART II

Deterministic Models

CHAPTER 3

Modelling Cascading Systems

In many systems of geographical interest, an important feature is that the interactions between system components – say between feeding levels in an ecosystem or age groups in a population – are characterized by *flows* of matter, energy, or people which can be resolved into *inputs* and *outputs*. Figure 3.1(*a*) depicts a forest ecosystem as three interacting components – plants, leaf litter lying on the ground, and soil. The three components are linked by a circulation of, among other things, carbon. Carbon enters the system as the gas carbon dioxide through the plant leaves. It moves through the system along the following route: from plants to litter by the processes of leaf fall and timber fall; from plants to soil by the process of root sloughage (the rotting of roots); and from litter to soil by the processes of decomposition and translocation (a downwards movement brought about by percolating water or burrowing animals, especially earthworms). It leaves the system as gaseous carbon dioxide produced by plants and organisms living in the litter and the soil in the process of respiration, and a little may be lost in drainage water. Figure 3.1(*b*) portrays the living portion of an ecosystem divided into a number of components, each one a feeding level – plants, herbivores, carnivores, and decomposers – which form a food chain. The components are linked by a flow of energy. Sunlight, or solar energy, is received by plants, turned into chemical energy of plant tissue, and passed along the food chain by the process of predation, the decomposers 'preying' on dead plant and animal tissue. Energy is lost from each feeding level as heat of respiration. Figure 3.1(*c*) shows a population as a three-component system, each component representing a different age group. The stock of population in each group is added to by births, in the case of the youngest age group, by survival from one age group to the next, except in the case of the youngest age group, and by immigration; and the stock of population in each age group is reduced by deaths and by emigration. The overall effect is a continual turnover of people in the system as a whole and in each age group, maintained by what are, in effect, inputs and outputs of people.

Notice that in passing through a system, mass, energy, and people may be *stored* in system components or may pass through them to emerge as output; and that the output of one component may form the input of another. So, carbon entering plants may be stored in plant tissue and may be lost from the plants in leaf fall and timber fall. The leaf fall and timber fall then form an input to the litter component which

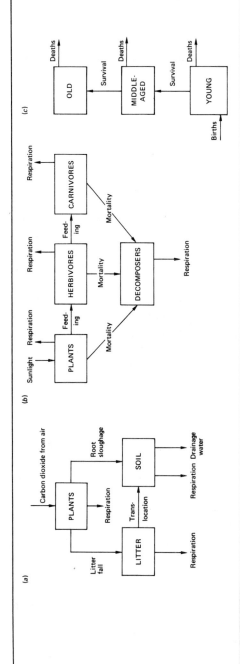

Figure 3.1. Some systems of interest to geographers. (*a*) Carbon flow in a forest ecosystem. (*b*) A food web. (*c*) A population system divided into three age groups.

may in turn store carbon and may lose carbon by the processes of decomposition and translocation. Systems of this ilk, following the definition given by Chorley and Kennedy (1971), are known as *cascading systems*; as well as the ones described, they include the hydrological cascade — the input, storage, and output of water in a catchment; the weathering and debris cascade — the input, storage, and output of sediment in a landscape; the glacier cascade — the input, storage, and output of ice in a valley glacier; and the wave cascade — the input, storage, and output of sediments along sections of coast.

RULES OF STORAGE AND FLOW

The passage of carbon, of energy, and of people through the systems pictured in figure 3.1, and indeed through all other cascading systems, is not a haphazard process: there are *rules* or *laws* which determine the rate and sometimes direction of transfer between system components such as the transfer of carbon from plants to litter, of energy from plants to herbivores, and the number of births in a population. These rules apply to outputs which pass into the environment but not to inputs from the environment: the loss of energy in respiration is partly under the system's control, but the amount of sunlight entering the system is determined by factors outside the system. The transfer and storage of mass, energy, and people in systems are subject to two groups of laws: *laws of conservation* and *transport laws*, also called *process laws* or *flow laws*.

The first group of laws, the laws of conservation, determine that each and every bit of matter and energy, and every person in a system must be recorded somewhere on an input—output—storage balance sheet: mass, energy and people cannot appear or disappear in an unaccountable manner: what goes into a system, or a system component, must be stored or come out. The law of energy conservation, known rather grandly as the first law of thermodynamics, states that energy may not be created nor destroyed, but may be transformed from one form to another. In practice, this means that the change in the amount of energy stored in a system over a time interval must equal inputs less outputs of energy over that period; so we may write an *energy balance* or *energy storage equation:*

$$\text{change in energy storage} = \left(\text{energy inputs} - \text{energy outputs} \right) \text{time interval.}$$

The law of mass conservation, sometimes referred to as the continuity condition, follows the same principle and we may write a *mass storage* or *continuity-of-mass equation*:

$$\text{change in mass storage} = \left(\text{mass inputs} - \text{mass outputs} \right) \text{time interval.}$$

The same logic is applied to the study of population changes. A change in population size may be brought about by inputs of people, either from births or immigration, and outputs of people, either from death or emigration. A *population storage equation* may be put

$$\text{change in population size} = \left(\begin{array}{c}\text{population} \\ \text{inputs}\end{array} - \begin{array}{c}\text{population} \\ \text{outputs}\end{array}\right)\begin{array}{c}\text{time} \\ \text{interval.}\end{array}$$

The second group of laws, the transport laws, serve to define the input and output rates which appear in the storage equations. One such law would set the output of carbon in leaf fall and timber fall from plants as proportional to the amount of carbon stored in the plants; this would be expressed as

$$\text{output of carbon} = \text{constant} \times \text{amount of carbon in store.}$$

Similarly, the output from a population could be set as proportional to population size and a population 'flow' law written as

$$\text{population output} = \text{constant} \times \text{population size.}$$

Many other transport laws have been established and we shall consider them in due course.

A storage equation in which the input and output rates are defined by suitable transport laws is a deterministic model. If the system should consist of just one component, then the model will consist of just one storage equation. More usually, the system will consist of several components and there will be one storage equation for each component. Because the output of one component may become the input of another, the equations relate to processes which proceed hand in hand; they are therefore termed *simultaneous equations*. And because the equations deal in storage change over discrete time intervals, they are *simultaneous, difference equations*. The set of equations will contain as yet unspecified constants and parameters which have been introduced through the transport laws. The model is *calibrated* by giving parameters values appropriate for the case being studied. When calibrated, the equations can, sometimes, be *solved* to reveal how the storages in each component change with time in response to the transport laws. The predicted changes then need to be matched against observed changes to see if the results are reasonable.

The procedure for modelling cascading systems in a deterministic way thus runs something like this:

1. State problem or hypothesis.
2. Make definitions and assumptions:
   (*a*) Define system of interest, including boundaries, components, inputs, and outputs.
   (*b*) Draw up storage equations, one for each component.
   (*c*) Define inputs and outputs in storage equations using transport laws.
3. Calibrate the set of equations.
4. Analyse the model and test the results.

Models of population change and models of mineral cycles in ecosystems will be used to show how all this works.

SIMPLE MODELS OF POPULATION CHANGE

Setting up a model

Change in the size of human populations may be studied by what in the ecological literature are termed *compartmental modelling techniques*. Populations may be regarded as storage compartments containing a stock of people which, during a specified interval of time, may be topped up by births and immigration, and depleted by deaths and emigration. The simplest models of population change deal with the growth of a single population in an isolated region. Such a population will change in response to births and deaths only — because the region is isolated, no migration takes place. The number of people in the population at a particular time — the stock — is a state variable and may be denoted by the letter x. The 'flows' of people entering and leaving the population as a result of births and deaths may be labelled as F_{births} and F_{deaths}, respectively. A storage equation may now be written for the population; it is

$$\Delta x/\Delta t = F_{births} - F_{deaths} \tag{3.1}$$

time rate   inputs   outputs
of change
in stock
of people

where Δx is the change in population size during a time interval, Δt.

The next step in setting up a model of population change is to define the flows of people — the F's in equation (3.1) — by population 'transport' rules. One possibility is to set births and deaths proportional to the size of the population. Doing this, the number of births per unit time is defined as

$$F_{births} = bx, \tag{3.2}$$

where b is the specific birth rate, that is, the number of births per unit of the population per unit time, say, 50 per thousand per year. So, for a population of 50 000, the number of births in a year would be 50 x 50 000/1000 = 2500. Similarly, the number of deaths per unit time is defined as

$$F_{deaths} = dx, \tag{3.3}$$

where d is the specific mortality rate, that is, the number of deaths per unit of the population per unit time, say 40 per thousand per year. So, for a population of 50 000, the number of deaths in a year would be 40 x 50 000/1000 = 2000.

Putting the definitions for births and deaths (equations 3.2 and 3.3) into the storage equation (3.1) gives

$$\frac{\Delta x}{\Delta t} = bx - dx, \tag{3.4}$$

which may be written

$$\frac{\Delta x}{\Delta t} = (b - d)x.$$

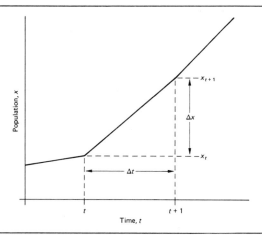

Figure 3.2. Defining the population increase, Δx, during a time interval, Δt.

The difference between the specific birth rate and the specific death rate, $(b - d)$, is called the specific growth rate of the population[1] and is commonly denoted by the letter r; so

$$r = b - d$$

and equation (3.4) reads

$$\frac{\Delta x}{\Delta t} = rx. \tag{3.5}$$

Δx is a small difference (see figure 3.2) and may be written

$$\Delta x = x_{t+1} - x_t,$$

where x_{t+1} is the population at time $t + 1$ and x_t is the population at time t. (The time interval, Δt, is equal to $(t + 1) - t$.) Therefore, equation (3.5) may be expressed as

$$x_{t+1} = x_t + (rx_t)\Delta t \tag{3.6}$$

which is the model, as yet uncalibrated.

Calibration and solution

If data are available from which to calibrate the model (equation 3.6) — if the r parameter can be given a value and the starting size of the population is known — it may be solved to show how the population will change with time.

As an example, let us calibrate the model for the population of England and Wales in 1751. At that time the population was 7·3 million, the birth rate was of the order of 0·036 per person per year, and the death rate was about 0·032 per person per year. Thus $r = b - d = 0·036 - 0·032 = 0·004$ per person per year. For convenience, the time interval used will be ten years; therefore $\Delta t = 1$ decade and r will be expressed as

0·04 per person per decade. In 1751, $t = 0$ and equation (3.6) is written

$$x_1 = x_0 + (rx_0)\Delta t.$$

As x_0 is 7·3, r is 0·04, and Δt is 1·0, we have

$$x_1 = 7·3 + (0·04 \times 7·3 \times 1·0)$$
$$= 7·59.$$

So the population of England and Wales in 1761, one decade after 1751, is predicted by the model to be 7·59 million. The population in 1771 may be found in a similar way; we may write

$$x_2 = x_1 + (rx_1)\Delta t,$$

and since x_1 was found to be 7·59 we have

$$x_2 = 7·59 + (0·04 \times 7·59 \times 1·0)$$
$$= 7·89 \text{ million.}$$

The population in 1781, x_3, could then be found; and so the process goes on for any number of time steps. The population of England and Wales predicted by the model up to the year 1981 is shown in table 3.1, column four.

Table 3.1. Results of the exponential population growth model.

Date	t (decades)	Observed population (millions)	Predicted population (millions) $r = 0·04$	$r = 0·14$
1751	0	7·3	7·3	7·3
1761	1	–	7·59	8·32
1771	2	–	7·89	9·48
1781	3	–	8·20	10·80
1791	4	–	8·52	12·31
1801	5	8·89	8·86	14·03
1811	6	10·16	9·21	15·99
1821	7	12·00	9·57	18·22
1831	8	13·89	9·95	20·77
1841	9	15·91	10·34	23·67
1851	10	17·92	10·75	26·98
1861	11	20·06	11·18	30·75
1871	12	22·71	11·62	35·05
1881	13	25·97	12·08	39·95
1891	14	29·00	12·56	45·54
1901	15	32·52	13·06	51·91
1911	16	36·07	13·58	59·17
1921	17	37·88	14·12	67·45
1931	18	39·95	14·68	76·89
1941	19	–	15·26	87·65
1951	20	43·75	15·87	99·92
1961	21	46·10	16·50	113·90
1971	22	48·85	17·16	129·84
1981	23	?	17·84	148·01

Testing the model

The success or failure of the model can in this case be assessed by comparing the predicted population changes with the known changes as recorded by the Registrar General. A very poor match between observed and predicted values is evident (table 3.1).

The question now is why should the model give poor predictions? The alternative reasons are, either that the assumptions of the model, and specifically the population 'transport' rules, are wrong; or the value of the parameter r is incorrect. Take the parameter r first; this was established as 0·04 per person per decade using the birth and death rates in 1751. A glance at figure 3.3 will shown that the birth and death rates however have not remained constant but fallen since 1751, the fall in the death rate preceding the fall in the birth rate. The pattern of change is of course the classic Demographic Transition, and it explains why the model under-predicts the population growth. As the death rate drops but the birth rate holds its 1751 level, as it does throughout the 'early expanding phase' (figure 3.3), the specific growth rate of the population, r, is for many decades of the order of 0·14 per person per decade. It would be interesting therefore to see what effect using the larger and more representative value of r has on the population predictions. The results of the model calibrated with a specific growth rate of 0·14 per person per decade are given in table 3.1 (column five). In this case the model over-predicts, by a long way, the growth of population. The indication is that the assumptions of the model are at fault.

It is not too difficult to see where the assumptions go wrong. If r is greater than 1·0, the model, because of the definitions that have been assumed for births and deaths, states that the growth of the population during a time interval is proportional to the size of the population at the start of the time interval. This kind of growth process was first seen to apply to the unlimited growth of populations by the

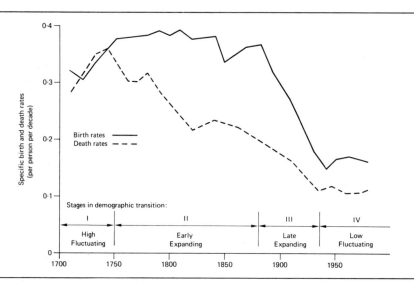

Figure 3.3. The demographic transition in England and Wales.

nineteenth-century Belgian astronomer, L A J Quetelet, who suggested that when a population is able to develop freely, without obstacles, it will grow according to a geometrical progression, that is, 1, 2, 4, 8, 16, and so on; this is termed exponential growth (p37). Because the population of England and Wales cannot grow indefinitely in an exponential manner — space and resources are too limited for this to happen — it is not surprising that the model, with its in-built assumptions generating exponential growth, fails to give trustworthy population predictions.

The next step is to re-examine the assumptions of the model, but before doing so consideration will be given to inaccuracies involved in the method used to solve the population growth equation.

Sources of error

Equation (3.6) is a difference equation. The method chosen to solve it, the Euler method, is mathematically primitive. A value for x_0 was given; then the value of x_1 was found by substituting in $x_1 = x_0 + 0 \cdot 14 x_0 \Delta t$; then x_2 was found from x_1; and so forth. More refined methods of the same type are available for solving difference equations; they all engage a process of what is called numerical approximation — estimating the value at one time from the value at a previous time. Errors of two sorts enter into these methods. First of all, because not even digital computers can work to an infinite number of decimal places, rounding errors appear which, though small at first, become magnified the more time steps that are taken. The second error arises because the function used to project from one time interval to another — a linear one in the Euler method but curvilinear in most — will only approximate the change during the time interval; this error increases with increasing duration of the time step, Δt. Table 3.2 shows calculations made on the equation $x_{t+1} = x_t + 0 \cdot 14 x_t \Delta t$ using three different values of Δt — one decade, two decades, and three decades: as the time interval becomes longer so the predicted population growth decreases. A danger here is that the longer the time interval, the more inaccurate the results, but the better the match with the observed data. The only way round this is to keep the time interval as brief as possible and not to predict too far into the future because this second type of error, like the first, increases the more time steps that are taken. In some instances, the error can lead to the solution's becoming unstable and the results are meaningless. For these reasons, numerical methods for solving difference equations should be treated with caution. The mathematician, W W Sawyer, related the story of a man who processed some meteorological data and forecast a typhoon; no typhoon came: the predicted typhoon was manufactured by an inappropriate numerical method.

If the time interval is made infinitesimally short, we may write dt instead of Δt (see p55). Equation (3.5) then reads

$$\frac{dx}{dt} = rx, \tag{3.7}$$

the increase in population, dx, also being infinitesimally small. Equation (3.7) is a differential equation and has an analytical solution in which the population at any time, not just at discrete time intervals, is given as some function of the initial population, x_0, the specific growth rate, r, and time, t. The analytical solution to

Table 3.2. The accuracy of the Euler method.

Date	t (decades)	Analytical solution $(x_t = 7\cdot3e^{0\cdot14t})$	Numerical solution (Euler method) with time interval, Δt, set at:		
			1 decade	2 decades	3 decades
1751	0	7·3	7·3	7·3	7·3
1761	1	8·39	8·32	–	–
1771	2	9·65	9·48	9·34	–
1781	3	11·11	10·80	–	10·36
1791	4	12·77	12·31	11·95	–
1801	5	14·70	14·03	–	–
1811	6	16·90	15·99	15·29	14·71
1821	7	19·45	18·22	–	–
1831	8	22·37	20·77	19·57	–
1841	9	25·73	23·67	–	20·88
1851	10	29·60	26·98	25·04	–
1861	11	34·05	30·75	–	–
1871	12	39·16	35·05	32·05	29·64
1881	13	45·05	39·95	–	–
1891	14	51·82	45·54	41·02	–
1901	15	59·61	51·91	–	42·08
1911	16	68·57	59·17	52·50	–
1921	17	78·87	67·45	–	–
1931	18	90·72	76·89	67·20	59·75
1941	19	104·36	87·65	–	–
1951	20	120·04	99·92	86·01	–
1961	21	138·08	113·90	–	84·84
1971	22	158·83	129·84	110·09	–
1981	23	241·74	148·01	–	–

equation (3.7) is

$$x_t = x_0 e^{rt} \tag{3.8}$$

which, unlike the solution to the difference equation, gives exact results. The derivation of the analytical solution involves difficult mathematical concepts; as far as this book is concerned, it will suffice to think of equation (3.8) as the function describing the population growth curve produced when the time interval is infinitesimally short.

Why seek numerical solutions, which give approximate results, when exact results can be obtained by analytical solutions? The reason is that, though analytical solutions are available for fairly simple differential equations, many others, especially those of interest to geographers, cannot be solved analytically and recourse to numerical solutions is necessary.

It will be instructive to see how exact results may be extracted from equation (3.8) and to compare these results with those from the Euler method. Calibrated for a base date of 1751 and with a specific growth rate of 0·14 per person per decade, equation (3.8) becomes

$$x_t = 7\cdot3e^{0\cdot14t}$$

and x_t, population size, can be found for any positive value of time, t. For instance, when $t = 0$,

$$x_0 = 7 \cdot 3 e^{(0 \cdot 14)(0 \cdot 0)}$$

$$= 7 \cdot 3 e^0$$

and as $e^0 = 1$

$$x_0 = 7 \cdot 3 \text{ million},$$

which is simply the initial population. When $t = 5$

$$x_5 = 7 \cdot 3 e^{(0 \cdot 14)(5 \cdot 0)}$$

$$= 7 \cdot 3 e^{(0 \cdot 7)}$$

$$= 13.08 \text{ million}.$$

Remembering that the specific growth rate is calibrated to operate over decades, this means that in 1801, five decades after 1751, the population of England and Wales, according to the exponential growth model, is precisely 13·08 million, rounding errors apart.

Table 3.2, column three, shows the population predictions for each decade as made by the analytical solution; compared with these exact results, the forecasts given by the Euler method are fairly accurate for about a century, after which time the error grows appreciably. A more accurate solution could be obtained from the Euler method by setting the time interval at one year or less.

Revising the assumptions

The exponential growth model failed to predict with good fidelity the growth of population in England and Wales since 1751. Faulty parameters did not seem to be the cause of this failure which suggests a re-examination of the assumptions of the model might be profitable.

The definitions of births and deaths assumed in equation (3.4) led to exponential growth. But, though early stages of population growth may follow an exponential path, as population density increases, space becomes more crowded and resources become scarcer. These obstacles arrest the growth of a population. It would seem worthwhile then, to add a crowding-cum-resource-limitation term, usually called an inhibiting factor, to the model. The common practice is to subtract from the exponential growth term, rx, a number which is proportional to, by a parameter c, the square of the population size, x^2. Thus equation (3.5) is recast as

$$\Delta x / \Delta t \qquad = rx \qquad\qquad - cx^2. \qquad\qquad (3.9)$$

| time rate of population change | exponential growth (the maximum possible rate of increase) | inhibiting factor (the unrealized increase resulting from obstacles to growth) |

Because the inhibiting factor brings down the population growth from its potential exponential rate, and because the degree to which it does so is proportional to the square of the population size, the effect of the inhibiting factor in reducing population growth is more pronounced when the population is big; this can be seen by writing equation (3.9) as

$$\frac{\Delta x}{\Delta t} = (r - cx)x. \tag{3.10}$$

The term $(r - cx)$ shows that the specific growth rate, r, is reduced in proportion to the population size: the bigger the population, x, gets, the slower it grows. The parameter c determines the reduction in the specific growth rate per unit of population and the product cx is the total reduction during a time interval brought about by inhibitory effects; cx is referred to as the environmental resistance. The parameter c is defined as

$$c = r/K,$$

where K is the maximum population the environment can support – the carrying capacity of the environment.

The revised model may be written

$$x_{t+1} = x_t + \{(r - cx_t)x_t\}\Delta t. \tag{3.11}$$

To calibrate this for the population growth in England and Wales since 1751, we need to specify the initial population, x_0, the specific growth rate, r, and the parameter c. As in the exponential model, x_0 is 7·3 million and r is 0·14 per person per decade. To define c, a carrying capacity for England and Wales must be established – 60 million is a not unreasonable figure. By definition, $c = r/K = 0.14/60 = 0.0023$. So the calibrated model is

$$x_{t+1} = x_t + \{(0.14 - 0.0023x_t)x_t\}\,1.0.$$

Starting at 1751 we have

$$x_1 = x_0 + \{(0.14 - 0.0023x_0)x_0\}\,1.0$$
$$= 7.3 + \{(0.14 - 0.0023 \times 7.3)7.3\}1.0$$
$$= 8.19 \text{ million.}$$

In other words, the population after one decade has grown to 8·19 million. And then to find the population after another decade we have

$$x_2 = x_1 + \{(0.14 - 0.0023x_1)x_1\}1.0$$
$$= 8.19 + \{(0.14 - 0.0023 \times 8.19)8.19\}1.0$$
$$= 9.18 \text{ million,}$$

and so on. The results are shown in table 3.3. The match between observed and predicted growth is very close, suggesting the model describes well the process of population growth in England and Wales since 1751.

The pattern of growth exhibits an early phase of exponential increase but the growth rate declines, little by little, until the size of the population comes to a standstill at the carrying capacity of 60 million. The analytical curve which describes this pattern of change is called a logistic, a term used by Edward Wright in 1599 to describe an S-shaped curve. The first attempt to derive an equation to describe the

Table 3.3. Results of the logistic population growth model.

Date	t (decades)	Observed population (millions)	Euler method with Δt set at 1 decade	Analytical solution, equation (3.13)
			Predicted population (millions)	
1751	0	7·3	7·3	7·3
1761	1	–	8·19	8·24
1771	2	–	9·18	9·29
1781	3	–	10·27	10·44
1791	4	–	11·46	11·70
1801	5	8·89	12·76	13·08
1811	6	10·16	14·17	14·57
1821	7	12·00	15·69	16·17
1831	8	13·89	17·32	17·87
1841	9	15·91	18·98	19·68
1851	10	17·92	20·80	21·57
1861	11	20·06	22 71	23·54
1871	12	22·71	24·70	25·57
1881	13	25·97	26·75	27·65
1891	14	29·00	28·84	29·74
1901	15	32·52	30·96	31·84
1911	16	36·07	33·08	33·92
1921	17	37·88	35·19	35·96
1931	18	39·95	37·26	37·95
1941	19	–	39·28	39·86
1951	20	43·75	41·23	41·69
1961	21	46·10	43·09	43·42
1971	22	48·85	44·85	45·05
1981	23	?	46·50	46·56

logistic growth process was made by Verhulst in 1844, and his work was extended by R Pearl and L J Reed in 1920. The analytical solution to the logistic growth equation is

$$x_t = \frac{K}{1 + Ce^{-rt}} \tag{3.12}$$

where the parameters are as previously defined and C is a constant of integration. Equation (3.12) describes an S-shaped or sigmoid curve for positive values of C which has to be evaluated in specific cases. The initial population, x_0, is defined by equation (3.12) when $t = 0$

$$x_0 = \frac{60}{1 + Ce^{-(r)(0)}}.$$

Remembering that $e^{-(r)(0)} = e^{-0} = 1$, this gives

$$x_0 = \frac{60}{1 + C}.$$

Since we know x_0 is 7·3 (million), we can solve equation (3.12) for C

$$C = \frac{60}{7\cdot3} - 1$$

$$= 7\cdot22.$$

The logistic equation (equation 3.12) can now be written for the case of population growth in England and Wales:

$$x_t = \frac{60}{1 + 7\cdot22e^{-0\cdot14t}} \cdot \tag{3.13}$$

The values of x for any positive value of time t can now be found. Thus the population after 1 decade ($t = 1$) is given by

$$x_1 = \frac{60}{1 + 7\cdot22e^{-0\cdot14 \times 1\cdot0}}$$

$$= \frac{60}{1 + (7\cdot22 \times 0\cdot86936)}$$

$$= 8\cdot24,$$

and the population after 20 decades is given by

$$x_{20} = \frac{60}{1 + 7\cdot22e^{-0\cdot14 \times 20}}$$

$$= \frac{60}{1 + (7\cdot22 \times 0\cdot06081)}$$

$$= 41\cdot69.$$

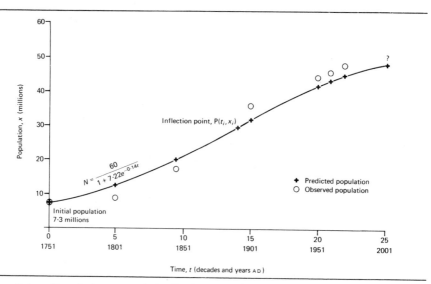

Figure 3.4.   Population growth in England and Wales since 1751 as predicted by a logistic growth model. Observed population growth is shown for comparison.

Table 3.3 (column five) shows several values predicted by the logistic model and a good fit with the values from the Euler method is apparent. The logistic curve described by equation (3.13) is drawn in figure 3.4. The coordinates of the point of inflexion, designated point P, are $P(t_i, x_i)$, where the following definitions hold

$$t_i = \frac{1}{r} \log_e C$$

$$x_i = \tfrac{1}{2}K.$$

Substituting the parameter values yields

$$t_i = \frac{1}{0 \cdot 14} \log_e 7 \cdot 22$$

$$= 14 \cdot 12$$

$$x_i = \tfrac{1}{2}60$$

$$= 30.$$

So the point of inflexion lies at 1892–3 with a population of 30 million.

Introducing population age structure

The study of population is an important facet of geography. From the early, and somewhat crude, models of the growth of single, isolated populations have evolved sophisticated models for forecasting inter-regional population change. With the continuing growth of population in a world of limited space and resources, it is useful to understand how, where, and why population changes, to be able to forecast where and in what numbers population will grow in another ten, twenty, or thirty years. The general principles of population growth were, as we have seen, appreciated in the early nineteenth century. Although simple exponential and logistic growth models of human populations now seem primitive, they can still provide fodder for the scare-mongers – a variant of the logistic model predicts that by AD 2026·87 the number of people on Earth will have become infinite; they also provide the basis of more recent and more elaborate models which disaggregate total population on the basis of age groups, sexes, and regions. Disaggregated population models are helpful to the planner who needs to know not just how many people there will be ten years hence, but where they will live, how many there will be in each age group, and what will be the proportion of women to men. A thorough treatment of these topics is given in the book by Rees and Wilson (1977); in this book, a simple example of an age-disaggregated population will be developed.

The first attempt to model changes in the age structure of a population was made by Leslie in 1945. Leslie's model predicts the age structure of a female population given the initial age structure and the birth and survival rates of each age group. The Leslie model could be set up by writing a storage equation for each age group but it is preferable to adopt Leslie's concise and efficient formulation which employed matrix algebra. The model, in matrix notation, is

$$\mathbf{x}_{t+1} = \mathbf{G}\mathbf{x}_t. \tag{3.14}$$

In equation (3.14), x_t is a column vector which defines the female population in each age class at time t. So, if there are three age classes then $x_t = \{x_1, x_2, x_3\}^T$; that is, a 3×1 column vector with x_1 the population of females in the first age class, x_2 the population in the second age class, and so on. x_{t+1} is an $n \times 1$ column vector, similar in structure to x_t, but defining the number of females in each age class at time $t + 1$. G is an $n \times n$ matrix which incorporates the birth rates and survival rates of individuals in each age class. The birth and survival components can, for the purposes of illustration, be separated. By way of example, take the female population in a Saxon village. Let us define three age groups — young, middle-aged and old, and an initial female population of twelve young Saxon women. This means $x_1, = 12$, $x_2 = 0$, $x_3 = 0$ so the initial population vector, x_t is

$$x_t = \{12, 0, 0\}^T.$$

Now, in a general case, the birth rate terms for each age class in the population may be written as

$$B = \begin{bmatrix} b_1 & b_2 & b_3 \\ 0 & 0 & 0 \\ 0 & 0 & 0 \end{bmatrix}$$

where b_1 is the birth rate for young Saxon women, b_2 is the birth rate for middle-aged Saxon women, and b_3 is the birth rate for old Saxon women. Specifically, let us use the following birth rates (in units of per person per decade)

$$B = \begin{bmatrix} 1 & 3 & 0 \\ 0 & 0 & 0 \\ 0 & 0 & 0 \end{bmatrix}$$

where $b_1 = 1$, so that each young woman on average produces one daughter before entering the middle-aged class or dying; $b_2 = 3$, so that each middle-aged woman in the village on average gives birth to three daughters before joining the old class or dying; and $b_3 = 0$, so that old women of the village on average give birth to no daughters. To see how these terms work we shall assume there is no survival over one time step. So assuming, the population after one time step can be found by pre-multiplying the initial population vector by the birth matrix, B:

$$x_{t+1} = \qquad B \qquad x_t$$

| population at time $t + 1$ | birth matrix | population at time t |

Substituting the appropriate values on the right-hand side

$$x_{t+1} = \begin{bmatrix} 1 & 3 & 0 \\ 0 & 0 & 0 \\ 0 & 0 & 0 \end{bmatrix} \begin{bmatrix} 12 \\ 0 \\ 0 \end{bmatrix}.$$

Following the procedure of matrix multiplication given on p57 this gives

$$x_{t+1} = \begin{bmatrix} (1 \times 12) + (3 \times 0) + (0 \times 0) \\ (0 \times 12) + (0 \times 0) + (0 \times 0) \\ (0 \times 12) + (0 \times 0) + (0 \times 0) \end{bmatrix} = \begin{bmatrix} 12 \\ 0 \\ 0 \end{bmatrix}$$

reflecting the fact that each young female on average produces one daughter before dying or entering the next age class.

The survival terms of the growth matrix can, in the general case, be written for each age class in the population as follows:

$$S = \begin{bmatrix} 0 & 0 & 0 \\ s_1 & 0 & 0 \\ 0 & s_2 & s_3 \end{bmatrix}$$

where s_1 is the survival rate of the young women in the village, s_2 in the survival rate of middle-aged females in the village, and s_3 is the survival rate of the old women of the village. Specifically, let us use the following survival rates (in units of per person per decade)

$$S = \begin{bmatrix} 0 & 0 & 0 \\ \frac{1}{3} & 0 & 0 \\ 0 & \frac{1}{4} & 0 \end{bmatrix}.$$

In this equation, $s_1 = \frac{1}{3}$, which is the probability that a young woman will survive to middle-age (the survival rate is $1-d$ where d is the death rate; this means that $1-\frac{1}{3} = \frac{2}{3}$ of the young women of the village die during each time step); $s_2 = \frac{1}{4}$, which is the probability of a middle-aged village woman surviving to old age; and $s_3 = 0$, which means that all old women will die during one time step (that is, one decade). To see how these terms work we shall assume there are no birth terms. So assuming the population after one time step can be found by pre-multiplying the initial population vector, x_t, by the survival matrix, S

$$x_{t+1} \quad = \quad S \quad \quad x_t.$$

population at   survival   population at
time $t + 1$ matrix   time t

Given an initial population of twelve individuals in each age class (the change in the initial population is made the better to illustrate the point) we get

$$x_{t+1} = \begin{bmatrix} 0 & 0 & 0 \\ \frac{1}{3} & 0 & 0 \\ 0 & \frac{1}{4} & 0 \end{bmatrix} \begin{bmatrix} 12 \\ 12 \\ 12 \end{bmatrix}.$$

Doing the matrix multiplication

$$x_{t+1} = \begin{bmatrix} (0 \times 12) + (0 \times 12) + (0 \times 12) \\ (\frac{1}{3} \times 12) + (0 \times 12) + (0 \times 12) \\ (0 \times 12) + (\frac{1}{4} \times 12) + (0 \times 12) \end{bmatrix} = \begin{bmatrix} 0 \\ 4 \\ 3 \end{bmatrix}.$$

The result indicates that, during one time step, all the young women of the village die or survive to middle-age: none remains after the time step; all the young women either die or join the old age class but the group is joined at time $t + 1$ by four survivors from the young age class; all the old women die and the three women in that class at time $t + 1$ are survivors from middle-age.

Putting the birth and survival matrices together we arrive at the growth matrix, **G**:

$$\mathbf{G} = \mathbf{B} + \mathbf{S}$$

$$= \begin{bmatrix} b_1 & b_2 & b_3 \\ 0 & 0 & 0 \\ 0 & 0 & 0 \end{bmatrix} + \begin{bmatrix} 0 & 0 & 0 \\ s_1 & 0 & 0 \\ 0 & s_2 & s_3 \end{bmatrix}$$

$$= \begin{bmatrix} b_1 & b_2 & b_3 \\ s_1 & 0 & 0 \\ 0 & s_2 & s_3 \end{bmatrix}.$$

For the village of Saxon women

$$\mathbf{G} = \begin{bmatrix} 1 & 3 & 0 \\ \frac{1}{3} & 0 & 0 \\ 0 & \frac{1}{4} & 0 \end{bmatrix}.$$

We have already noted (equation 3.14) that a basic model of population change is given by the matrix equation:

$$\mathbf{x}_{t+1} = \mathbf{G}\,\mathbf{x}_t.$$

Thus to find the population age structure at time $t + 1$; which in our example will be a decade later than time t, the population age structure at time t is pre-multiplied by the growth matrix, **G**. In the example of the Saxon village, with the initial female population given as $\mathbf{x}_t = \{12, 0, 0\}^T$ we get

$$\begin{bmatrix} 1 & 3 & 0 \\ \frac{1}{3} & 0 & 0 \\ 0 & \frac{1}{4} & 0 \end{bmatrix} \begin{bmatrix} 12 \\ 0 \\ 0 \end{bmatrix} = \begin{bmatrix} 12 \\ 4 \\ 0 \end{bmatrix}$$

G	$\mathbf{x}_t$	$\mathbf{x}_{t+1}$
(0 x 3)	(1 x 3)	(1 x 3)
growth matrix	initial population	population at time $t + 1$

The vector $\{12, 4, 0\}^T$ is the number of Saxon women in each age class after one step (one decade). There are twelve young women ($x_1 = 12$), four middle-aged women ($x_2 = 4$), and no old women ($x_3 = 0$). To find the age structure after a further decade, the new column vector $\mathbf{x}_{t+1}$, is pre-multiplied by the growth matrix, **G**:

$$\begin{bmatrix} 1 & 3 & 0 \\ \frac{1}{3} & 0 & 0 \\ 0 & \frac{1}{4} & 0 \end{bmatrix} \begin{bmatrix} 12 \\ 4 \\ 0 \end{bmatrix} = \begin{bmatrix} 24 \\ 4 \\ 1 \end{bmatrix}.$$

G	x_{t+1}	x_{t+2}
(3 x 3)	(1 x 3)	(1 x 3)
growth matrix	population at time $t+1$	population at time $t+2$

The pre-multiplication process can be repeated for any number of time steps, though this would assume that birth rates and survival rates do not change with time. In each case, the age structure at time $t+i$ is found by pre-multiplying the column vector describing the population age structure at time $t+(i-1)$ by the growth matrix. Continuing the example,

$$\begin{bmatrix} 1 & 3 & 0 \\ \frac{1}{3} & 0 & 0 \\ 0 & \frac{1}{4} & 0 \end{bmatrix} \begin{bmatrix} 24 \\ 4 \\ 0 \end{bmatrix} = \begin{bmatrix} 36 \\ 8 \\ 1 \end{bmatrix}.$$

G	x_{t+2}	x_{t+3}
(3 x 3)	(1 x 3)	(1 x 3)
growth matrix	population at time $t+2$	population at time $t+3$

Table 3.4 shows the female population age structure of the Saxon village calculated by this iterative method for ten time steps — a span of 90 years. The results are plotted in figure 3.5. Notice that the age structure of the population is stable, each class growing at the same rate.

The basic model of change in an age-structured population can be modified in several ways. For instance, allowance can be made for emigration and immigration. A full treatment migration is far too difficult to be dealt with in this book, and we shall give a very simplified example of how migration can be included in the model for the Saxon village.

Let us assume that a nearby village of bellicose Celts make occasional raids on the Saxon females. This population loss can be regarded as an emigration component, albeit a forced one. Assuming that the pusillanimous Saxons make no reprisal raids on the Celtic village so there are no immigrants, we can write the migration vector, **m**, as follows:

$$\mathbf{m} = \begin{bmatrix} -12 \\ 0 \\ 0 \end{bmatrix}$$

To see how this migration vector may be used in the model we shall take the female population of the Saxon village at time $t+1$, that is, $x_{t+1} = \{12, 4, 0\}^T$. Now, to find the population at x_{t+2} this population has to be pre-multiplied by **G**. Let us assume that during this decade the Celts first start to make their rapacious excursions. In this case we must add the migration vector to the model:

$$\mathbf{G}x_{t+1} + \mathbf{m} = x_{t+2}.$$

Table 3.4. Female population of Saxon village: age structure.

Age class	Time step, t (decades)									
	0	1	2	3	4	5	6	7	8	9
Young	12	12	24	36	60	96	156	252	408	660
Middle-aged	0	4	4	8	12	20	32	52	84	136
Old	0	0	1	1	2	3	5	8	13	21

Substituting our example values yields

$$\mathbf{x}_{t+2} = \begin{bmatrix} 1 & 3 & 0 \\ \frac{1}{3} & 0 & 0 \\ 0 & \frac{1}{4} & 0 \end{bmatrix} \begin{bmatrix} 12 \\ 4 \\ 0 \end{bmatrix} + \begin{bmatrix} -12 \\ 0 \\ 0 \end{bmatrix}$$

$$= \begin{bmatrix} 24 \\ 4 \\ 1 \end{bmatrix} + \begin{bmatrix} -12 \\ 0 \\ 0 \end{bmatrix}$$

$$= \begin{bmatrix} 12 \\ 4 \\ 1 \end{bmatrix}.$$

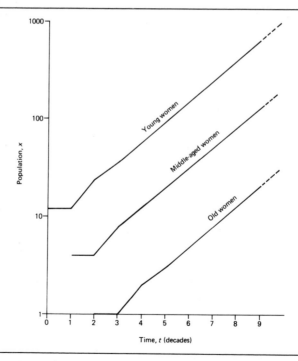

Figure 3.5. The growth of the female population in a Saxon village.

Carrying out the same procedure for the next decade we get

$$\mathbf{x}_{t+3} = \begin{bmatrix} 1 & 3 & 0 \\ \frac{1}{3} & 0 & 0 \\ 0 & \frac{1}{4} & 0 \end{bmatrix} \begin{bmatrix} 12 \\ 4 \\ 1 \end{bmatrix} + \begin{bmatrix} -12 \\ 0 \\ 0 \end{bmatrix}$$

$$= \begin{bmatrix} 24 \\ 4 \\ 1 \end{bmatrix} + \begin{bmatrix} -12 \\ 0 \\ 0 \end{bmatrix}$$

$$= \begin{bmatrix} 12 \\ 4 \\ 1 \end{bmatrix}$$

It will be found that for all future time steps the new population vector is $\{12, 4, 1\}^T$; this simply means that in this example the 'harvesting' of twelve young females leads to a stable population which remains constant in size.

Another modification of the basic model allows the inclusion of both sexes. Thus far we have not incorporated the male contingent of the Saxon village: we have tacitly assumed the men have successfully been doing their bit in the growth process. The inclusion of both sexes makes the growth matrix more complex. Alternate rows and columns are given to males and females, respectively. (Some workers use separate matrices for males and females.) The general growth matrix formulation for the two-sex, three age class case looks like this:

$$\mathbf{G} = \begin{bmatrix} 0 & b_{m1} & 0 & b_{m2} & 0 & b_{m3} \\ 0 & b_{f1} & 0 & b_{f2} & 0 & b_{f3} \\ s_{m1} & 0 & 0 & 0 & 0 & 0 \\ 0 & s_{f1} & 0 & 0 & 0 & 0 \\ 0 & 0 & s_{m2} & 0 & s_{m3} & 0 \\ 0 & 0 & 0 & s_{f2} & 0 & s_{f3} \end{bmatrix}$$

Number of males born to a female in each of the three age groups

Number of females born to a female in each of the three age groups

Survival rate of young males

Survival rate of young females

Survival rates of middle-aged and old males

Survival rates of middle-aged and old females

The column vector of the initial population is also divided into male and female subgroups

$$
\mathbf{x}_t = \begin{bmatrix} x_{m1} \\ x_{f1} \\ x_{m2} \\ x_{f2} \\ x_{m3} \\ x_{f3} \end{bmatrix}
\begin{array}{l}
\text{number of males in age group 1} \\
\text{number of females in age group 1} \\
\text{number of males in age group 2} \\
\text{number of females in age group 2} \\
\text{number of males in age group 3} \\
\text{number of females in age group 3}
\end{array}
$$

To see how this two-sex case would work we shall assume that the Saxon village has birth and survival rates that are the same for both men and women. Thus young women produce one son and one daughter per decade and the middle-aged women produce three sons and three daughters. With an initial population of twelve young men and twelve young women, the population after a decade is given by

$$
\mathbf{x}_{t+1} = \begin{bmatrix}
0 & 1 & 0 & 3 & 0 & 0 \\
0 & 1 & 0 & 3 & 0 & 0 \\
\frac{1}{3} & 0 & 0 & 0 & 0 & 0 \\
0 & \frac{1}{3} & 0 & 0 & 0 & 0 \\
0 & 0 & \frac{1}{4} & 0 & 0 & 0 \\
0 & 0 & 0 & \frac{1}{4} & 0 & 0
\end{bmatrix}
\begin{bmatrix} 12 \\ 12 \\ 0 \\ 0 \\ 0 \\ 0 \end{bmatrix}
=
\begin{bmatrix} 12 \\ 12 \\ 4 \\ 4 \\ 0 \\ 0 \end{bmatrix}
$$

$$\qquad\qquad\quad\ \mathbf{G}\qquad\qquad\qquad \mathbf{x}_t\qquad \mathbf{x}_{t+1}$$

After another decade

$$
\mathbf{x}_{t+2} = \begin{bmatrix}
0 & 1 & 0 & 3 & 0 & 0 \\
0 & 1 & 0 & 3 & 0 & 0 \\
\frac{1}{3} & 0 & 0 & 0 & 0 & 0 \\
0 & \frac{1}{3} & 0 & 0 & 0 & 0 \\
0 & 0 & \frac{1}{4} & 0 & 0 & 0 \\
0 & 0 & 0 & \frac{1}{4} & 0 & 0
\end{bmatrix}
\begin{bmatrix} 12 \\ 12 \\ 4 \\ 4 \\ 0 \\ 0 \end{bmatrix}
=
\begin{bmatrix} 24 \\ 24 \\ 4 \\ 4 \\ 1 \\ 1 \end{bmatrix}
$$

$$\qquad\qquad\quad\ \mathbf{G}\qquad\qquad\qquad \mathbf{x}_{t+1}\qquad \mathbf{x}_{t+2}$$

Notice the population changes in the same manner as the female population did when considered by itself (table 3.4); this is to be expected as the same birth and survival rates are used for the male population.

MODELS OF MINERAL CYCLES

Mineral cycles in ecosystems can be profitably studied with compartment model techniques. Whether viewed as global circulations, or circulations at local scales, say within a small river catchment, mineral cycles may be represented as a set of storage compartments which are linked by inputs and outputs of minerals. The simplest of mineral cycle models might consider just a few storage units — plants and animals, litter, and soil. Even with so few compartments, differences in mineral circulation patterns within the world's major biomes can be explored (see Gersmehl 1976); this

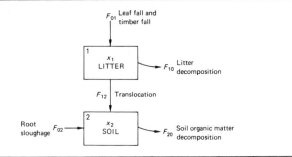

Figure 3.6. Inputs, outputs, and storages of carbon for litter and soil compartments.

will be demonstrated by developing a model of carbon flow in an ecosystem as described earlier (p61). For clarity of exposition, the system of interest will be reduced to two compartments only — litter and soil. Thus, in effect, the flow and storage of carbon in a soil profile will be modelled.

The model is sketched in figure 3.6. The litter compartment is labelled as compartment 1 and the soil compartment is labelled as compartment 2. The system consists of two state variables: x_1 is the carbon content of the litter store and x_2 is the carbon content of the soil store. Notice that the subscript refers to compartments and not time. Where reference to time is called for a second subscript is added: for example, $x_{1,t}$ would be the carbon content of the litter at time t; $x_{2,t+1}$ would be the soil carbon content at time $t + 1$; and $x_{1,9}$ would be the litter carbon content nine time steps from the initial state. For an input–output link between any pair of compartments, a source or donor compartment and a terminal or receptor compartment may be defined. If the source of a flow is compartment 1 and the destination is compartment 2, then the flow may be denoted as F_{12}, the subscripts indicating the flow is from compartment 1 to compartment 2. The environment, which lies outside the boundary of the system of interest, is regarded as compartment 0. So a flow from compartment 2 to the environment would be labelled F_{20}. All the other flows may be labelled according the same logic (figure 3.6).

Of the two inputs from outside the system of interest, F_{01} is leaf fall and timber fall, and F_{02} is root sloughage. Carbon stored in litter is subject to decomposition by soil organisms which release carbon dioxide in respiration, shown as output F_{10}, and also to translocation by water and burrowing animals, output F_{12}, which feeds into the soil store. The carbon stored in the soil is depleted by the respiration of soil organisms and this is shown as output F_{20}.

Carbon storage equations

Having defined the system of interest, the next step is to draw up a storage equation for each compartment. The change of storage in the two carbon stores is brought about by inputs and outputs of carbon. Denoting the change in carbon stored in litter during a small interval of time, Δt, by Δx_1, and the change in carbon stored in the soil during a small interval of time by Δx_2, storage equations may be expressed thus

$$\Delta x_1 = \quad (F_{01} \qquad\qquad - F_{10}) \qquad \Delta t$$
$$\Delta x_2 = \quad (F_{02} + F_{12} - F_{20}) \qquad \Delta t. \tag{3.15}$$

change in	inputs	outputs	time
carbon	of	of	interval
storage	carbon	carbon	

The change in carbon storage in the litter compartment, Δx_1, is the difference between the amount of carbon in store at the start of the time interval, time t, and the amount of carbon in store at the end of the time interval, time $t + 1$; the change in storage in the soil compartment can be stated in like manner. So the following definitions obtain

$$\Delta x_1 = x_{1,t+1} - x_{1,t}$$
$$\Delta x_2 = x_{2,t+1} - x_{2,t},$$

equations (3.15) may be written

$$x_{1,t+1} - x_{1,t} = (F_{01} - F_{10})\Delta t$$
$$x_{2,t+1} - x_{2,t} = (F_{02} + F_{12} - F_{20})\Delta t,$$

and rearranged to read

$$x_{1,t+1} = x_{1,t} + (F_{01} - F_{10})\Delta t$$
$$x_{2,t+1} = x_{2,t} + (F_{02} + F_{12} - F_{20})\Delta t. \tag{3.16}$$

Equations (3.16), which form the model in its uncalibrated state, state that the amount of carbon stored in each compartment after a time interval depends upon the amount of carbon stored at the start of the time interval, and the inputs and outputs of carbon during the time interval; they convey the same information as equations (3.15) and both sets of equations may be put as follows

$$\Delta x_1 / \Delta t = (F_{01} - F_{10})$$
$$\Delta x_2 / \Delta t = (F_{02} + F_{12} + F_{20}) \tag{3.17}$$

where the *rate* of change of carbon storage in each compartment is equated with the difference between inputs and outputs of carbon.

Carbon transport laws

Inputs and outputs of carbon – the F's in equations (3.16) – now need defining by transport laws. On an annual basis, inputs of carbon from leaf fall and timber fall, and from root sloughage, can be regarded as constant values which will be specified when the model is calibrated. The outputs of carbon may, as a first approximation anyway, be defined by a linear transport law[2]. A linear transport law states that the flow of water, or nutrients, or energy between two compartments is some function of the storage in the donor (or, more rarely, receptor) compartment. In general

$$F_{ij} = f(x_i)$$

and in a specific case, the function might be

$$F_{ij} = k_{ij}x_i,$$

where k_{ij} is the specific rate of transfer or turnover rate (this will be explained shortly). The transfer is said to be donor-controlled because the donor compartment determines both maximum rate and actual rate of transfer.

Applying the linear transport law to the model, the output from the litter compartment resulting from respiration may be defined as

$$F_{10} \qquad = \quad k_{10} \qquad \qquad x_1$$

output from litter to air	turnover rate from litter to air	carbon storage in litter

which states that the output of carbon in respiration from litter is proportional, by a turnover rate from litter to air, to the storage of carbon in the litter. The turnover rate, k_{10}, is the fraction of carbon stored in the litter which is lost in respiration in a unit interval of time; it may be defined as the output, F_{10}, divided by the storage of carbon in litter when inputs and outputs for the litter compartment are equal, that is, when the system is in a steady state (storage values do not change with time). For instance, if in the steady state the litter compartment contained 1000 kilograms of carbon per hectare, and if the output of carbon in respiration were 10 kilograms of carbon per hectare per day, the turnover rate would be

$$k_{10} = 10/1000,$$

which may be expressed as a fraction, 0·01 per day, or as a percentage, 1 per cent per day; this would mean it would take 100 days to displace all the carbon stored in the litter by respiration. The turnover rate is also referred to as the transfer coefficient and the rate constant. The reciprocal of the turnover rate is the turnover time and this represents the time taken to displace all the stored carbon in respiration under steady-state conditions – as has been stated, this is 100 days. In the physical sciences, the term time constant is used in preference to turnover time, and in the geological literature the term residence time is widely used; all three terms have the same physical significance and broadly represent the ratio of throughput to storage; they thus indicate how much of the input is stored and how much emerges as output during a unit time interval.

Other flows in the carbon model are defined as

$$F_{20} \quad = \quad k_{20} \qquad \qquad x_2$$

output from soil to air	turnover rate from soil to air	carbon storage in soil

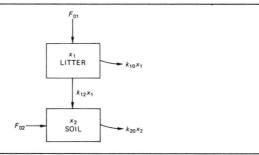

Figure 3.7.   Carbon flows defined by transport laws.

and

$$F_{12} = k_{12} \qquad x_1.$$

output	turnover	carbon
from	rate from	storage
litter	litter	in
to soil	to soil	litter

All input and output definitions are shown in figure 3.7.

Calibration and solution of the carbon model

Substituting the definitions of input and output terms into equations (3.16) gives

$$
\begin{aligned}
x_{1,t+1} &= x_{1,t} + (F_{01} - k_{10}x_1 - k_{12}x_1)\Delta t \\
x_{2,t+1} &= x_{2,t} + (F_{02} + k_{12}x_1 - k_{20}x_2)\Delta t.
\end{aligned}
\tag{3.18}
$$

This set of simultaneous, difference equations, which forms the basic model, describes the flow and storage of carbon in the system. To use the model it must be calibrated for particular cases; in other words, the inputs from outside the system, F_{01} and F_{02}, and the turnover rates, k_{10}, k_{12}, and k_{20} must be supplied.

Figure 3.8 lists parameter values typical of a prairie grassland soil. Inputs are in units of kilograms of carbon per hectare per year (kg C/ha/yr); turnover rates are in units of

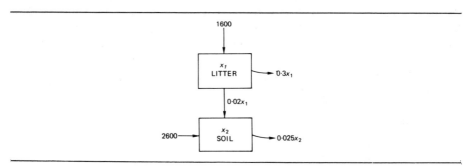

Figure 3.8.   The carbon cycle model calibrated for a prairie grassland.

per year. Substituting the parameter values into equations (3.18) gives

$$x_{1,t+1} = x_{1,t} + (1600 - 0.30x_{1,t} - 0.02x_{1,t})\Delta t$$
$$x_{2,t+1} = x_{2,t} + (2600 + 0.02x_{1,t} - 0.025x_{2,t})\Delta t.$$

(3.19)

The dynamics of the prairie grassland system — how the values of carbon storage in each compartment change with time — is uniquely defined by equations (3.19). To find the storage in each compartment at any time, the equations must be solved simultaneously, the Euler method being suitable for doing this.

Assume the compartments start out empty. The initial state of the prairie grassland is thus $x_{1,0} = 0.0$ and $x_{2,0} = 0.0$. The amount of carbon in each of the two stores after a year has elapsed can be found by substituting $x_{1,0}$ and $x_{2,0}$ into equations (3.19); this produces

$$x_{1,1} = 0.0 + \{1600 - (0.32)(0.0)\}1.0 \qquad\qquad = 1600$$
$$x_{2,1} = 0.0 + \{2600 + (0.02)(0.0) - (0.025)(0.0)\}1.0 = 2600.$$

So, after one year, 1600 kilograms of carbon per hectare are stored in the litter compartment and the soil compartment contains 2600 kilograms of carbon per hectare. The storages after a second year has passed are then worked out like this

$$x_{1,2} = x_{1,1} + (1600 - 0.32x_{1,1})1.0$$
$$x_{2,2} = x_{2,1} + (2600 - 0.02x_{1,1} + 0.025x_{2,1})1.0$$
$$= 1600 + \{1600 - (0.32)(1600)\}1.0$$
$$= 2600 + \{2600 + (0.02)(1600) - (0.025)(2600)\}1.0$$
$$= 2688 \text{ kg C/ha}$$
$$= 5230 \text{ kg C/ha}.$$

These values can then be used to find the storages after three years and so on. Calculations for several time steps are plotted in figure 3.9. Notice that in a steady

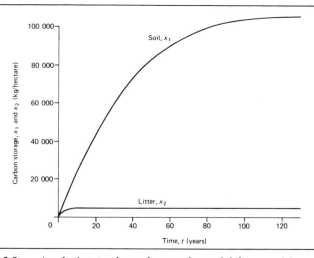

Figure 3.9. A solution to the carbon cycle model for a prairie grassland.

state, that is, when inputs and outputs balance, the litter contains 5000 kg C/ha and the soil 108 000 kg C/ha, values which accord well with the observed storage of carbon in prairie grassland soils.

The carbon cycle model could be calibrated for any environment. Different parameters produce different system dynamics and thus different steady-state storages of carbon. Using the model, it should be possible to predict the steady-state contents of litter and soil compartments in any environment. A quick method of deriving steady-state storages is founded on the following logic.

In a steady state, $\Delta x_1 / \Delta t$ and $\Delta x_2 / \Delta t$ in equations (3.17) will both be zero as storage will not change with time. This being so, equations (3.17), when calibrated for a prairie grassland, read

$$0 \cdot 0 = 1600 - 0 \cdot 32 x_1$$
$$0 \cdot 0 = 2600 + 0 \cdot 02 x_1 - 0 \cdot 025 x_2.$$

Rearranging to put the constants 1600 and 2600 on the left-hand side, and to line up the x_1's and x_2's produces

$$-1600 = -0 \cdot 32 x_1$$
$$-2600 = +0 \cdot 02 x_1 - 0 \cdot 025 x_2. \tag{3.20}$$

Equations (3.20) may be put into matrix form as

$$\begin{bmatrix} -1600 \\ -2600 \end{bmatrix} = \begin{bmatrix} -0 \cdot 32 & 0 \cdot 0 \\ +0 \cdot 02 & -0 \cdot 025 \end{bmatrix} \begin{bmatrix} x_1 \\ x_2 \end{bmatrix} \tag{3.21}$$

$$\mathbf{z} \qquad\qquad \mathbf{K} \qquad\qquad \mathbf{x}$$

(the reader should verify this by multiplying out the right-hand side) where $\mathbf{z}$ is an input vector, $\mathbf{K}$ is a turnover rate matrix, and $\mathbf{x}$ is a state vector. To find steady-state values of x_1 and x_2, two steps are taken. Firstly, the inverse of the turnover rate matrix is found; it is[3]

$$\mathbf{K}^{-1} = \begin{bmatrix} -3 \cdot 125 & 0 \cdot 0 \\ -2 \cdot 5 & -40 \cdot 0 \end{bmatrix}.$$

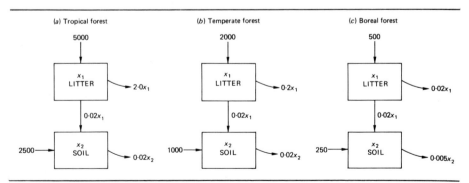

Figure 3.10. The carbon cycle model calibrated for: (*a*) A tropical forest; (*b*) A temperate forest; (*c*) A boreal forest.

Table 3.5. Steady-state carbon storage in three types of forest.

	Steady-state carbon storage (kg C/ha)			
	Litter, x_1		Soil, x_2	
Ecosystem	Observed	Predicted	Observed	Predicted
Tropical forest	1500	2475	102 600	127 475
Temperate forest	12 100	9080	106 300	108 880
Boreal forest	20 000	12 500	129 200	100 000

Secondly, the left-hand side of equations (3.21) is multiplied by $\mathbf{K}^{-1}$

$$\begin{bmatrix} -3\cdot125 & 0\cdot0 \\ -2\cdot5 & -40\cdot0 \end{bmatrix} \begin{bmatrix} -1600 \\ -2600 \end{bmatrix} = \begin{bmatrix} 5000 \\ 108\ 000 \end{bmatrix}$$

$$\mathbf{K}^{-1} \qquad \mathbf{a} \qquad \mathbf{y}^s$$

the resulting vector, $\mathbf{x}^s$, containing the steady-state values of carbon stored in litter, x_1^s (5000 kg/ha) and soil, x_2^s (108 000 kg/ha).

This method can be used to examine the effect of altering the parameters on the steady-state storages of carbon. Figure 3.10 depicts values of parameters suitable for tropical, temperate, and boreal forests. Solving for steady-state storages of carbon in soil and litter in each of the three cases gives the results shown in table 3.5. Notice the predicted storages are in agreement with typical observed storages in the three forest ecosystems.

APPLICATIONS

Compartment models have been successfully applied to a variety of ecological systems, including those where man plays an important role. Using a compartment model, Hett and O'Neill (1974) were able to show that one reason the natives of the Aleutian Islands had developed a dependence upon the sea for food, clothing, and building materials before contact with civilization was that the marine food supply is 300 times less sensitive to changes in the environment than is the terrestrial food supply – the sea seems to be a more 'dependable' source of food; it was also found that the Aleut population played a negligible role in maintaining the stability of the whole ecosystem. Park *et al* (1974) constructed a general model of a lake ecosystem consisting of 16 compartments including attached aquatic plants, phytoplankton, zooplankton, bottom-dwelling aquatic insects, fish, suspended organic matter, decomposers, sediment, and nutrients. The model has given insight into the effects of nutrient enrichment on the Lake George ecosystem, New York: doubling the input of phosphates into the southern end of the lake leads to an increase in the spring of large diatoms, including those which lower the water quality by imparting to it a bad smell and taste. Nutrient enrichment in Lake Erie has been studied with a spatial compartment model by Di Toro *et al* (1975).

The management of grazing ecosystems has been aided by the use of compartment models: Goodall (1970) has built a model to simulate sheep grazing in Australia with rainfall treated as a random variable; Milner (1972) simulated the dynamics of the Soay sheep population on the St Kilda Nature Reserve; and Swartzman and Singh (1974) considered the value of a simulation model in assessing the worth of various management schemes in a successional tropical grassland. Several compartment models are described in Huggett (1980).

As a case study of a compartment model, the one built by O'Neill and Burke (1971) to study the movement of DDT and its breakdown product, DDE, in human food chains

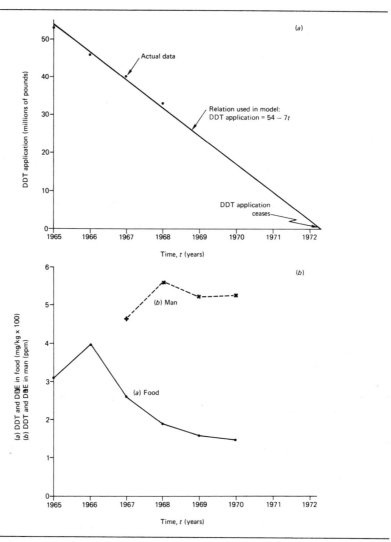

Figure 3.11.   Data used in modelling DDT and DDE movement through the human food chain in the USA. Based, with permission, on Table 1 in R V O'Neill and O W Burke (1971).

will be described. The model was specifically designed to draw implications about future pesticide concentrations in humans in the USA from the scanty data shown in figure 3.11. Bearing in mind the paucity of data, a simple model was built in which pesticide concentrations in three compartments — food directly contaminated by spraying; food indirectly contaminated (by recycling of pesticide through the soil for instance); and man — were linked by flows of pesticides. The time rate of change in storage of pesticide in each compartment was expressed as a function of inputs less outputs

$$\frac{\Delta \text{FOOD}(a)}{\Delta t} = \begin{array}{c} \text{inputs by direct} \\ \text{spraying} \end{array} - \begin{array}{c} \text{elimination rate} \\ \text{from FOOD}(a) \end{array}$$

$$\frac{\Delta \text{FOOD}(b)}{\Delta t} = \begin{array}{c} \text{inputs by indirect} \\ \text{contamination} \end{array} - \begin{array}{c} \text{elimination rate} \\ \text{from FOOD}(b) \end{array}$$

$$\frac{\Delta \text{MAN}}{\Delta t} = \begin{array}{c} \text{inputs from FOOD}(a) \\ \text{and FOOD}(b) \end{array} - \begin{array}{c} \text{elimination rate} \\ \text{from MAN} \end{array}$$

Or, in symbols

$$\frac{\Delta x_1}{\Delta t} = a_1 F(t) - b_1 x_1$$

$$\frac{\Delta x_2}{\Delta t} = a_2 F(t) - b_2 x_2 \qquad\qquad (3.22)$$

$$\frac{\Delta x_3}{\Delta t} = a_3 (x_1 + x_2) - b_3 x_3,$$

where x_1 is the food directly contaminated by spraying, x_2 is the food indirectly contaminated, and x_3 is the concentration of pesticide in man. The a parameters control the rate of input, and the b parameters control the elimination rate, of DDT and DDE in food and man. The function $F(t)$ represents DDT usage and is the actual application rate in the USA.

To render the model operational, the parameters were derived for the US case. The method used was to try several different values until the model output successfully matched the observed data shown in figure 3.11(b). The parameters for man, x_3, are $a_3 = 0 \cdot 1430$ and $b_3 = 0 \cdot 4740$. The function $F(t)$ was assumed to be linear and of the form

$$F(t) = 54 - 7t,$$

which follows the application rate shown in figure 3.11(a) and implies that DDT usage ceases in 1973. The equations were then solved to give the concentrations of DDT in x_1, x_2, and x_3 up to the year 2022 (figure 3.12).

A number of other cases were also simulated with the model and these are also shown in figure 3.12. The first case is the one we have described — DDT usage reduced at a rate of about 7 million pounds per year. The results show that levels of DDT in man continue to drop but nevertheless measurable amounts are still present half a century hence. In the second case it was assumed DDT usage was maintained at the

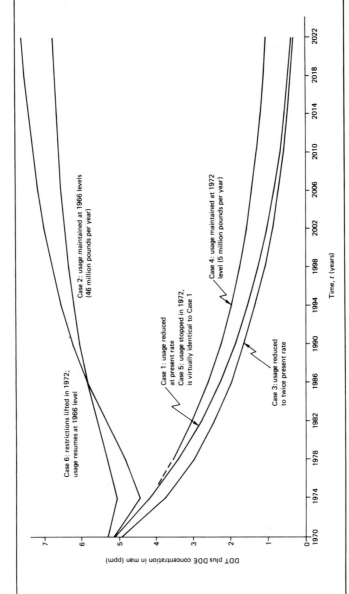

Figure 3.12. DDT plus DDE concentration in man as predicted by O'Neill and Burke's model with varying assumptions about DDT application rate. Based, with permission, on Table 3 in R V O'Neill and O W Burke (1971).

1966 level of 46 million pounds per year. The predictions show that in this case DDT levels in man start to rise around 1978 and continue to rise (though at a decreasing rate) to a final concentration in excess of 6·7 ppm in 2022. The third case assumed a doubling of the present rate of decrease in DDT usage (14 million pounds per year rather than 7) effective from 1966. The results show a more rapid decrease of DDT concentration in man than in case one, a concentration of 1 ppm being reached four years earlier. Case four investigated DDT usage maintained at 1972 levels (5 million pounds per year); this retards the rate of decrease and concentrations exceed 1 ppm in 2022. The results of case 5, in which DDT application is stopped in 1972, produces almost identical results to case one. Finally, in case 6, the resumption of 1966 levels of DDT usage after 1972 was assumed. Not surprisingly, this leads to rising DDT levels in man, being 7·6 ppm and still climbing in 2022.

The implications of the results are self-evident. Cases 1, 3, and 5 show the persistence of DDT in man arising from its slow turnover in the environmental pool. Despite its limitations (limited data, for example), the model is a method for projecting future DDT levels on the basis of present data and a knowledge of major mechanisms which are thought to dominate exchange of DDT in food chains leading to man.

CONCLUDING REMARKS

This chapter has shown how systems of geographical interest characterized by storages and flows of matter or energy can be modelled mathematically on a sound theoretical basis. The models have been kept as simple as possible. Relatively simple models are not without value, as perusal of Gersmehl's (1976) paper and chapter six of Huggett's (1980) book will show; but many models found in the literature are more refined, both in terms of transport laws and calibration, than those described in this chapter. For example, it was noted that sunlight is received by plants and turned into chemical energy of plant tissue. Actual models of plant production are sophisticated and take into account environmental factors other than sunlight intensity in the computation of plant production, notably leaf area, nitrogen supply, and temperature (see Park *et al* 1974, Titus *et al* 1975). Likewise, the decay of organic matter was considered a simple process in which the loss of organic matter by decay during a time interval was set proportional to the amount of organic matter in store. More realistic models of organic decay processes include various organic substances separately — humic material, faeces, dead plant material, and so on — and the effect of water and temperature regimes (see Hunt 1977, Schlesinger 1977). Population models have reached a very high level of refinement and have peculiar problems of calibration (see Pressat 1978, Rees and Wilson 1977).

Storage equations of sorts may be written for socio-economic systems, though their formulation is commonly based on thinner theoretical grounds, and their calibration is more problematical and more objectionable, than is the case with physical systems. The most infamous models of socio-economic systems are the ones built by Forrester (1969, 1971) to simulate what he called the world system and an urban system. Another model dealing with societary problems is that constructed by Watt *et al*

(1974) to predict the likely effects of the 1973 oil price rise on population, transport, and other parts of society in the United States. Details of these socio-economic models and others like them may be found in Bennett and Chorley (1978) and Huggett (1980).

Notes

1   It is also known as the Malthusian parameter after the Reverend Thomas Malthus.

2   Processes which seem to conform to a linear transport law include water release from soil and rock to rivers, the energy loss in respiration of organisms, and the loss of nutrients from ecosystem components, especially where the loss is to an abiotic compartment — litter, soil, atmosphere, and so on.
    Many ecologists argue that transfers of matter and energy between compartments which represent living components of ecosystems are described by a function of the general form

$$F_{ij} = f(x_i, x_j),$$

where x_i is the storage in the donor compartment and x_j is the storage in the receptor compartment. Specific instances of the function usually take the form

$$F_{ij} = k_{ij} x_i x_j.$$

The product of the two storages rendering it nonlinear, this function describes certain ecological phenomena which could not be expressed by a linear transport law: the asymptotic approach of flow rates to a maximum or minimum level; the sudden change in inputs and outputs when a threshold value is passed; and the very existence of flow depending on *both* compartments having something in store.

3   As was seen earlier (p28), this may be done for a 2 x 2 matrix by swapping the position of the elements on the main diagonal, changing the signs of the other two elements, and multiplying each element by the reciprocal of the matrix determinant. Now swapping positions and changing signs leaves matrix **K** looking like this

$$\begin{bmatrix} -0 \cdot 025 & 0 \cdot 0 \\ -0 \cdot 02 & -0 \cdot 32 \end{bmatrix}.$$

The determinant is $(-0 \cdot 32 \times -0 \cdot 025) - (0 \cdot 0 \times 0 \cdot 2) = 0 \cdot 008$ and the reciprocal of this is $1/0 \cdot 008 = 125$. Multiplying each element of the rearranged matrix by 125 yields the inverse, $\mathbf{K}^{-1}$

$$\mathbf{K}^{-1} = \begin{bmatrix} -3 \cdot 125 & 0 \cdot 0 \\ -2 \cdot 5 & -40 \cdot 0 \end{bmatrix}$$

This may be checked by pre-multiplying **K** by its inverse $\mathbf{K}^{-1}$

$$\underset{\mathbf{K}^{-1}}{\begin{bmatrix} -3 \cdot 125 & 0 \cdot 0 \\ -2 \cdot 5 & -40 \cdot 0 \end{bmatrix}} \underset{\mathbf{K}}{\begin{bmatrix} -0 \cdot 32 & 0 \cdot 0 \\ +0 \cdot 02 & -0 \cdot 025 \end{bmatrix}} = \underset{\mathbf{I}}{\begin{bmatrix} 1 & 0 \\ 0 & 1 \end{bmatrix}}$$

which, as it should do, results in the formation of an identity matrix.

CHAPTER 4

Space–time Deterministic Models

The models dealt with in the previous chapter considered the behaviour of a system at a particular place. The system measures were representative of the system at that place, but locational differences in system measures were excluded from the models. It was discovered, for instance, how carbon storage in litter and soil changes with time, but nothing was learned about how the stored carbon was distributed within a soil profile. A vast number of problems in geography involve *location* as an important facet – some might even say that without a locational facet, a model is not geographical – and interest centres on how system measures are distributed over a system of interest or spatial domain: how does the depth of overland flow vary in a catchment? How is sulphur dioxide distributed in urban air? How does the concentration of suspended sediment vary in a lake? How is population growth influenced by migration between regions? To answer questions like these, the system of interest must be studied in some sort of regional or spatial framework.

STORAGE, FLOW, AND SPATIAL FRAMEWORKS

One of two spatial frameworks is commonly used, depending on the kind of system being studied and the sort of data available. In the first type, the spatial domain is regarded as *continuous* in all directions and is defined, usually, in Cartesian coordinates. In this framework, sulphur dioxide concentrations would be assumed to vary throughout a three-dimensional parcel of air overlying an urban area; the concentration of sulphur dioxide could be defined at any point in the parcel of air, the value at a particular point being denoted by, say, $C(x, y, z)$, where x and y are horizontal coordinates and z is a vertical coordinate (figure 4.1a). The value of sulphur dioxide concentration may also vary with time, t, and the full space–time designation would be $C(x, y, z, t)$. Similarly, the height of the land surface within a spatial domain could be assumed to vary continuously in the space defined by the x and y coordinates and with time. So to specify the height of the land surface, h, at any particular spot and time within a two-dimensional spatial domain, the general notation $h(x, y, t)$ could be used (figure 4.1b). Considering sediment concentration, S, along a river, a one-dimensional spatial domain is set up, the sediment concentration at any point in

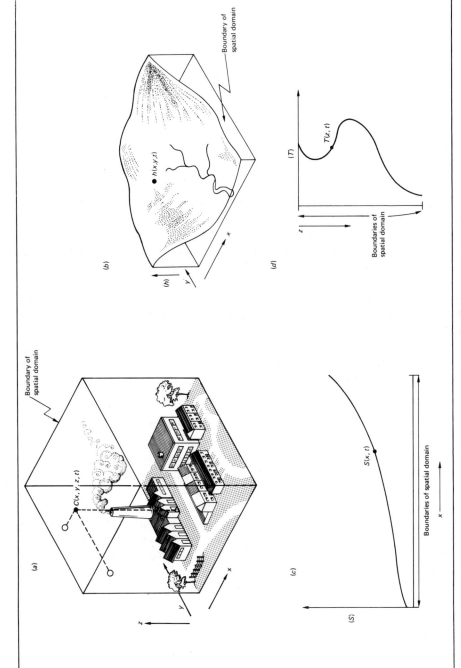

Figure 4.1. Spatial domains in continuous space.

which, and at any time, being written $S(x,t)$ (figure 4.1c). Looking at heat distribution in a soil profile, a vertical spatial domain is of interest, the temperature at any point in which could be designated $T(z,t)$ (figure 4.1d).

In the second type of spatial framework, the spatial domain comprises either a set of *regular* or *irregular regions,* each of which is a spatial cell of finite size, or simply a set of sampling points. The size and shape of the spatial cell is often arbitrary and can be geared to the problem being considered, the availability of data, and the level of spatial detail required in different parts of the spatial domain. In physical applications, the simplest kind of spatial cell is a cuboid or block. Blocks can be built up to fill the space under scrutiny. To study processes in a soil profile, the blocks may be stacked; to model flow in a river they may be placed end to end in a line; to study landscape processes in two dimensions they may be placed side by side; to study three-dimensional domains they may be placed side by side as well as stacked (figure 4.2). Other shapes may also be employed: in studying lakes it is common to use slices of water stacked one upon another. In human applications, which will be dealt with in Chapter 5, the spatial cells correspond to a set of discrete regions such as wards, enumeration districts, or counties.

Having established a spatial problem or hypothesis and having erected a suitable spatial domain in which to study it, the next step is to draw up storage equations, one for each spatial cell. Take the example of a soluble pollutant dumped in a canal in which the water is not flowing. To model the subsequent spread of pollutant by the process of diffusion, the canal could be divided into a line of spatial cells. The pollutant will tend to diffuse from cell to cell, upstream and downstream of the cell into which it was dumped. A pollutant *storage equation* of the form

$$\begin{matrix} \text{change in pollutant} \\ \text{storage in cell} \end{matrix} = \left(\begin{matrix} \text{pollutant} \\ \text{inputs} \end{matrix} - \begin{matrix} \text{pollutant} \\ \text{outputs} \end{matrix} \right) \begin{matrix} \text{time} \\ \text{interval} \end{matrix}$$

could be drawn up for each cell in the canal. The pollutant input and output terms then need defining by a *transport law* which represents the diffusive process — *Fick's first law of diffusion* fits the bill; it states that the diffusive flow rate of a solute (in the example the pollutant) is proportional, by a solute diffusion coefficient, to the negative gradient of solute concentration

$$\begin{matrix} \text{solute flow} \\ \text{rate} \end{matrix} = \begin{matrix} \text{solute diffusion} \\ \text{coefficient} \end{matrix} \times \begin{matrix} \text{negative solute} \\ \text{concentration gradient.} \end{matrix}$$

This transport law would apply to all cells save those at the ends of the canal section of interest where special flow conditions may need to be specified — these are known as the *boundary conditions.* The negative gradient of pollutant concentration could be expressed as the negative difference between pollutant concentration in two adjacent cells divided by the distance between the cell centres. (As will be seen later, the negative sign simply ensures that the pollutant always moves from the cell with the higher pollutant concentration to the cell with the lower pollutant concentration and never the other way — the process of solute diffusion is said to be irreversible; if the cells have identical concentrations of pollutant, no flow will take place.) The significance of all this is that the inputs and outputs depend on the spatial distribution of pollutant along the canal; hence the model deals explicitly with locational aspects of flow and storage.

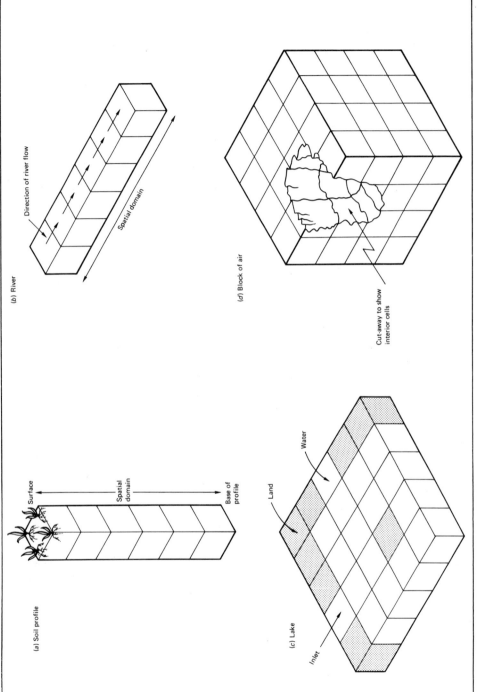

Figure 4.2. Spatial domains filled by discrete spatial units (cells).

(a) Soil profile

Surface

Spatial domain

Base of profile

(b) River

Direction of river flow

Spatial domain

(c) Lake

Inlet

Land

Water

(d) Block of air

Cut-away to show interior cells

Two other important spatial transport laws define the flow of heat and water. *Fourier's law of heat conduction* defines the flow rate of heat through any system. The law permits heat to pass from hot parts of the system to cool parts, but never from cool regions to hot ones: heat flow is an irreversible process. The flow of heat takes place along a temperature gradient. In the absence of temperature gradients in a system (that is, the temperature is the same at all points within the system), and assuming that heat is not being transferred by convention, no flow of heat can occur. Specifically, Fourier's law of heat conduction states that the flow rate of heat is proportional, by a parameter known as a heat conduction coefficient, to the steepness of the negative temperature gradient

$$\frac{\text{heat flow}}{\text{rate}} = \frac{\text{heat conduction}}{\text{coefficient}} \times \frac{\text{negative temperature}}{\text{gradient}}$$

the negative sign ensuring that heat flows 'downhill' from hot to cold and not the other way. The greater the difference in temperature between two parts of a system, the steeper the temperature gradient and the faster the heat flow rate.

Darcy's law defines the flow rate of water in a porous medium: the flow rate of water is proportional, by a water conduction coefficient, to the negative moisture gradient

$$\frac{\text{water flow}}{\text{rate}} = \frac{\text{water conduction}}{\text{coefficient}} \times \frac{\text{negative moisture}}{\text{gradient}}.$$

The greater the difference in moisture content between two parts of, say, the soil, the steeper the moisture gradient and the more rapid the flow rate of water.

Once a storage equation has been written for each cell in a spatial domain, and once all inputs and outputs have been suitably defined, the model is complete. All that remains is to *calibrate* it, to obtain a *solution*, and to *test* the results — three steps which in practice are often difficult and time-consuming to implement.

To summarize, the procedure for modelling a physical space—time system runs something like this:

1. Define problem: state hypothesis.
2. Define system of interest in terms of physical components, boundaries, and spatial domain.
3. Describe the system components by suitable measures (state variables) — these will usually be storages of mass or energy in the spatial cells which constitute the system.
4. Draw up storage equations for each of the cells which lie inside the spatial domain, but not for those on the boundaries.
5. Define by physical transport laws the inputs and outputs between cells as described in the storage equations.
6. Specify (*a*) the storages at the start of the time period over which the model will run — these are the initial conditions, (*b*) what goes on at the boundaries of the spatial domain — these are the boundary conditions, and (*c*) the parameter values.
7. Solve the storage equations for specific initial and boundary conditions and so reveal how the distribution of mass or energy within the spatial domain of interest changes with time.
8. Test the predicted changes against observed changes and, in the light of the test,

accept or reject the model. If the model is rejected, assumptions or parameters may be changed and the model re-run.

The next three sections of the chapter will show how to build, calibrate, solve, and test models of three different, physical, space—time processes — heat flow in a soil profile, water movement in soil or rock, and salt movement in an enclosed sea.

MODELLING HEAT FLOW IN A SOIL PROFILE

Assumptions

The assumptions made in modelling heat transfer in soils are twofold:

(*a*) Heat flow, that is the flow of thermal energy, obeys the law of energy conservation.

(*b*) Heat flows from warm areas to cool areas, the rate of flow being defined by Fourier's law of heat conduction.

Formulation

(*a*) We shall formulate the model for a block of soil, a cube in shape, whose volume is 10 cm^3. In formulating the model we shall also need to consider a block of soil of the same volume lying above the block of interest in the soil profile and a similar block lying under the block of interest.

The block of interest will consist of soil particles, water, and air spaces, all of which have the capacity to store heat. The amount of heat a substance can store per unit volume at a given temperature is known as heat capacity. Air stores 0·0003 calories per cubic centimetre per degree Centigrade (cal/cm^3/°C); for instance, water stores 1·0 cal/cm^3/°C. All soil components have different heat capacities, but formulae are available for evaluating an overall heat capacity for a block of soil. We shall assume our block is a sandy soil and has a heat capacity of 0·5 cal/cm^3/°C.

(*b*) Heat may enter the block and heat may leave the block. For simplicity, we shall consider the case in which heat passes to and from the block via its upper and lower faces only: heat transfer to blocks of soil positioned to the side of the block of interest is not permitted. The problem is thus reduced to one-dimensional heat flow in a soil profile.

(*c*) Storage, input, and output of heat is, we are assuming, subject to the law of energy conservation. Consequently, a heat storage equation may be drawn up which equates the change in the heat storage in the block during a given time interval to the numerical difference between heat inputs and heat outputs

$$\begin{bmatrix} \text{change in heat storage in} \\ \text{the block during a given} \\ \text{time interval} \end{bmatrix} = \begin{bmatrix} \text{heat} \\ \text{inputs} \end{bmatrix} - \begin{bmatrix} \text{heat} \\ \text{outputs} \end{bmatrix}. \qquad (4.1)$$

(*d*) The heat inputs and heat outputs will be expressed as heat flow rates as defined by Fourier's law of heat conduction. Remember that Fourier's law defines heat flow rates as proportional, by a heat conduction coefficient, to the negative gradient of temperature between the points that heat is flowing

$$\begin{bmatrix} \text{heat flow} \\ \text{rate} \end{bmatrix} = \begin{bmatrix} \text{heat conduction} \\ \text{coefficient} \end{bmatrix} \times \begin{bmatrix} \text{negative gradient} \\ \text{of temperature} \end{bmatrix}. \tag{4.2}$$

In our block of soil there are two heat flows to be considered: one through the top of the block and one through the bottom of the block; so we have to define two heat flow rates. For heat flow through the top face of the block, the negative gradient of temperature may be expressed as the negative difference between the temperature of the block of interest and the temperature of the block lying above it, divided by the distance between the centres of the blocks

$$\begin{bmatrix} \text{negative gradient} \\ \text{of temperature} \end{bmatrix} = - \left[\frac{\left(\begin{array}{cc} \text{temperature of} & \text{temperature of} \\ \text{block of interest} & \text{overlying block} \end{array} \right)}{\text{distance between centres of blocks}} \right] \tag{4.3}$$

So if the temperature of the block of interest is 10 °C, the temperature of the block above it is 8 °C, and the distance between the centres of the two blocks is 10 cm, then the negative temperature gradient through the top face is

$$-(10 - 8)/10 = -0.2\ ^\circ\text{C/cm}.$$

To define the heat flow rate from this we need to know the value of the heat conduction coefficient. Assume we have measured this and found it to be 0·01 cal/cm/s/°C. The heat flow rate will then be

$$0.01\ \text{cal/cm/s/}^\circ\text{C} \times -0.2\ ^\circ\text{C/cm} = -0.002\ \text{cal/cm}^2\text{/s}.$$

This then is the rate at which heat leaves the block per unit area (cm^2) of the top face. The area of the top face is 100 cm² so the total heat outflow per unit time is 0·2 cal/s.

For heat flow through the bottom face of the block, the negative gradient of temperature may be expressed as the negative difference between the temperature of the block of interest and the temperature of the block lying beneath it, divided by the distance between the centres of the two blocks

$$\begin{bmatrix} \text{negative gradient} \\ \text{of temperature} \end{bmatrix} = - \left[\frac{\left(\begin{array}{cc} \text{temperature of} & \text{temperature of} \\ \text{block of interest} & \text{block below} \end{array} \right)}{\begin{array}{c} \text{distance between the centres of} \\ \text{blocks} \end{array}} \right]. \tag{4.4}$$

So if the temperature of the block of interest is 10 °C, the temperature of the block below is 15 °C, and the distance between the centres of the blocks is 10 cm, then the negative gradient of temperature through the bottom face is

$$-(10 - 15)/10 = +0.5\ ^\circ\text{C/cm}.$$

To define the heat flow rate from this the conduction coefficient through the bottom of the block must be measured: we shall assume, as with the top face, a value of 0·01 cal/cm/s/°C. The heat flow into the block through the bottom face will thus be +0·005 cal/cm²/s; and the total heat inflow through the bottom face, which has an area of 100 cm², is 0·5 cal/s.

(e) In the case we have considered, heat flows into the block through the bottom face and out of the block through the top face. To find the change in heat storage in the block during a given time interval, according to the storage equation, we must find the difference between heat inputs and heat outputs during the time interval. We have found the heat input rate is 0·005 cal/cm²/s, and the heat output rate is 0·002 cal/cm²/s. Because the surface areas of the top and bottom faces are 100 cm², we have to multiply these flow rates by 100 cm² (as we have already seen) as well as by the time interval. The general equation is

$$
\begin{bmatrix} \text{change in heat storage} \\ \text{in the block during a} \\ \text{given time interval} \end{bmatrix} = \left[\begin{pmatrix} \text{heat} \\ \text{inputs} \end{pmatrix} - \begin{pmatrix} \text{heat} \\ \text{outputs} \end{pmatrix} \times \begin{array}{l} \text{area} \\ \text{of} \\ \text{face} \end{array} \times \begin{array}{l} \text{length} \\ \text{of time} \\ \text{interval} \end{array} \right]. \qquad (4.5)
$$

In the example, for a time interval of one minute,

$$= (0·005 - 0·002) \times 100 \times 60$$

$$= 18·0 \text{ cal.}$$

Notice that whether or not heat moves into or out of a face depends on temperature differences between adjacent blocks. Thus, if the block of interest were cooler than the two adjacent blocks, heat would move into it through both top and bottom faces: there would be no outputs and the block would gain heat. Should the block of interest be hotter than the two adjacent blocks, heat will move out of it through both top and bottom faces: there will be no inputs and the block will lose heat. Note also the case in which heat inputs and heat outputs are equal; the difference between inputs and outputs is then zero and this means the amount of heat stored in the block remains constant during the time interval. This, the steady state, will arise when the temperature gradients through the top and bottom faces of the block are equal (assuming the heat conduction coefficients are the same).

To find the difference in heat inputs and heat outputs, we have compared the negative temperature gradients (multiplied by their respective heat conduction coefficients) through the two faces. The storage equation can thus be written in terms of temperature

$$
\begin{bmatrix} \text{change in storage} \\ \text{in the block} \\ \text{during a given} \\ \text{time interval} \end{bmatrix} = \begin{bmatrix} \text{difference between product of} \\ \text{heat conduction coefficient and} \\ \text{negative temperature gradient} \\ \text{through top and bottom faces} \\ \text{of block} \end{bmatrix} \begin{array}{l} \text{area} \\ \times \text{of} \\ \text{face} \end{array} \times \begin{array}{l} \text{time} \\ \text{interval} \end{array} . \qquad (4.6)
$$

We will find this expression useful a little later.

(f) Now, we may, again applying the principle of energy conservation, work out the change in heat storage of the block over a given time interval in another way. Say the heat content of the block is at a particular time 10 000 cal; and the heat storage one minute later is 15 000 cal. Clearly, the change in heat storage over this time interval is +5000 cal; this is calculated by subtracting the heat storage at the start of the time interval from the heat storage at the end of the time interval.

At this juncture, we shall use an equation which relates heat storage in calories, heat capacity in cal/cm^3/°C, and temperature in °C; it is

heat storage = heat capacity × temperature

or

temperature = heat storage/heat capacity. (4.7)

Using this relation, the temperature of the block at the start of the time interval is, remembering that its heat capacity is 0·5 cal/cm^3/°C and its volume is 10 000 cm^3, 10 000/500 = 20 °C. And at the end of the time interval the temperature is 15 000/500 = 30 °C.

From here we can see that the change in heat storage in the block can be found by multiplying the heat capacity of the block, the volume of the block, and the difference in temperature of the block between the end and start of the time interval

$$
\begin{bmatrix} \text{change in heat} \\ \text{storage in the} \\ \text{block during a} \\ \text{given time interval} \end{bmatrix} = \begin{bmatrix} \text{heat} \\ \text{capacity} \end{bmatrix} \times \text{volume} \times \begin{bmatrix} \text{temperature} \\ \text{difference} \\ \text{between end} \\ \text{and start of} \\ \text{time interval} \end{bmatrix} \quad (4.8)
$$

(g) We now have two expressions for the change in heat storage in the block during a given time interval:

(i) The difference between heat inputs and heat outputs which we have defined more rigorously as the difference between the product of heat conduction coefficient and negative temperature gradient through the top and bottom faces of the block, times the area of a face, times the time interval (equation 4.6);

(ii) The product of heat capacity, volume, and temperature difference between end and start of the time interval (equation 4.8).

Equating these two expression yields

$$
\begin{bmatrix} \text{heat} \\ \text{capacity} \end{bmatrix} \times \text{volume} \times \begin{bmatrix} \text{temperature} \\ \text{difference} \\ \text{between end} \\ \text{and start of} \\ \text{time interval} \end{bmatrix}
$$

$$
= \begin{bmatrix} \text{difference between} \\ \text{product of heat} \\ \text{conduction coefficient} \\ \text{and negative} \\ \text{temperature gradient} \\ \text{through top and} \\ \text{bottom faces of block} \end{bmatrix} \times \begin{bmatrix} \text{area} \\ \text{of} \\ \text{face} \end{bmatrix} \times \begin{bmatrix} \text{time} \\ \text{interval} \end{bmatrix} \quad (4.9)
$$

We can simplify this equation by writing 'time temperature difference' and 'space temperature difference' instead of the rather wordy counterparts; and, more importantly, dividing both sides of the equation by volume times time interval: this gives

$$\left[\frac{\text{heat capacity} \times \text{volume} \times \begin{array}{c}\text{time temperature}\\\text{difference}\end{array}}{\text{volume} \times \text{time interval}} \right]$$

$$= - \left[\frac{\begin{array}{c}\text{space temperature}\\\text{difference}\end{array} \times \text{area} \times \text{time interval}}{\text{volume} \quad \text{length} \times \text{time interval}} \right]$$

$$\left[\frac{\text{heat capacity} \times \begin{array}{c}\text{time temperature}\\\text{difference}\end{array}}{\text{time interval}} \right] = - \left[\frac{\begin{array}{c}\text{space temperature}\\\text{difference}\end{array}}{\text{unit length}} \right] \tag{4.10}$$

Equation (4.10) is what we have been seeking: it is a model of heat transfer through the block which, as we shall see, enables us to calculate the temperature of the block at any time providing we have a starting temperature in all three blocks and the temperature in the two adjacent blocks at all subsequent times. The starting temperatures are called the initial conditions and the temperature of the adjacent blocks at all times are the boundary conditions.

Before showing how the model will work for a particular case, we shall switch from verbal expressions to symbols. The following symbols will be used

C = heat capacity
λ = heat conduction coefficient (usually called thermal conductivity)
T = temperature
Δ = a difference
Δt = a time interval
Δz = a distance interval in the z direction (vertical).

Putting these in the place of verbal terms in equation (4.10) we have

$$C \frac{\Delta T}{\Delta t} = - \left[\frac{\Delta}{\Delta z} \left(-\lambda \frac{\Delta T}{\Delta z} \right) \right] . \tag{4.11}$$

Or, if the thermal conductivity, λ, is constant with depth

$$\frac{\Delta T}{\Delta t} = \frac{\lambda}{C} \frac{\Delta}{\Delta z} \left(\frac{\Delta T}{\Delta z} \right) \tag{4.12}$$

which may be written

$$\frac{\Delta T}{\Delta t} = \frac{\lambda}{C} \frac{\Delta^2 T}{\Delta z^2} . \tag{4.13}$$

The term λ/C is the thermal diffusivity, κ, and has units of cm^2/s. Equation (4.13), the basic model, relates the warming or cooling of the block of soil, $\Delta T/\Delta t$, to the curvature of the temperature profile, $\Delta^2 T/\Delta z^2$, and can be used to predict temperature changes in the block; it could equally well be applied to a soil profile represented as a set of blocks lying one upon the other. Providing the initial temperatures of all blocks in the soil profile are given, and the temperatures of the uppermost and lowermost blocks specified at all times, equation (4.13) may be solved numerically by a procedure akin to the Euler method which was used in the previous chapter.

A numerical solution

The first step in obtaining a numerical solution to equation (4.13) is to write in full the gradients and curvatures. The time temperature gradient, $\Delta T/\Delta t$, can be expressed as the difference between the temperature at time $t + 1$ and the temperature at time t, divided by the time interval, Δt, which is defined by $(t + 1) - t$; we may write

$$\frac{\Delta T}{\Delta t} = \frac{T_{t+1} - T_t}{\Delta t}. \tag{4.14}$$

This temperature change is to be studied at different points along the soil profile so a reference system for points in the soil profile is needed. Let the total soil profile depth be Z, and let Z be divided into a set of discrete points, $z_1, z_2, z_3, \ldots, z_n$. For convenience, let the spacing between points, Δz, be constant and let $\Delta z = z_i - z_{i-1}$. Using this system, subscripts may be added to T in equation (4.14) to indicate the depth at which the temperature gradient is taken

$$\left(\frac{\Delta T}{\Delta t}\right)_i = \frac{T_{i,t+1} - T_{i,t}}{\Delta t} \tag{4.15}$$

where the i's denote depth z_i (figure 4.3).

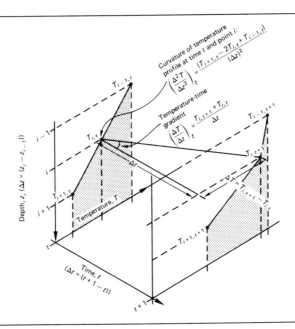

Figure 4.3. The geometry of the rate of temperature change, $\Delta T/\Delta t$, and the curvature of the temperature profile, $\Delta^2 T/\Delta z^2$.

Now the curvature of the temperature profile is expressed as the temperature at depth $i - 1$, less twice the temperature at depth i, plus the temperature at depth $i + 1$, all divided by Δz squared (cf. p46)

$$\frac{\Delta^2 T}{\Delta z^2} = \frac{T_{i-1} - 2T_i + T_{i+1}}{(\Delta z)^2} . \tag{4.16}$$

As the curvature varies with time we must add subscripts to specify to which time the equation refers. So, as explained in figure 4.3,

$$\left(\frac{\Delta^2 T}{\Delta z^2}\right)_t = \frac{T_{i-1,t} - 2T_{i,t} + T_{i+1,t}}{(\Delta z)^2} . \tag{4.17}$$

Combining the finite difference expressions in equations (4.15) and (4.17) and adding the term for thermal diffusivity, which is assumed to be constant, we find that

$$\frac{T_{i,t+1} - T_{i,t}}{\Delta t} = \kappa \left(\frac{T_{i-1,t} - 2T_{i,t} + T_{i+1,t}}{(\Delta z)^2}\right). \tag{4.18}$$

Multiplying through by Δt and adding $T_{i,t}$ to both sides we get

$$T_{i,t+1} - T_{i,t} + T_{i,t} = T_{i,t} + \kappa \Delta t \left(\frac{T_{i-1,t} - 2T_{i,t} + T_{i+1,t}}{(\Delta z)^2}\right), \tag{4.19}$$

which may be rearranged to yield

$$T_{i,t+1} = T_{i,t} + \frac{\kappa \Delta t}{(\Delta z)^2} (T_{i-1,t} - 2T_{i,t} + T_{i+1,t}). \tag{4.20}$$

This is our basic working equation.

We have one equation of the same form as equation (4.20) for every point in the soil profile, i. There are n such points so there are n equations which act at the same time — they are simultaneous equations. To solve these simultaneous equations by what is called an explicit method is a straightforward process, though very tedious, if not utterly impracticable, without a computer. Nevertheless, the principles can be learned by working through a simple case by hand.

Let us say there are five points in the soil: z_0, z_1, z_2, z_3, z_4. The initial conditions define the temperature at these depths at $t = 0$. Substituting these values into the right-hand side of equation (4.20) we can calculate the values for each of the five depths at time $t + 1$.

Given $T_{0,t} = 10$, $T_{1,t} = 0$, $T_{2,t} = 0$, $T_{3,t} = 0$ and $T_{4,t} = 0$, we find that

$$T_{1,t+1} = T_{1,t} + \kappa \frac{\Delta t}{(\Delta z)^2} (T_{1-1,t} + -2T_{1,t} + T_{1+1,t})$$

$$= 0 + \kappa \frac{\Delta t}{(\Delta z)^2} (10 - 0 + 0).$$

If we set $\kappa = 0 \cdot 5 \text{ cm}^2/\text{min}$, $\Delta t = 1$ min, and $\Delta z = 1$ cm

$$= 0 \cdot 5 \times 10$$

$$= 5 \cdot 0 \,^\circ C.$$

At all other depths we get zero. So the values of temperature for $t = 1$ are $T_{0,1} = ?$, $T_{1,1} = 5{\cdot}0$, $T_{2,1} = 0$, $T_{3,1} = 0$, $T_{4,1} = ?$. The question mark lies by $T_{0,1}$ since this is a boundary value and has yet to be specied. $T_{4,1}$ is also a boundary value and it also needs specifying. If we assume that the temperature at the surface is maintained at $10\,^{\circ}C$ and that at the lower boundary is fixed at $0\,^{\circ}C$, we can proceed with the solution of the equation. It is worth mentioning here though that what we are in fact doing is, given initial conditions and boundary conditions for all future time steps, filling in the missing values in what is called the region of integration (figure 4.4).

It is because the values of T are found directly from the values at the end of the previous time step that the method is called explicit.

Having found the values for $t = 1$, we can now use these new values in equation (4.20) to solve for $t = 2$. Firstly, taking depth $i = 1$ we get

$$T_{1,t+1} = T_{1,t} + \kappa \, \frac{\Delta t}{(\Delta z)^2} \, (T_{1-1,t} - 2T_{1,t} + T_{1+1,t})$$

$$T_{1,2} - T_{1,1} + \kappa \, \frac{\Delta t}{(\Delta z)^2} (T_{0,1} - 2T_{1,1} + T_{3,1})$$

$$= 5{\cdot}0 + 0{\cdot}5(10{\cdot}0 - (2{\cdot}0)(5{\cdot}0) + 0{\cdot}0)$$

$$= 5{\cdot}0\,^{\circ}C.$$

Secondly, at depth $i = 2$,

$$T_{2,2} = T_{2,1} + \kappa \, \frac{\Delta t}{(\Delta z)^2} \, (T_{1,1} - 2T_{2,1} + T_{3,1})$$

$$= 0{\cdot}0 + 0{\cdot}5(5{\cdot}0 - (2)(0{\cdot}0) + 0{\cdot}0)$$

$$= 2{\cdot}5\,^{\circ}C.$$

The temperature at all other depths for time $t = 2$ will be zero. The reader may check this. We have now calculated the temperature profile after two minutes. These new

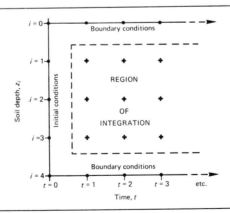

Figure 4.4. Region of integration. The solid dots are nodes in the region of integration whose values are specified by the initial conditions or boundary conditions. The crosses are nodes in the region of integration whose values are to be solved.

Table 4.1. Hand calculations of heat flow in soil profile
(figures rounded).

$t = 0$, $z_0 = 10.0$, all other i's $= 0.0$

$t = 1$,	$i = 1$	$T = 0.0 + 0.5(10.0 - 0.0 + 0.0) = 5.0$
	$i = 2$	$T = 0.0 + 0.5(0.0 - 0.0 + 0.0) = 0.0$
	$i = 3$	$T = 0.0 + 0.5(0.0 - 0.0 + 0.0) = 0.0$

$t = 2$,	$i = 1$	$T = 5.0 + 0.5(10.0 - 10.0 + 0.0) = 5.0$
	$i = 2$	$T = 0.0 + 0.5(5.0 - 0.0 + 0.0) = 2.5$
	$i = 3$	$T = 0.0 + 0.5(0.0 - 0.0 + 0.0) = 0.0$

$t = 3$,	$i = 1$	$T = 5.0 + 0.5(10.0 - 10.0 + 2.5) = 6.25$
	$i = 2$	$T = 2.5 + 0.5(5.0 - 5.0 + 0.0) = 2.5$
	$i = 3$	$T = 0.0 + 0.5(2.5 - 0.0 + 0.0) = 1.25$

$t = 4$,	$i = 1$	$T = 6.25 + 0.5(10.0 - 12.25 + 2.5) = 6.5$
	$i = 2$	$T = 2.5 + 0.5(6.25 - 5.0 + 0.0) = 3.125$
	$i = 3$	$T = 1.25 + 0.5(2.5 - 2.5 + 0.0) = 1.25$

$t = 5$,	$i = 1$	$T = 6.5 + 0.5(10.0 - 13.0 + 3.125) = 6.625$
	$i = 2$	$T = 3.125 + 0.5(6.5 - 6.25 + 1.25) = 3.875$
	$i = 3$	$T = 1.25 + 0.5(3.125 - 2.5 + 0.0) = 1.562$

$t = 6$,	$i = 1$	$T = 6.625 + 0.5(10.0 - 13.25 + 3.875) = 6.937$
	$i = 2$	$T = 3.875 + 0.5(6.625 - 7.75 + 1.562) = 4.093$
	$i = 3$	$T = 1.562 + 0.5(3.875 - 3.125 + 0.0) = 1.937$

$t = 7$,	$i = 1$	$T = 6.937 + 0.5(10.0 - 13.875 + 4.093) = 7.046$
	$i = 2$	$T = 4.093 + 0.5(6.937 - 8.817 + 1.937) = 4.437$
	$i = 3$	$T = 1.937 + 0.5(4.093 - 3.875 + 0.0) = 2.046$

$t = 8$,	$i = 1$	$T = 7.046 + 0.5(10.0 - 14.093 + 4.437) = 7.218$
	$i = 2$	$T = 4.437 + 0.5(7.046 - 8.875 + 2.046) = 4.546$
	$i = 3$	$T = 2.046 + 0.5(4.437 - 4.093 + 0.0) = 2.218$

$t = 9$,	$i = 1$	$T = 7.218 + 0.5(10.0 - 14.437 + 4.546) = 7.273$
	$i = 2$	$T = 4.564 + 0.5(7.218 - 9.093 + 2.218) = 4.718$
	$i = 3$	$T = 2.218 + 0.5(4.546 - 4.437 + 0.0) = 2.273$

$t = 10$,	$i = 1$	$T = 7.273 + 0.5(10.0 - 14.546 + 4.718) = 7.359$
	$i = 2$	$T = 4.718 + 0.5(7.273 - 9.437 + 2.273) = 4.773$
	$i = 3$	$T = 2.273 + 0.5(4.718 - 4.546 + 0.0) = 2.359$

$t = 11$,	$i = 1$	$T = 7.359 + 0.5(10.0 - 14.718 + 4.773) = 7.386$
	$i = 2$	$T = 4.773 + 0.5(7.359 - 9.546 + 2.359) = 4.859$
	$i = 3$	$T = 2.359 + 0.5(4.773 - 4.718 + 0.0) = 2.386$

---

values can in turn be substituted into equation (4.20) to solve the temperature profile after three minutes, that is, at $t = 3$. Starting with depth $i = 1$

$$T_{1,3} = T_{1,2} + \kappa \frac{\Delta t}{(\Delta z)^2} (T_{0,2} - 2T_{1,2} + T_{2,2})$$

$$= 5 \cdot 0 + 0 \cdot 5(10 \cdot 0 - (2)(5 \cdot 0) + 2 \cdot 5)$$

$$= 6 \cdot 25 \,^\circ\text{C}.$$

At depth $i = 2$,

$$T_{2,3} = T_{2,2} + \kappa \frac{\Delta t}{(\Delta z)^2} (T_{1,2} - 2T_{2,2} + T_{3,2})$$

$$= 2 \cdot 5 + 0 \cdot 5(5 \cdot 0 - (2)(2 \cdot 5) + 0 \cdot 0)$$

$$= 2 \cdot 5 \,^\circ\text{C}.$$

At depth $i = 3$,

$$T_{3,3} = T_{3,2} + \kappa \frac{\Delta t}{(\Delta z)^2} (T_{2,2} - 2T_{3,2} + T_{4,2})$$

$$= 0 \cdot 0 + 0 \cdot 5(2 \cdot 5 - (2)(0 \cdot 0) + 0 \cdot 0)$$

$$= 1 \cdot 25 \,^\circ\text{C}.$$

The calculations up to $t = 11$ are given in table 4.1 and the results are shown in figures 4.5 and 4.6. The values in table 4.1 slowly converge on the steady-state solution for the given boundary conditions, that is, $T_1 = 7 \cdot 5$, $T_2 = 5 \cdot 0$ and $T_3 = 2 \cdot 5$. We can verify that this is the steady state by substituting the values in equation (4.20).

$i = 1$ $7 \cdot 5 + 0 \cdot 5(10 \cdot 0 - 15 \cdot 0 + 5 \cdot 0) = 7 \cdot 5$

$i = 2$ $5 \cdot 0 + 0 \cdot 5(7 \cdot 5 - 10 \cdot 0 + 2 \cdot 5) = 5 \cdot 0$

$i = 3$ $2 \cdot 5 + 0 \cdot 5(5 \cdot 0 - 5 \cdot 0 + 0 \cdot 0) = 2 \cdot 5$.

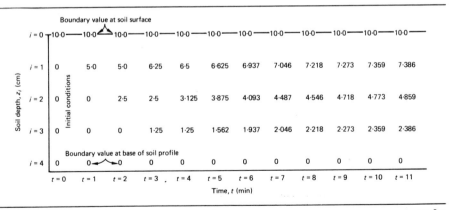

Figure 4.5. Solution to soil heat flow problem. The boundary values are set at 10 $^\circ$C for the soil surface and 0 $^\circ$C for the base of the profile. As time progresses, heat diffuses into the profile and approaches the steady-state solution for the given boundary conditions. All temperatures are in $^\circ$C and the figures are rounded.

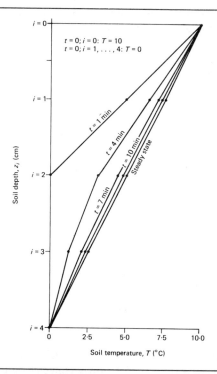

Figure 4.6.   Change in the soil temperature profile through numerical solution to soil heat flow problem. Notice how the solution converges on the steady-state profile for the given boundary conditions.

Clearly $T_{i+1} = T_i$ and the temperature profile does not change with time, and therefore inputs balance outputs — the system is in a steady state.

Testing the model

The case just solved, because it is hypothetical and has highly simplified boundary conditions, cannot be readily tested by comparing predicted change with change observed in the field. However, many researchers have solved the soil heat transfer model for realistic boundary conditions. For instance, Wierenga and de Wit (1970) used the one-dimensional form of the soil heat transfer model to predict the temperature fluctuation in subsoil from the temperature variation at the soil surface, taking into account changes in the apparent thermal conductivity with depth and soil temperature. To test the model, predicted soil temperatures were compared with soil temperatures observed below the surface of the bare field which was being modelled. In wet soils, observed and predicted temperatures were in close agreement, but in dry soils differences occurred during part of the day, probably because heat transfer in water vapour, though important in dry soils, was not incorporated in the model.

MODELLING WATER FLOW IN ROCK AND SOIL

Assumptions

The assumptions we shall make in order to model water movement in rock and soil are:

(*a*) Water flow obeys the law of mass conservation and so is subject to a mass storage equation.

(*b*) Water, in a porous medium such as rock and soil, moves from areas of high hydraulic head to areas of low hydraulic head, the rate of flow being defined by Darcy's law of water flow. (It might help to think of hydraulic head as a measure of water storage).

Formulation

We do not need to go into quite the same detail of formulation as we did for the case of heat flow. Here is an outline of the steps taken to derive a storage equation for water flow which will be our model.

(*a*) In this model, the spatial domain is two-dimensional. It is an extensive area, say 25 km^2, divided into blocks of 1 km^3 each. The surface of the area will thus be divided into 1 km^2 cells and for convenience we shall let the coordinate lines which intersect to form these cells be at right angles to one another, one set orientated in a north—south direction, the other set in an east—west direction. The cells extend for 1 km below the land surface so forming 1 km^3 blocks.

Consider just one of the blocks in the interior of the region; this is the block of interest. Adjacent to this block are four others which lie one to the north, one to the east, one to the south, and one to the west; these blocks are linked to the block of interest by faces of area 1 km^2 which we shall refer to as the north face, east face, south face, and west face.

The water storage in any one of the blocks will be defined as the product of water density (usually 1 g/cm^3), volumetric water content of the block, and volume of the block. So the volume of the block is 1 km^3, water density is 1 g/cm^3; assuming the volumetric water content is 10 per cent, this means that there are 0·1 km^3 of water in the block and this has a mass of 10^{12} g.

(*b*) Water moving through pores in rock and soil may pass from one block to another across any of the faces.

(*c*) Storage, inputs, and outputs of water in the block are subject to the mass conservation principle. Therefore, we may write a general water storage equation for the block as

$$\begin{bmatrix} \text{change in water} \\ \text{storage in block} \\ \text{over time interval} \end{bmatrix} = \begin{bmatrix} \text{water} \\ \text{inputs} \end{bmatrix} - \begin{bmatrix} \text{water} \\ \text{outputs} \end{bmatrix}. \tag{4.21}$$

(*d*) The water inputs and water outputs may, we are assuming, be defined by Darcy's law which states that water flow rate is proportional, by a water conduction coefficient, to the negative gradient of hydraulic head. As was said in the assumptions at the start of this section, hydraulic head can be thought of as a measure of the water content of the block; it might be even more helpful to think of it as the height of the water table above the base of the block. In this case, if the water table is at 100 m in the block of interest, and at 200 m in the block to the north, the gradient of hydraulic head is 100 m/km. The flow rate of water between the blocks would normally be measured in units of volume per unit area per unit time, say $cm^3/cm^2/s$, and is defined as

$$\begin{bmatrix} \text{water flow} \\ \text{rate} \end{bmatrix} = \begin{bmatrix} \text{water conduction} \\ \text{coefficient} \end{bmatrix} \times \begin{bmatrix} \text{negative gradient} \\ \text{of hydraulic head} \end{bmatrix}. \tag{4.22}$$

(*e*) The change in water storage in the block over a time interval is given by the equation

$$\begin{bmatrix} \text{change in water} \\ \text{storage in block} \\ \text{over time interval} \end{bmatrix} = \begin{bmatrix} \left(\begin{matrix} \text{water} \\ \text{inputs} \end{matrix} - \begin{matrix} \text{water} \\ \text{outputs} \end{matrix} \right) \times \begin{matrix} \text{area} \\ \text{of} \\ \text{face} \end{matrix} \times \begin{matrix} \text{time} \\ \text{interval} \end{matrix} \end{bmatrix}. \tag{4.23}$$

As with heat flow, this equation can be put in a more useful form by defining water inputs and outputs as the difference between the product of water conduction coefficient and negative gradient of hydraulic head through the north and south faces, and the east face and west faces:

$$\begin{bmatrix} \text{change in water} \\ \text{storage in block} \\ \text{over time interval} \end{bmatrix} = \begin{bmatrix} \text{difference between} \\ \text{product of water conduction} \\ \text{coefficient and negative} \\ \text{hydraulic head through} \\ \text{opposite faces of block} \end{bmatrix} \times \begin{matrix} \text{area} \\ \text{of} \\ \text{face} \end{matrix} \times \begin{matrix} \text{time} \\ \text{interval} \end{matrix} \tag{4.24}$$

(*f*) The change in water storage may also be expressed as water content at the end of the time interval minus the water content at the start of the time interval, times the volume of the block

$$\begin{bmatrix} \text{change in water} \\ \text{storage in block} \\ \text{over time interval} \end{bmatrix} = \begin{bmatrix} \text{volume} \\ \text{of} \\ \text{block} \end{bmatrix} \times \begin{bmatrix} \text{difference in water storage} \\ \text{between end and start} \\ \text{of time interval} \end{bmatrix}. \tag{4.25}$$

(*g*) Equating the two expressions for change in water storage (equations 4.24 and 4.25), we get

$$\begin{bmatrix} \text{volume} \times \begin{matrix} \text{difference in} \\ \text{water storage} \\ \text{between end and} \\ \text{start of time} \\ \text{interval} \end{matrix} \end{bmatrix} = \begin{bmatrix} \text{difference between} \\ \text{product of water} \\ \text{conduction coefficient} \\ \text{and negative hydraulic} \\ \text{head through opposite} \\ \text{faces of block} \end{bmatrix} \times \text{area} \times \begin{matrix} \text{time} \\ \text{interval} \end{matrix}.$$

$$\tag{4.26}$$

Dividing equation (4.26) by volume and time interval and simplifying the phrases

$$\left[\frac{\text{volume} \times \begin{array}{c}\text{time difference}\\\text{in water storage}\end{array}}{\text{volume} \times \text{time interval}} \right] = - \left[\frac{\begin{array}{c}\text{space difference in}\\\text{water storage in N–S}\\\text{and E–W directions}\end{array} \times \text{area} \times \begin{array}{c}\text{time}\\\text{interval}\end{array}}{\text{volume}_{[\text{length}]} \times \text{time interval}} \right]$$

$$\left[\frac{\begin{array}{c}\text{time difference}\\\text{in water storage}\end{array}}{\text{time interval}} \right] = - \left[\frac{\begin{array}{c}\text{space difference in water}\\\text{storage in N–S and E–W directions}\end{array}}{\text{length}} \right] \qquad (4.27)$$

Next we change to symbols with

θ = water content
H = hydraulic head
K = water conduction coefficient or, as it is usually called hydraulic conductivity
Δ = a difference
Δt = a time interval
Δx = distance interval in x direction (north–south),
Δy = distance interval in y direction (west–east)

to yield, remembering we are interested in flows in two directions,

$$\frac{\Delta \theta}{\Delta t} = - \left[\frac{\Delta}{\Delta x} \left(-K \frac{\Delta H}{\Delta x} \right) + \frac{\Delta}{\Delta y} \left(-K \frac{\Delta H}{\Delta y} \right) \right]. \qquad (4.28)$$

If the hydraulic conductivity, K, is constant, as it would be if the rock were homogeneous, equation (4.28) becomes

$$\frac{\Delta \theta}{\Delta t} = K \left[\frac{\Delta^2 H}{\Delta x^2} + \frac{\Delta^2 H}{\Delta y^2} \right]. \qquad (4.29)$$

Equation (4.29) shows that the time rate of change of water content of the block (or any other block in the region) is proportional to the curvature of the height of the water table (hydraulic head) in the x direction and in the y direction. So 'hills' in the water table will shed water, whereas 'valleys' will receive water; 'passes' will receive water from one direction and shed it in another, and so on. As a storage equation for water movement in porous media, equation (4.29) is of immense importance in hydrology.

Using the model

An interesting case of the water storage equation is the one in which $\Delta \theta / \Delta t = 0$; in other words, the steady-state case

$$K \left(\frac{\Delta^2 H}{\Delta x^2} + \frac{\Delta^2 H}{\Delta y^2} \right) = 0. \qquad (4.30)$$

Equation (4.30) is called the Laplace equation for ground-water flow; it can be used, for instance, to find the height of the water table within an area, given the height of the

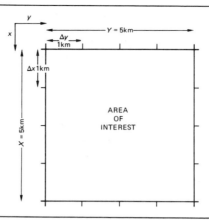

Figure 4.7.   Area of interest. x and y are spatial coordinates.

water table on the boundaries of the area. We shall see how this is done for a simple case. The region of interest may be defined by spatial coordinates x and y. Let the total distance of interest in the x direction be X and the distance of interest in the y direction Y; let X be divided into m increments each of length Δx and let Y be divided into n increments each of length Δy; and let $\Delta y = \Delta x$ be a constant spacing where $\Delta x = x_i - x_{i-1}$ and $\Delta y = y_j - y_{j-1}$. In our example, $X = Y = 5$ km and $\Delta x = \Delta y = 1$ km (figure 4.7).

By superimposing a grid over the area of interest, we find a number of nodes at the intersections of the coordinate lines of the grid. The result is called a mesh-point or node map. We have met this before (see p48). At this juncture, however, we shall amend the node reference system so that it will be more suitable for computer manipulations. The changes are simply that the map origin is in the top left-hand corner, rather than in the bottom left-hand corner, and that the x direction runs, so to speak, from top to bottom and the y direction runs from side to side. Thus amended, the node references are as in figure 4.8. Nodes within the inner rectangle of figure 4.8 are called interior nodes and the nodes around the edge are exterior or boundary nodes.

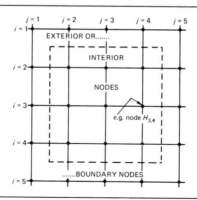

Figure 4.8.   Mesh-map of area of interest. Interior nodes lie within the dashed lines.

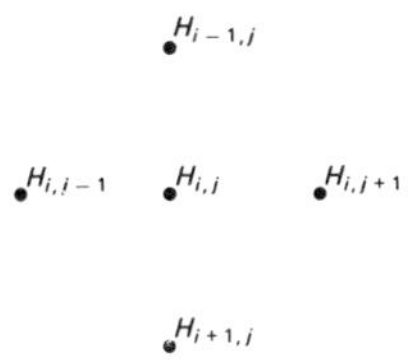

Figure 4.9. Node reference scheme.

The procedure is now to write equation (4.30), in its full difference form, for each node. With the nodes around node (i,j), written as shown in figure 4.9, we get

$$0 = K \left[\frac{H_{i+1,j} - 2H_{i,j} + H_{i-1,j}}{(\Delta x)^2} + \frac{H_{i,j+1} - 2H_{i,j} + H_{i,j-1}}{(\Delta y)^2} \right]. \tag{4.31}$$

Setting K at 1 km per month and with $\Delta x = \Delta y = 1$ km, equation (4.31) reduces to

$$0 = H_{i+1,j} + H_{i-1,j} + H_{i,\,j+1} + H_{i,j-1} - 4H_{i,j} \tag{4.32}$$

where the hydraulic head, H, is measured in kilometres. The right-hand side of equation (4.32) is simply the sum of the values of hydraulic head at the four nearest nodes to (i,j), less four times the value of hydraulic head at node (i,j). To solve the equation numerically, the arithmetic must be systematically carried out for all points of the grid. This is a straightforward process except for the boundary nodes. Boundary nodes are of two sorts: edges and corners. Clearly, an edge node has just three points adjacent to it and a corner node just two. This is why the boundary nodes must be specified in advance. The method we shall look at for solving the equation is known as the Gauss–Seidel method. It requires not only that boundary values should be specified, but also that arbitrary initial values should be given for the interior nodes. It matters not what these initial values are and it is convenient to assign a value of zero to them. Let us take an example of a 5 x 5 grid with all the boundary values set at zero except $H_{1,2}$ $H_{1,3}$ and $H_{1,4}$, which we shall set thus: $H_{1,2} = H_{1,4} = 0.1$ km (100 m) and $H_{1,3} = 0.3$ km (300 m). The initial conditions are shown in figure 4.10. We now need to find the values of the nine interior nodes. Equation (4.32) may be applied in turn to each interior node to give nine simultaneous equations:

$$
\begin{aligned}
H_{3,2} + H_{1,2} + H_{2,3} + H_{2,1} - 4H_{2,2} &= 0 \\
H_{3,3} + H_{1,3} + H_{2,4} + H_{2,2} - 4H_{2,3} &= 0 \\
H_{3,4} + H_{1,4} + H_{2,5} + H_{2,3} - 4H_{2,4} &= 0 \\
H_{4,2} + H_{2,2} + H_{3,3} + H_{3,1} - 4H_{3,2} &= 0 \\
H_{4,3} + H_{2,3} + H_{3,4} + H_{3,2} - 4H_{3,3} &= 0 \qquad (4.33) \\
H_{4,4} + H_{2,4} + H_{3,5} + H_{3,3} - 4H_{3,4} &= 0 \\
H_{5,2} + H_{3,2} + H_{4,3} + H_{4,1} - 4H_{4,2} &= 0 \\
H_{5,3} + H_{3,3} + H_{4,4} + H_{4,2} - 4H_{4,3} &= 0 \\
H_{5,4} + H_{3,4} + H_{4,5} + H_{4,5} - 4H_{4,4} &= 0
\end{aligned}
$$

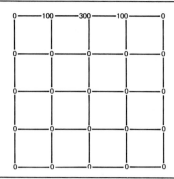

Figure 4.10.   Initial conditions: height of water table in metres.

In the Gauss–Seidel method, each of equations (4.33) is rearranged so that the unknowns (interior nodes) appear on the right-hand side. Thus

$$H_{2,2} = \frac{(H_{3,2} + H_{1,2} + H_{2,3} + H_{2,1})}{4 \cdot 0} \qquad (4.34)$$

$$H_{2,3} = \frac{(H_{3,3} + H_{1,3} + H_{2,4} + H_{2,2})}{4 \cdot 0},$$

and so on for each equation. The procedure then entails a series of iterations for each of which an approximation to the value of the interior nodes is given. These values converge on final, stable values which depend on the boundary values and the characteristics of the equation which is being solved. To compute the values by hand is out of the question even in a simple case; recourse to a computer is essential. Nevertheless, the way in which the equations operate can be demonstrated by considering a few steps for a few nodes.

Take the node (2,2). The equation for this node is

$$H_{2,2} = \frac{(H_{3,2} + H_{1,2} + H_{2,3} + H_{2,1})}{4 \cdot 0}.$$

$H_{2,1}$ and $H_{1,2}$ are boundary nodes having the values 0·0 and 0·1 km, respectively, $H_{2,3}$ and $H_{3,2}$ are interior nodes which have initially been set at 0·0. Thus for the first iteration

$$H_{2,2} = \frac{(0 \cdot 0 + 0 \cdot 1 + 0 \cdot 0 + 0 \cdot 0)}{4 \cdot 0}$$

$$= 0 \cdot 025 \text{ km}.$$

The process is repeated for the next node which is node (2,3), and the equation for it is

$$H_{2,3} = \frac{(H_{3,3} + H_{1,3} + H_{2,4} + H_{2,2})}{4 \cdot 0}.$$

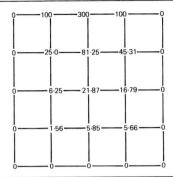

Figure 4.11. First iteration: height of water table in metres.

$H_{1,3}$ is a boundary node set at 0·3 km. $H_{3,3}$, $H_{2,2}$ and $H_{2,4}$, are all interior nodes which were initially set at zero before the first iteration began. Remember though that $H_{2,2}$ has now a value of 0·025 km. The appropriate equation for point (2,3) in the first iteration is thus

$$H_{2,3} = \frac{(0·0 + 0·3 + 0·0 + 0·025)}{4·0}$$

$$= 0·08125 \text{ km}.$$

This process is repeated for all interior nodes and the resulting values of hydraulic head are shown in figure 4.11. A second iteration is then applied, starting at point 2,2 and running through all nine interior nodes but this time using values from the first iteration. The values of hydraulic head after iterations are shown in figure 4.12. After several iterations, stable values of hydraulic head are approached. In our example this is achieved at the end of seventeenth iteration and the stable values are shown in figure 4.13. In real terms, these values are the stable values of hydraulic head within the area of interest, given the particular boundary values of hydraulic head. It might

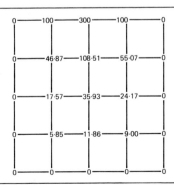

Figure 4.12. Second iteration: height of water table in metres.

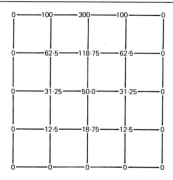

Figure 4.13.   Stable heights of water table (17 iterations).

help to visualize what we have done using a simple analogy. Imagine that the boundary values are the heights of pegs which have been driven into the ground. A net formed of string is fitted to the tops of the pegs. The stable values in this case would be the heights of the 'interior nodes' of the stretched string net.

The maps of hydraulic head may be contoured, lines of equal hydraulic head being known as equipotentials. The map shows the height of the water table above a datum. Water movement across the surface is at right angles to the equipotentials and may be represented as streamlines (figure 4.14).

Testing the model

The model should be put to the test by comparing the predicted pattern of hydraulic head over the region of interest with the observed pattern over the area, but this cannot be done for the case given as it is an hypothetical case.

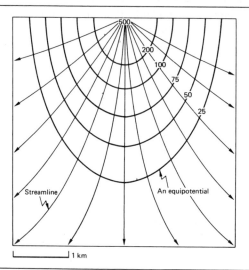

Figure 4.14.   Streamlines on the water table surface. Contour heights are in metres.

MODELLING SALT MOVEMENT IN AN ENCLOSED SEA

Assumptions

We shall assume that

(*a*) Salt movement is subject to the law of mass conservation.

(*b*) Salt may be moved within the enclosed sea by two processes. Firstly, by convection, which is bodily transport by sea currents; and secondly, by diffusion along salt concentration gradients in the sea at a rate defined by Fick's law of diffusion.

Formulation

(*a*) Imagine an enclosed sea divided into a series of blocks, the absolute dimensions of which need not yet concern us, which, as well as lying side by side, are also stacked one upon the other. We shall formulate a salt storage equation for a block occupying a central position within the sea. As in the case of water movement, the block has blocks adjacent to it in the north, south, east, and west directions; it also has a block on top and a block below. The block of interest is joined to the other six blocks by six faces which we shall call the north face, east face, south face, west face, top face, and bottom face.

The salt is stored in the block in solution. The units of salt storage would be, say, mg/l. An added complication arises here, for if the salt concentrations should exceed 311 mg/l, solid salt would precipitate and accumulate on the bottom of the sea. For simplicity, we shall assume that saturation is not attained.

(*b*) Salt may enter the block of interest through any of the six faces.

(*c*) Storage, inputs, and outputs of salt are subject to the law of mass conservation.

Therefore, we may write a general salt storage equation for the block as

$$\begin{bmatrix} \text{change in salt} \\ \text{storage in block} \\ \text{over time interval} \end{bmatrix} = \begin{bmatrix} \text{salt} \\ \text{inputs} \end{bmatrix} - \begin{bmatrix} \text{salt} \\ \text{outputs} \end{bmatrix}. \qquad (4.35)$$

(*d*) The inputs and outputs of salt, we are assuming, may be due to two things.

   (i) Convection – salt carried bodily in moving water. The salt flow rate in or out of the block through any of the faces can be given as the product of salt concentration in the block from which the water is coming and the velocity of the moving water

$$\begin{bmatrix} \text{salt flow rate due} \\ \text{to convection} \end{bmatrix} = \begin{bmatrix} \text{salt} \\ \text{concentration} \times \begin{matrix} \text{current} \\ \text{velocity} \end{matrix} \end{bmatrix}. \qquad (4.36)$$

   (ii) Diffusion – movement of salt along a concentration gradient between two blocks. The salt flow rate due to diffusion is, according to Fick's law, proportional to the negative gradient of salt concentration through the face of the block.

$$\begin{bmatrix} \text{salt flow rate} \\ \text{due to diffusion} \end{bmatrix} = \begin{bmatrix} \text{salt diffusion} \\ \text{coefficient} \end{bmatrix} \times \begin{bmatrix} \text{negative salt} \\ \text{concentration} \\ \text{gradient across} \\ \text{a face} \end{bmatrix}. \tag{4.37}$$

Combining convective and diffusive terms gives the total salt flow rate across any one face

$$\begin{bmatrix} \text{salt} \\ \text{flow} \\ \text{rate} \end{bmatrix} = \begin{bmatrix} \text{salt} \\ \text{concentration} \end{bmatrix} \times \begin{bmatrix} \text{current} \\ \text{velocity} \end{bmatrix} + \begin{bmatrix} \text{salt} \\ \text{diffusion} \\ \text{coefficient} \end{bmatrix} \times \begin{bmatrix} \text{negative salt} \\ \text{concentration} \\ \text{gradient across} \\ \text{a face} \end{bmatrix}. \tag{4.38}$$

(*e*) The change in salt storage in the block of water over a time interval is given by the equation

$$\begin{bmatrix} \text{change in salt} \\ \text{storage in block} \\ \text{over time interval} \end{bmatrix} = \begin{bmatrix} \left(\begin{matrix} \text{salt} \\ \text{inputs} \end{matrix} - \begin{matrix} \text{salt} \\ \text{outputs} \end{matrix} \right) \times \begin{matrix} \text{area} \\ \text{of} \\ \text{face} \end{matrix} \times \begin{matrix} \text{time} \\ \text{interval} \end{matrix} \end{bmatrix}, \tag{4.39}$$

a more useful version of which has the input and output terms defined as

$$\begin{bmatrix} \text{change in salt} \\ \text{storage in block} \\ \text{over time interval} \end{bmatrix} = \begin{bmatrix} \text{difference between} \\ \text{sum of convective} \\ \text{and diffusive terms} \\ \text{through opposite} \\ \text{pairs of faces of} \\ \text{the block} \end{bmatrix} \times \begin{matrix} \text{area} \\ \text{of} \\ \text{face} \end{matrix} \times \begin{matrix} \text{time} \\ \text{interval} \end{matrix}. \tag{4.40}$$

(*f*) Changes in salt storage may also be given by the salt concentration at the end of the time interval, less the salt concentration at the start of the time interval, times the volume of the block

$$\begin{bmatrix} \text{change in salt} \\ \text{storage in block} \\ \text{over time interval} \end{bmatrix} = \begin{bmatrix} \text{volume} \\ \text{of} \\ \text{block} \end{bmatrix} \times \begin{matrix} \text{difference in salt concentration} \\ \text{at end and start of time} \\ \text{interval} \end{matrix}. \tag{4.41}$$

(*g*) Equating the two changes of storage expressions yields

$$\begin{bmatrix} \text{volume} \times \begin{matrix} \text{time difference} \\ \text{in salt} \\ \text{concentration} \end{matrix} \end{bmatrix} = \begin{bmatrix} \text{space difference in} \\ \text{diffusive and} \\ \text{convective terms} \\ \text{across opposite} \\ \text{pairs of faces} \end{bmatrix} \times \text{area} \times \begin{matrix} \text{time} \\ \text{interval} \end{matrix}. \tag{4.42}$$

Dividing by volume times time interval gives

$$
\left[\frac{\begin{array}{c} \text{time difference} \\ \text{volume} \times \text{in salt} \\ \text{concentration} \end{array}}{\text{volume} \times \text{time interval}} \right] = - \left[\frac{\begin{array}{c} \text{space difference in} \\ \text{diffusive and} \\ \text{convective terms} \quad \times \text{area} \times \frac{\text{time}}{\text{interval}} \\ \text{across opposite} \\ \text{pairs of faces} \end{array}}{\dfrac{\text{volume} \times \text{time interval}}{\text{length}}} \right]
$$

$$
\left[\frac{\begin{array}{c} \text{time difference} \\ \text{in salt} \\ \text{concentration} \end{array}}{\text{time interval}} \right] = - \left[\frac{\begin{array}{c} \text{space difference in diffusive} \\ \text{and convective terms across} \\ \text{opposite pairs of faces} \end{array}}{\text{length}} \right]. \tag{4.43}
$$

Converting to symbols, using

C = salt concentration
D = salt diffusion coefficient
v_x = current velocity in x direction (north–south)
v_y = current velocity in y direction (west–east)
v_z = current velocity in z direction (top–bottom)
Δ = a difference
Δt = time interval
Δx = distance interval in x direction
Δy = distance interval in y direction
Δz = distance interval in z direction

we have

$$
\frac{\Delta C}{\Delta t} = - \underbrace{\left[\frac{\Delta}{\Delta x} \left(Cv_x - D\frac{\Delta C}{\Delta x} \right) \right]}_{\underbrace{\text{convection} \quad \text{diffusion}}_{x \text{ direction}}} + \underbrace{\left[\frac{\Delta}{\Delta y} \left(Cv_y - D\frac{\Delta C}{\Delta y} \right) \right]}_{\underbrace{\text{convection} \quad \text{diffusion}}_{y \text{ direction}}} + \underbrace{\left[\frac{\Delta}{\Delta z} \left(Cv_z - D\frac{\Delta C}{\Delta z} \right) \right]}_{\underbrace{\text{convection} \quad \text{diffusion}}_{z \text{ direction}}}.
$$

$$\tag{4.44}$$

Assuming D is constant, expansion of brackets and the separation of diffusive and convective terms gives

$$
\frac{\Delta C}{\Delta t} = - \underbrace{\frac{\Delta(Cv_x)}{\Delta x} - \frac{\Delta(Cv_y)}{\Delta y} - \frac{\Delta(Cv_z)}{\Delta z}}_{\text{convection}} + \underbrace{D\frac{\Delta^2 C}{\Delta x^2} + \frac{\Delta^2 C}{\Delta y^2} + \frac{\Delta^2 C}{\Delta z^2}}_{\text{diffusion}}. \tag{4.45}
$$

The salt storage equation (4.45) shows that in the enclosed sea, salt storage will be determined by (1) the salt concentration field, that is, the pattern of salt concentration in the blocks into which the sea has been divided: salt diffuses from blocks with high concentrations to blocks with low concentrations of salt. And (2) the water current velocity field, that is, the pattern of sea current movements between the blocks. The use of the equation will be demonstrated in the next section by the classic work of Briggs and Pollack (1967).

APPLICATIONS

Many processes of interest to physical geographers can be represented as a flow through a series of linked spatial system units. For instance, flood flow in streams has been studied by looking at the fate of a unit input of rainfall to a stream channel as it is passed from one downstream storage cell to another. The general kind of pattern found in flood routing studies is a discharge wave which, in passing downstream, becomes more attenuated and less peaked (figure 4.15). Throughflow along a slope produces a similar pattern, rainfall progressing down-slope in a wave-like fashion. The down-slope movement of soil solutes and colloids has been modelled by Huggett (1975), who found a peak concentration develops at the junction of convex and concave slope sections and thence wanders down-slope. Solute transfer in rivers, notably the transfer of pollutants, has been widely modelled. Thomann (1973) built a model incorporating pollutant input, diffusion, advection, and decay. For the case in which pollutant input changes with a period of seven days, it was found that pollutant concentrations diminish downstream but vary cyclically in response to the input variations.

Atmospherical pollution models usually consider pollutant levels and other relevant variables at a number of points in a three-dimensional parcel of air. The parcel is usually divided into a number of spatial zones and storage equations set up which describe and define the transport of pollutants from one spatial zone to another under given meteorological conditions and for given inputs of pollutants (Seinfeld and Kyan 1971). The same kind of modelling strategy can be applied to any portion of the atmosphere. The aim of these models is to predict the space—time variation in atmospherical properties such as moisture, pressure, and wind fields in the planetary

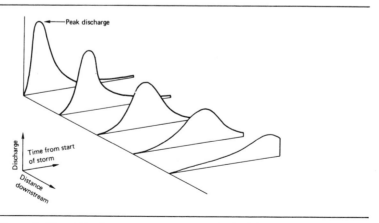

Figure 4.15.   The downstream progress of a flood wave.

boundary layer. The models involve complex sets of partial difference equations governing the energy, mass, and momentum fluxes in the atmosphere and can only be solved on high-speed, digital computers. The most difficult step in building large-scale atmospherical models is commonly finding suitable parameters to fit in the equations. Some models of the general circulation are discussed and used by Mintz (1968), Smagorinsky *et al* (1970) and Barry (1975), to select but a few references from a long list. At a smaller scale, models of the soil—plant—atmosphere continuum are discussed by Terjung (1976).

To conclude this chapter, two case studies will be discussed in detail, the first a model of an urban heat island, the second a model of water and salt movement in a marine basin.

A model of an urban heat island

Myrup (1969) used a deterministic model of the Earth's surface energy budget to study the urban heat island problem; he hoped the model would throw some light on questions relating to the causes of the urban temperature excess, the possibility of using engineering techniques to reduce the heat island in existing cities, and the possibility of building future cities in such a way as to eliminate the heat island effect. To keep the model manageable, a number of assumptions were made but they need not concern us. Essentially, the model considered inputs and outputs of energy to and from a surface block of soil or urban fabric:

$$\frac{\Delta G}{\Delta t} \quad = \quad Q \quad - \quad I \quad - \quad H \quad - \quad LE \quad - \quad S \; .$$

time rate of heat storage in block of soil or urban fabric	solar energy flux	net infrared energy flux	sensible heat flux	latent heat flux	soil heat flux

To operationalize the model, Myrup used known physical laws to define the input and output terms. This having been done, the solution of the model required the specification of parameters in the input and output formulae. The value of Q, the incoming solar energy flux, required latitude, date, and atmospheric transmission coefficient (to determine the amount of solar energy at the top of the atmosphere which actually reaches the urban air), and surface albedo (reflectivity); the sensible heat flux or the transfer of warm air, H, and the latent heat flux or the transfer of energy stored in water vapour, LE, required wind speed, temperature, and humidity (the wetness of the air) at the upper bounding surface of the urban atmosphere, and the roughness of the 'canopy' (trees, buildings, and so on); and the soil heat flux, S, required thermal conductivity, atmospheric relative humidity near the soil surface, and the soil temperature at the lower boundary in the soil.

The model was run for (1) parameters measured at a rural site in the vicinity of Davis, California; and (2) arbitrary but realistic parameters for an hypothetical city near to the rural site occupying 100 km^2 and having a population of one million people. The equations which comprise the urban heat island model were solved by procedures similar to those which were developed for the soil heat flow model. Several

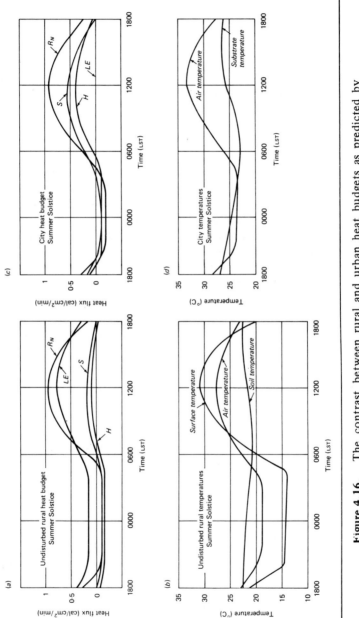

Figure 4.16. The contrast between rural and urban heat budgets as predicted by Myrup's (1969) model. (*a*) The rural heat budget calculated for the summer solstice. (*b*) Surface, air, and soil temperatures in the rural area for the summer solstice. (*c*) The surface heat budget calculated for an hypothetical city in the vicinity of Davis, California, for the summer solstice. (*d*) Air and substrate temperatures for an hypothetical city in the summer solstice. (All from L O Myrup 1969, with permission of the author and the American Meteorological Society.)

different experiments were carried out. The first considered the differences between city and countryside. The results, shown in figure 4.16, indicate a contrast, in heat budget terms, between urban and rural areas. In the rural heat budget most of the available energy is used to evaporate water, whereas in the city case most of available energy is used to heat concrete, no energy being used in evaporation. The absence of evaporation in the city results from the city's having just a 10 per cent evaporating area, which is too low for the specific humidity at street level to attain levels high enough to induce the upwards diffusion of water vapour. By comparing the maximum and minimum temperatures of the urban and rural cases, it will be seen that the model predicts a day-time heat island of 6·0 °C and a night-time heat island of 4·1 °C.

The second experiment was for the city park case, the case with the largest observed heat island effect. With suitable modifications of parameters for a city park, the model produced results which gave a day-time heat island of 11·5 °C and a night-time heat island of 8·1 °C at dawn. The third experiment was for the case in which the evaporating area of the countryside was reduced to 50 per cent, which is probably more realistic, as the value of 100 per cent in the other cases would require the entire area to be irrigated; and in which the evaporating area of the city was increased to 30 per cent, a realistic change because most cities are not entirely concrete jungles. The results show that during the day the city is 2·9 °C cooler than the countryside, but at night it is slightly warmer.

The fourth experiment was for a transect from urban area to countryside and was made to obtain a clearer picture of how rural urban climates merge into one another. For this experiment, all parameters were held same as in the first experiment, except conductivity, diffusivity, roughness length, and evaporating area which were varied along the transect in an appropriate manner. Figure 4.17 shows the predicted

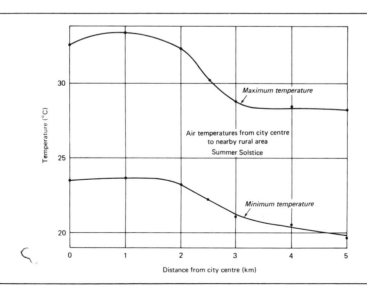

Figure 4.17.   Air temperature calculated at various distances from the city centre for the summer solstice. (From L O Myrup 1969, with permission of the author and the American Meteorological Society.)

maximum and minimum temperatures at various distances from the city centre. The peak maximum temperature is at 1 km from the city centre, rather than at the city centre itself, because the city centre has the greater roughness length, corresponding to the presence of three-storey buildings there, which produces a more efficient upwards diffusion of heat. The steepest gradient of temperature change, both maximum and minimum, occurs between 2 and 3 km from the city centre and seems attributable mainly to changes in evaporating area.

The fifth experiment was run to assess the sensitivity of the predicted temperature to changes in the model parameters. This was done by holding all parameters for the city case constant except the one whose effect was being tested. The parameters studied in this way were wind speed, albedo, roughness, and evaporating area. The results for wind speed showed that city temperatures are most sensitive to low wind speeds and that wind speeds tend to increase minimum and decrease maximum temperatures thus counteracting the heat island effect. The results for albedo showed that city temperatures are fairly insensitive to albedo, at least in the albedo range 0·1 to 0·3 which is typical for most cities. Maximum city temperature was found to be sensitive to changes in roughness length, the sensitivity figure being $-2\cdot3\ ^\circ$C per 10 cm increase in roughness length. The effect of evaporating area on city temperature was interesting because it backed up the results of the city—countryside traverse, which suggested that the greatest sensitivity occurs when the evaporating area is between 20 and 30 per cent. In general, a 10 per cent increase in evaporating area leads to a $3\cdot5\ ^\circ$C rise in maximum city temperature.

The predictions of this simple model are clearly promising in so far as they seem reasonable and help to explain many of the observed features of the urban heat island phenomenon. In connection with the questions the model was built to answer, it is evident that one way of avoiding or at least alleviating the sweltering summer nights that regularly occur in cities of the eastern USA is to provide a greater evaporating area, certainly up to the seemingly critical level of 20 per cent; such provision could be made by creating more roof-top gardens and extra parks.

Water and salt movement in a marine basin

Briggs and Pollack (1967) built a space—time deterministic model to study the pattern of salt concentration and the formation of halite (rock salt) in an inland sea. The specific case they modelled was an evaporite basin which existed in Late Silurian times in the Michigan basin area of the USA (figure 4.18). The basin was bounded by reef banks which partially isolated it from open sea. Briggs and Pollack assumed that, because of evaporative loss of water in the basin, sea water would have been drawn in to make good the loss. The sea water carried dissolved salt into the basin over the reefs and through a limited number of distinct inlets; but the saltier water inside the basin was prevented by the reefs from returning to the open sea; sea water which entered the basin would have been subject to evaporation and salt would have precipitated when the salt concentration in the water reached saturation point. Briggs and Pollack expressed these ideas mathematically using three sub-models, each of which contains simplifying assumptions.

(1) A model of fluid flow to describe the steady-state, potential flow of water in

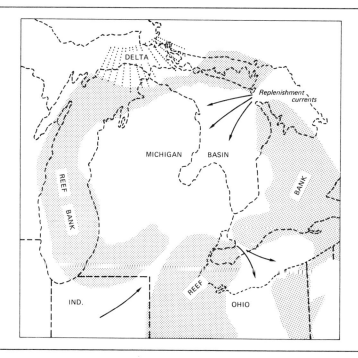

Figure 4.18.   Inferred geography in, and adjacent to, the Michigan Basin during Late Silurian time. (From L I Briggs and H N Pollack, Digital model of evaporite sedimentation, *Science*, Vol. 155, pp. 453—456, 27 January 1967). Copyright 1967 by the American Association for the Advancement of Science.

the basin. The chief assumption here was that flow could be adequately defined by the Laplace equation of flow (p113) for the case of an incompressible, density-constant fluid. The loss of water by evaporation from the surface water of the basin was included in this sub-model.

(2) A model to describe the transport of salt in the basin. Since salt tends to become concentrated in the basin with time, a time-dependent model based on a mass storage equation was needed. The assumption was made that salt was transported by currents of water in the basin and by diffusion along salt concentration gradients.

(3) A model to incorporate the precipitation of salt when salt concentrations in the sea water reach saturation point (311 g/l).

To simplify the geographical situation, Briggs and Pollack assumed initially that just two inlets admitted sea water to the basin and one outlet existed in the south-east; reefs acted as a barrier to the influx of sea water. The area of interest was divided into square, two-dimensional cells by superimposing a meshwork. Initial conditions and boundary conditions having been specified, the equations which comprise each sub-model were applied to each cell to find: the direction and velocity of water movement (figure 4.19*a*); the steady-state concentration of salt in the sea water (figure 4.19*b*); and the area of salt precipitation. To test the results, the predicted pattern of halite distribution was compared with the observed thickness pattern of halite in the area

(a)

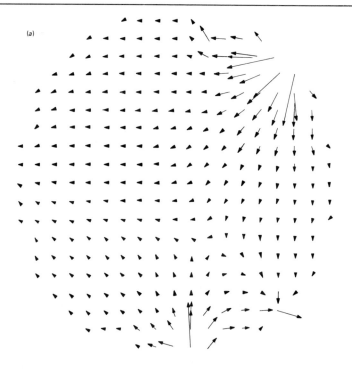

(b)

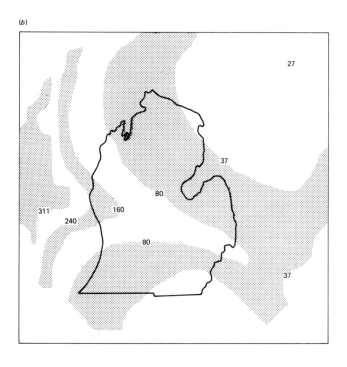

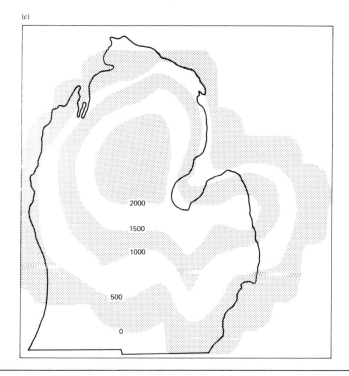

Figure 4.19.   (*a*) Flow diagram of sea currents from a simulation run in which two inlets and one outlet to the Michigan basin were assumed. (*b*) Isosalinity contours produced after a steady state has been attained in simulation run assuming two inlets and one outlet to the Michigan basin. Values are in grams of salt per litre. Saturation values of 311 g salt per litre occur in the western part of the area. (*c*) Thickness contour map based on bore-hole data showing distribution of halite in Cayugan series (Upper Silurian) in the Michigan basin. (All from L I Briggs and H N Pollack 1967, Digital model of evaporite sedimentation, *Science*, Vol. 155, pp. 453–456, 27 January 1967.) Copyright 1967 by the American Association for the Advancement of Science.

(figure 4.19*c*). The two patterns did not match, the predicted pattern suggesting halite would be found in the western part of the basin (the areas in figure 4.19*b* with concentrations in excess of 311 g/l), the actual pattern showing halite in the centre of the basin. No amount of modification to the relative flow volumes of sea water through the two inlets and one outlet could shift the site of salt deposition to the central area. However, when, as well as the two inlets and one outlet, a chain of 'leaks' through the reefs were incorporated in the model, the salt concentrations reached saturation point in the central area of the basin (figure 4.20*a*) and the predicted salt thickness pattern after 12 000 years of deposition accorded remarkably well with the observed pattern (figure 4.20*b*).

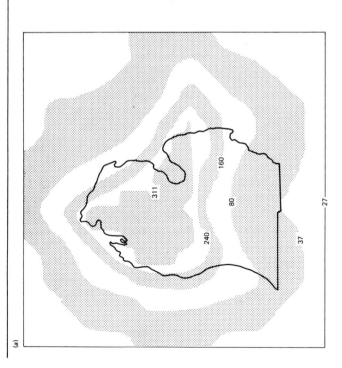

Figure 4.20. (*a*) Isosalinity map showing salt concentrations (grams of salt per litre) after 4000 years of operation of a model which, as well as two inlets and one outlet, allows for 'leaky' reefs. (*b*) Simulated salt thickness pattern employing 'leaky reef' hypothesis. Thickness values are in feet. (Both from L I Briggs and H N Pollack 1967, Digital model of evaporite sedimentation, *Science*, Vol. 155, pp. 453–456, 27 January 1967.) Copyright 1967 by the American Association for the Advancement of Science.

CONCLUDING REMARKS

The space—time deterministic models given in the text have been kept simple. Actual models are usually far more complex in so far as they consider a large number of spatial units; boundary conditions which change with time; the transfer and storage of more than one material or energy form; or situations in which parameters in the transport laws depend on location in space or the amount of material or energy in store (or even both). For instance, models of mineral transfer in soil profiles may involve a large number of soil layers, say up to 50; may look at different phases of a mineral such as water-soluble, absorbed, immobilized, and precipitated; and may consider sources, such as weathering, and sinks, such as plant uptake. In these cases, calibration will entail specifying parameters in the transport laws for each phase of the mineral (these parameters will vary from one soil layer to another if the soil is heterogeneous), and parameters in equations which describe the transformation of the mineral from one phase to another (for example, see Mansell *et al* 1977). Similar complexities of model structure and calibration are found in mathematical models of slope profile development and atmospheric processes, some of the simpler examples of which are explored in Huggett (1980, Chapter 7).

Examples of space—time deterministic models are legion. None of them is easy for the mathematical novice to play with; but this does not mean that the principles upon which they are built, the problems involved in their calibration, and the validity and interpretation of the results they generate, cannot be appreciated by the geographer who has received little or no special training in mathematics. Terjung (1976) searched all major western atmospheric science journals, and found a handful of articles which use mathematical modelling techniques contributed by geographers; the same is true in journals of geomorphology, hydrology, and soil science. Does this mean that geographers should stick with their well tried statistical forms of analysis? Terjung (1976) argued that climatic processes — and to this could be added virtually all Earth-surface processes — are too complicated and too interconnected to be amenable to statistical analysis: geographers of the future who wish to work in the environmental sciences must be willing to learn the basic methods of environmental science — calculus, physics, chemistry, engineering, modern biology, and computer programming; and future physical geography courses should have at their core basic thermodynamics and hydrodynamics as they affect man and his environmental envelope.

This chapter has tried to put across the principles of the approach espoused by Terjung. It was not intended to enable readers to become proficient modellers of space—time systems; rather, it has provided a grounding in the procedures and logic of deterministic space—time modelling that should open up a large body of literature of immense relevance to many geographical studies.

CHAPTER 5

Spatial Interaction Models

With the exception of population models, our discussion of deterministic models has been concerned with the prediction of changes that occur in the physical landscape. Most of these models are founded on the notion that the rate of flow of matter and energy can be used to construct model equations which predict how the structure of the physical landscape evolves. For instance, the amount of organic litter on the ground (the structure) is dependent upon the rate of leaf and timber fall and the rates of litter decomposition and translocation. In this sense such models use rates of movement as 'independent' variables to predict the value of a dependent variable representing some structural aspect of the physical landscape. In this chapter, where we will consider the flows of people, goods and income across the human landscape, the modelling relationship between flow and structure is reversed. Flows in human geography are often termed *spatial interactions*, and a spatial interaction model is an equation which predicts the size and direction of some flow (the dependent variable) using independent variables which measure some structural property of the human landscape. For instance, the daily pattern of journey-to-work flows in a city can be predicted using structural variables such as the distribution of workers, the distribution of employment and the costs of travelling to work. Similarly, flows of shopping expenditure in a city can be predicted using knowledge about consumer expenditure levels and shopping centre sizes.

The original spatial interaction models are called gravity models because their mathematical assumptions are similar to those embodied in Isaac Newton's law of gravitational attraction. Indeed, gravity models have a long geographical pedigree. The ideas contained in the simplest gravity model originate from the work of the English demographer Ravenstein (1885, 1889) who observed, from an analysis of migration data, that a population centre, j, attracted migrants from a sending centre, i, in direct proportion to the population of the sending centre and in inverse proportion to the distance, d_{ij}, between the two centres. A similar notion was expressed by Reilly (1931) in a book entitled *The Law of Retail Gravitation* which summarized much of his earlier work. Like Ravenstein, Reilly used the variables population size and distance, in his case to predict the amount of retail trade attracted by a city. Reilly's law states that 'two cities attract retail trade from an intermediate town in direct proportion to the population of the two cities, and in inverse proportion to the square of the distance between each city and the intermediate town.' Reilly's law was designed primarily to delimit the trading area of a city and later in this chapter we will describe a shopping expenditure model which is an extension of Reilly's original idea.

The ideas of Ravenstein and Reilly are partial expressions of the simple gravity model which is a direct analogy with the law of gravity propounded by Newton in 1687. Newton asserted that the force of attraction, F, between two bodies is the product of their masses, m_1 and m_2, divided by the square of the distance between them, d_{12}^2, which gives the familiar formula

$$F = G.m_1 m_2 / d_{12}^2 \qquad (5.1)$$

where G is a universal constant, the pull of gravity. The translation of Newton's law into a geographical flow model is usually attributed to the astronomer Stewart (1941, 1950). In geography, force is equated with the numbers of movements (or trips) between two regions; mass is a variable such as population size and measures a region's capacity either to generate or to attract trips; and distance is measured either in physical terms or some surrogate such as travel cost or travel time.

During the 1950s and early 1960s geographers tested the gravity model as an explanation for migration, information and freight flows, and Olsson (1965) has reviewed developments during this period. Much of the work in this period was concerned with the construction of statistical tests designed to assess the degree of correspondence between the model predictions and observed patterns of flows. However, mathematical refinements to the structure of the simple gravity model were generated by developments in the field of urban and regional planning. Beginning in 1959 with the Chicago Area Transportation Study, large-scale planning projects which evaluated future land-use and transportation requirements of metropolitan conurbations became a popular instrument of urban policy formulation. The research carried out in conjunction with such studies produced numerous theoretical refinements to the gravity model which, in particular, improved the model's ability to predict journey-to-work trips and flows of shopping expenditure. Today, geographers recognize a family of gravity models (Wilson 1971) whose mathematical structures each add more realism to the assumptions of Stewart's original model. This chapter begins by describing mathematical structure of the four kinsmen in the gravity model family; the simple gravity model, the production-constrained model, the attraction-constrained model, and the production–attraction-constrained model. We then consider some further work by Wilson (1967, 1970) which uses entropy-maximizing principles to establish a more elegant theoretical framework for the analysis of spatial interactions than that provided by the gravity analogy.

THE SIMPLE GRAVITY MODEL

Notation and assumptions

All spatial interaction models predict the size of the flow between a pair of places. A flow is denoted by the variable T_{ij}, where T measures the size of the predicted flow, i is a subscript which identifies the place where the flow begins, and j is a subscript identifying the place where the flow ends. For example, suppose we wish to represent the number of migrants who move *from i* = London *to j* = Manchester in a given year. We can represent this flow verbally by the term

$T_{\text{London, Manchester}}$.

Similarly, the flow of people *from i* = Manchester *to j* = London is given by

$T_{\text{Manchester, London}}$.

Clearly, it is a tedious procedure to write place names under T to denote the direction of flow. It is far more economical to give each place a number and then label flows according to our place numbers. For instance, if London is numbered 1 and Manchester 2, then our previous examples become T_{12} and T_{21}, respectively.

Having established a notation to identify individual flows we now consider the three assumptions of the simple gravity model. Firstly, it is assumed that the size of any flow is proportional to ($\propto$) a variable W_i which measures the trip-generation capacity of the region where the flow begins. For studies where the flows are numbers of people, W_i is often defined as the population of the origin region. This assumption may be written symbolically as the proposition

$$T_{ij} \propto W_i, \tag{5.2}$$

which asserts that the magnitude of the flow leaving any region i will increase linearly with the population size of the region (figure 5.1*a*). Secondly, it is assumed that the size of T_{ij} is proportional to a variable W_j which measures the trip-attraction capacity of the region where the flow ends. Again, attractiveness is often measured by the population size of the destination region. This assumption is written symbolically as

$$T_{ij} \propto W_j \tag{5.3}$$

and asserts that the magnitude of the flow arriving in any zone j will increase linearly with the population size of the destination region (figure 5.1*b*).

The third assumption concerns the distance, d_{ij}, between the origin region, i, and the destination region, j. The proposition asserts that the amount of interaction, T_{ij}, *declines* in proportion with the square of the distance, d_{ij}^2, between the two regions, and which we write as

$$T_{ij} \propto 1/d_{ij}^2,$$

or more economically in the form

$$T_{ij} \propto d_{ij}^{-2}. \tag{5.4}$$

The validity of this proposition is often justified by the evidence of data for different types of interaction which shows that short-distance flows occur more frequently than long-distance flows. This empirical characteristic of spatial interactions is usually described by the term *distance decay*. However, apart from maintaining the strict analogy with Newton's law, there is no theoretical justification for expecting flows to decline exactly with the square of the distance between regions. For this reason it makes more geographical sense to allow distance to be raised to some power α and write proposition (5.4) more generally as

$$T_{ij} \propto d_{ij}^{-\alpha}. \tag{5.5}$$

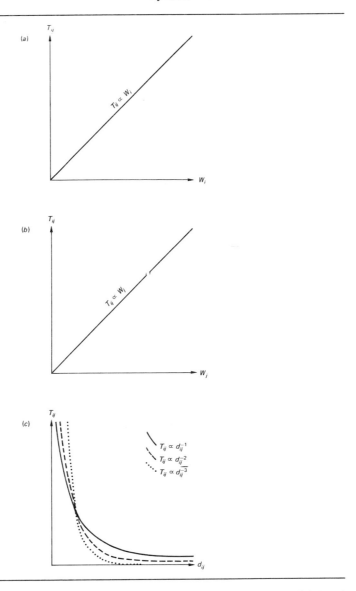

Figure 5.1 The gravity model: basic propositions. (*a*) Origin generation. (*b*) Destination attraction. (*c*) Distance decay.

The exact value given to α will depend on the empirical evidence available to the research worker about the particular flows he is studying. The effect of varying the value of the distance exponent is shown in figure 5.1(*c*), which illustrates the relationship between T_{ij} and d_{ij} for the values $\alpha = 1$, 2 and 3. When $\alpha = 1$, the decline of interaction is relatively gentle. However, raising α to progressively higher powers steepens the gradient of the curve such that the number of short-distance interactions are increased relative to the number of long-distance interactions. For this reason the value of α is said to measure the *frictional effect of distance*.

When the three gravity propositions are combined together the basic gravity model formula is obtained as

$$T_{ij} = k \cdot W_i W_j / d_{ij}^{\alpha}. \tag{5.6}$$

Notice that the attraction and generation propositions are incorporated by the multiplication of the terms W_i and W_j, while division by some power of distance produces the distance-decay effect. The only undefined term in formula (5.6) is k, which is a scaling constant. The need to include this constant arises because the independent variables W_i, W_j and d_{ij} are not measured in units of flow. The method of calculating k will be demonstrated in the following section.

The calculation of gravity model predictions

To illustrate the sequence of calculations needed to solve the gravity formula we will consider the problem of predicting migration flows among the $n = 3$ regions of the study area shown in figure 5.2. Beginning with purely notational aspects of this problem, it can be seen that each region has been given an identity number which is used to give values to the subscripts i and j. With $n = 3$ regions, there are $n^2 = 9$ possible directions of flow from an origin region to a destination region, and these

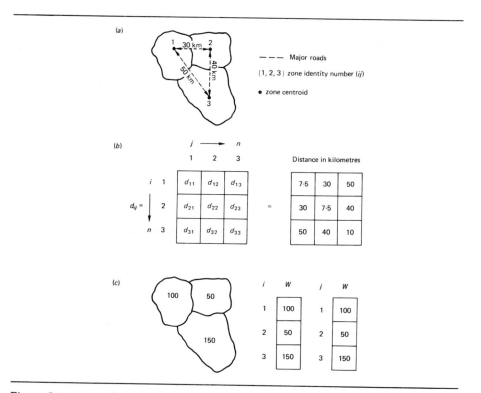

Figure 5.2. Data for the gravity model migration problem. (*a*) Zones and networks. (*b*) Distance matrix. (*c*) Zone populations W_i and W_j (thousands).

different flows are recorded in a matrix of the form

$$T_{ij} = \begin{array}{c} \\ i \\ \\ n \end{array} \begin{array}{c} j \xrightarrow{\hspace{2cm}} n \\ \begin{array}{cccc} & 1 & 2 & 3 \\ 1 & \boxed{\begin{array}{ccc} T_{11} & T_{12} & T_{13} \\ T_{21} & T_{22} & T_{23} \\ T_{31} & T_{32} & T_{33} \end{array}} \\ 2 & & \\ 3 & & \end{array} \end{array}$$

Notice that this matrix records flows that take place *within* the boundaries of a region, in addition to flows *between* pairs of regions. These within-region flows are denoted by the elements $T_{(i=j)}$ listed in the principal diagonal of the matrix; for example, T_{22} will record the gravity prediction for the number of migrations that begin and end within the boundaries of region 2.

It is known that in a given year a total $T = 15\,000$ people change their residence within the boundaries of the study area. The values of the independent variables W_i, W_j, and d_{ij}, which are needed to make the gravity model predictions for each element T_{ij} in the flow matrix, are listed in figure 5.2. The population of each region has been used to represent the capacity of a region to generate, W_i, and attract, W_j, migrations. The distances between each region are measured as the shortest road distance between the centre of each region. Because migrations within regions are to be calculated, the principal diagonal of the distance matrix (figure 5.2b) contains the elements $d_{(i=j)}$ which measure the average length of migrations within each region. It is a common practice to define a within-region distance as one quarter of the distance between the centre of the region and the nearest neighbouring region. For our example, this convention gives the distance within region 1 as

$$d_{11} = d_{12}/4 = 30/4 = 7 \cdot 5.$$

Finally, we must specify the value of the distance exponent, and in this example we will assume $\alpha = 2$.

The calculations needed to obtain the gravity predictions $\{T_{ij}\}$ take place in three stages. The first step is to calculate an initial trip matrix $\{T_{ij}^{(1)}\}$, where the superscript (1) simply denotes that this is the first set of calculations. This matrix $\{T_{ij}^{(1)}\}$ is obtained from the equation

$$T_{ij}^{(1)} = W_i W_j / d_{ij}, \tag{5.7}$$

which is the gravity formula with the scaling constant omitted. Using the data listed in figure 5.2, the first two elements of the matrix $\{T_{ij}^{(1)}\}$ are calculated as

$$T_{11}^{(1)} = W_1 W_1 \, d_{11}^{-2}$$
$$= 100 \times 100 \times (1/7 \cdot 5^2) = 177 \cdot 8$$
$$T_{12}^{(1)} = W_1 W_2 \, d_{12}^{-2}$$
$$= 100 \times 50 \times (1/30^2) = 5 \cdot 6.$$

Repetition of these procedures for the remaining elements gives the completed initial[1] trip matrix as

$$
T_{ij}^{(1)} = \begin{array}{c|ccc} & & j \longrightarrow & n \\ & 1 & 2 & 3 \\ \hline i \ 1 & 177\text{·}8 & 5\text{·}6 & 6\text{·}0 \\ 2 & 5\text{·}6 & 44\text{·}4 & 4\text{·}7 \\ n \ 3 & 6\text{·}0 & 4\text{·}7 & 225\text{·}0 \end{array}
$$

The initial interactions $T_{ij}^{(1)}$ are each proportional to the predicted number of migrations between region i and region j, but they are not expressed in units of numbers of migrants. Therefore, we need to find a method of transforming the initial interactions into final predictions, T_{ij}, which satisfy the constraint

$$
\sum_{i}^{n} \sum_{j}^{n} T_{ij} = T. \tag{5.8}
$$

This constraint simply asserts that the sum of the predicted migrations must be equal to the known number of migrations, which in this example is $T = 15\,000$. To satisfy this condition we calculate the scaling constant k which converts each of the unscaled predictions $T_{ij}^{(1)}$ into numbers of migrants. Thus the second stage of the calculation is to evaluate the scaling constant from the formula

$$
k = T \Big/ \sum_{i}^{n} \sum_{j}^{n} T_{ij}^{(1)}, \tag{5.9}
$$

which is the ratio between the known total number of flows, T, and the sum of all the initial unscaled predictions. For our example problem k is evaluated as

$$
\begin{aligned}
k &= T/[T_{11} + T_{12} + T_{13} +_{21} +, \ldots, + T_{33}] \\
&= 15\,000/[177\text{·}8 + 5\text{·}6 + 6\text{·}0 + 5\text{·}6 +, \ldots, + 225] \\
&= 15\,000/479\text{·}8 = 31\text{·}263.
\end{aligned}
$$

Finally, to obtain the predicted trip matrix which satisfies the constraint (equation 5.8) we simply calculate the transformation

$$
T_{ij} = kT_{ij}^{(1)}, \tag{5.10}
$$

for each value of $T_{ij}^{(1)}$. The matrix of scaled values is listed in figure 5.3(*a*) and the reader may wish to check that the sum of these values is equal to $T = 15\,000$ migrations.

Goodness-of-fit

The gravity model assumptions are gross simplifications of aggregate human behaviour and therefore our predictions should be checked against reality. Figure 5.3 shows some simple methods for assessing the degree of correspondence, or goodness-of-fit, between the predicted migrations (figure 5.3*a*) and the observed migrations (T_{ij}^{*}, figure 5.3*b*) in our study area. Clearly, if the prediction were perfect, then all the values T_{ij} would be equal to their respective observed migrations, T_{ij}^{*}. Therefore the

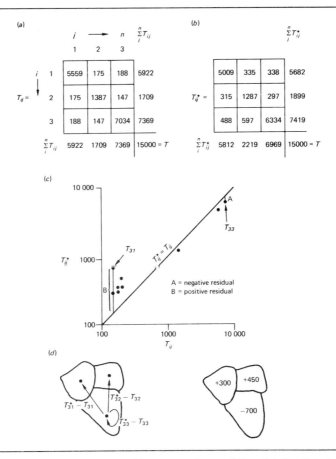

Figure 5.3. The migration problem: goodness-of-fit. (*a*) Predicted migrations. (*b*) Observed migrations. (*c*) Graph of errors. (*d*) Zonal errors — migrations leaving zone 3, $T^*_{3j} - T_{3j}$

graph of the line $T_{ij} = T^*_{ij}$ (figure 5.3*c*) depicts perfect goodness-of-fit between prediction and observation. However, by plotting each point (T_{ij}, T^*_{ij}) on the graph it can be seen that errors in prediction have occurred. The individual values $T^*_{ij} - T_{ij}$ are termed residuals and may be either positive or negative. When the residual is negative the model has over-predicted reality and, conversely, positive residuals are indicative of under-prediction. For the migration problem it turns out that all the intra-zonal (short-distance) residuals are negative, while all the inter-zonal (relatively long-distance) residuals are positive. This pattern of error suggests we have chosen the wrong distance exponent to calculate the prediction. If the exponent were to be increased, to say $\alpha = 2 \cdot 5$, the goodness-of-fit would be improved because short-distance interactions would be increased at the expense of long-distance interactions.

A further aspect of goodness-of-fit is the mapping of residual values for flows either entering or leaving a particular zone. The pattern of residual signs provides information about the observed attractiveness of zones as origins or destinations for flows. For example, the map of residual flows leaving zone 3 (figure 5.3d) shows that zones 1 and 2 (positive residuals) are favoured by migrants from zone 3 in preference to relocation within their home zone.

THE PRODUCTION-CONSTRAINED MODEL

The need for constraints

Before we proceed to a discussion of more complex gravity models it is necessary to demonstrate one of the fundamental weaknesses of the simple gravity model. This weakness can be illustrated by comparing properties of the matrices $\{T_{ij}\}$ and $\{T_{ij}^*\}$ in figures 5.3(a) and (b), respectively. If the elements T_{ij}^* in any row i of the observed matrix are added together $\Sigma_j^n T_{ij}^*$, this total is equal to the observed number of migrations that originated in region i. For example, 5682 migrations originated in region 1 during the study period. Similarly, the sum over the elements in any row j of the observed matrix $\Sigma_i^n T_{ij}^*$ is equal to the total number of migrations which ended with a location in region j. A comparison of the predicted row and column totals (figure 5.3a) with the corresponding observed row and column totals (figure 5.3b) will reveal that the respective sums are not equal. These inequalities highlight the weakness of the simple gravity model. The model contains no mathematical mechanisms for ensuring that the predicted number of flows originating in a region are made equal to the observed number of migrations originating in that region, or alternatively, that predicted arrivals are equal to observed arrivals. Therefore, if the number of flows either beginning or ending in regions is known prior to prediction, the simple gravity model prediction will not take account of this available information.

The need to modify the gravity model to balance predicted row and column totals with observed row and column totals was first recognized in the development of models to predict flows of shopping expenditure between regions in a city, or between city regions. In this section we examine the properties of a model which is described in Wilson (1974). This model has a virtually identical mathematical structure to two earlier models, namely, Huff's (1964) model to predict consumer patronage of a set of shopping centres, and the market potential model of Lakshmanan and Hansen (1965). The latter model was designed as part of a planning project in Baltimore which attempted to predict the impact of future shopping centre developments on patterns of shopping expenditure.

A production-constrained gravity model is the general term used to describe types of interaction where it can be assumed that the total number of flows leaving each zone i is known prior to making a prediction and therefore can be incorporated into the model design. The simple gravity model used the variable W_i, usually measured by zonal population, to represent a zone's capacity to generate trips. In the production-constrained model W_i is replaced by the variable O_i which represents the known total number of flows beginning in each zone i. While the value O_i is still assumed to be proportional to a zone's ability to generate flows, the production-constrained model is constructed so that the following constraint is satisfied

$$\sum_j^n T_{ij} = O_i. \tag{5.11}$$

This constraint asserts that the sum of the flows in any row i of the predicted flow matrix must be equal to the total number of flows originating in zone i, that is

$$T_{11} + T_{12} + T_{13} + , \ldots , T_{1n} = O_1$$
$$T_{21} + T_{22} + T_{23} + , \ldots , T_{2n} = O_2, \text{etc.}$$

The model equations

To explain how the constraint (equation 5.11) is built into the production-constrained gravity model we will consider the problem of predicting flows of shopping expenditure between the three-region city illustrated in figure 5.4. The shopping model is defined by the equation

$$T_{ij} = A_i O_i W_j c_{ij}^{-\alpha} \tag{5.12}$$

where T_{ij} is the flow of shopping expenditure from the residents of zone i to shops in zone j in some unit time period, usually a week or a month; O_i is the total shopping expenditure of the residents of zone i in the unit time period[2] ; W_j is a generalized measure of the attractiveness of zone j for shopping expenditure, usually the amount of retail floorspace in zone j; c_{ij} is the cost of travelling between zone i and zone j[3] ; and α is the exponent of distance decay. Notice that the only fundamental difference

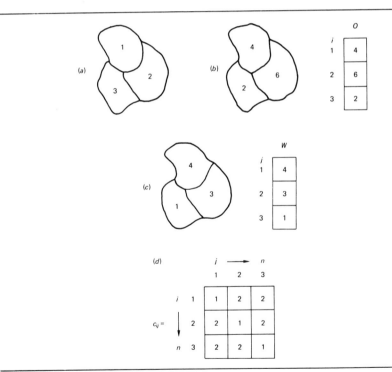

Figure 5.4. Shopping expenditure data for a three-region city. (a) Zone numbers, (i,j). (b) Weekly shopping expenditure, O_i (£ x 10^5). (c) Zonal retail floorspace, W_j (m^2 x 10^6). (d) Costs of travel between zones i and j (£).

between equation (5.12) and the simple gravity model (formula 5.6) is that a variable A_i has replaced the scaling constant k. Recall that in the simple gravity model k was calculated to ensure that the total predicted number of flows was made equal to the total observed number of flows. However, because the constraint (equation 5.11) has now been added to the model specifications, it is necessary to calculate a scaling constant A for each row i to ensure that the sum of the expenditure leaving zone i for each region j sum to the known total zonal expenditure O_i. A_i is simply the ratio between the known zonal shopping expenditure, O_i, and the sum of the unscaled predicted shopping expenditures leaving zone i for each zone j, that is,

$$A_i = O_i \bigg/ \left[\sum_j^n O_i W_j c_{ij}^{-\alpha} \right]. \tag{5.13}$$

Because the value of O_i remains constant during the summation over j in the denominator of equation (5.13), we can rewrite the expression as

$$A_i = O_i \bigg/ \left[O_i \sum_j^n W_j c_{ij}^{-\alpha} \right].$$

Cancelling out the O_i, we obtain a simpler formula for A_i as

$$A_i = 1 \bigg/ \left[\sum_j^n W_j c_{ij}^{-\alpha} \right]. \tag{5.14}$$

To obtain the predicted trip matrix for shopping expenditure flows in the three-region city (figure 5.4) we begin by calculating the values of A_i from formula (5.14). In this example, we will assume that the exponent of distance (cost) decay is $\alpha = 1$. The scalar for region 1 is

$$\begin{aligned} A_1 &= 1/[W_1/c_{11} + W_2/c_{12} + W_3/c_{13}] \\ &= 1/[4/1 + 3/2 + 1/2] \\ &= 1/6 = 0.167. \end{aligned}$$

Similarly, the scalar for region 2 is

$$\begin{aligned} A_2 &= 1/[W_1/c_{21} + W_2/c_{22} + W_3/c_{23}] \\ &= 1/[4/2 + 3/1 + 1/2] \\ &= 1/5.5 = 0.182. \end{aligned}$$

The reader may wish to check that the value of the scalar for region 3 is

$$A_3 = 0.222.$$

We now possess all the information necessary to calculate the individual flows of shopping expenditure from equation (5.12), for example,

$$\begin{aligned} T_{11} &= A_1 O_1 W_1/c_{11} \\ &= 0.167 \times 4 \times 4/1 = 2.67 \\ T_{12} &= A_1 O_1 W_2/c_{12} \\ &= 0.167 \times 4 \times 3/2 = 1.00. \end{aligned}$$

Table 5.1. Flows of shopping expenditure
in the three-region city.

$$T_{ij} = \begin{array}{c} \\ i \end{array} \downarrow \begin{array}{c|ccc|c} & \multicolumn{3}{c}{j \longrightarrow n} & \\ & 1 & 2 & 3 & O_i \\ \hline 1 & 2\cdot67 & 1\cdot00 & 0\cdot33 & 4\cdot00 \\ 2 & 2\cdot18 & 3\cdot28 & 0\cdot54 & 6\cdot00 \\ n\ 3 & 0\cdot89 & 0\cdot67 & 0\cdot44 & 2\cdot00 \\ \hline T_{\cdot j} & 5\cdot74 & 4\cdot95 & 1\cdot31 & \end{array}$$

The full set of shopping expenditure flows is listed in table 5.1. Notice that the sum of the individual interactions in each row i is equal to the known weekly zonal shopping expenditure O_i, indicating the constraint imposed by the introduction of the scalars A_i.

Impact Analysis

Although the shopping model predicts flows of expenditure between each pair of regions, its major geographical application is based on the model's ability to predict the amount of expenditure attracted to each shopping centre j[4]. The values in each column of the interaction matrix are the predicted flow of expenditure from each region i to the shopping centre located in region j. For example, figure 5.5 maps the flow of expenditure from each origin region i to the shopping centre in region $j = 1$. Therefore, the sum over the interactions in each column j, that is

$$T_{\cdot j} = \sum_{i}^{n} T_{ij} \tag{5.15}$$

yields a prediction, $T_{\cdot j}$, for the total amount of expenditure at each shopping centre j (table 5.1). If each value $T_{\cdot j}$ is divided by the shopping centre size, W_j, we obtain the predicted sales per square metre at each centre. For example, sales per square metre of centre 1 are obtained as

$$T_{\cdot 1}/W_1 = 574\ 000/4\ 000\ 000 = £0\cdot14 \text{ per } m^2.$$

Now, although shopping centre size influences the flow from i to j, the shopping

Figure 5.5. Predicted flows of shopping expenditure to shopping centre $j = 1$ ($£ \times 10^5$).

model does not constrain the flow to each centre, $T._{j}$, in proportion to centre size. Therefore, depending on the values of other variables in the model, shopping centres are free to compete for sales per square metre. Figure 5.6(a) graphs the sales per square metre against centre size for the three centres in our city. On this graph the dotted line represents the average sales per square metre $(T/\Sigma_{j}^{n} W_{j}$ = £1 200 000/8 000 000 = £0·15 per m^2) for the whole city. Notice that centre 2 is the most competitive centre and attracts relatively more expenditure in proportion to its size than either centre 1 or centre 3.

In urban planning the shopping model is often used to asses the *impact* of planned changes to shopping centre size on the pattern of shopping expenditure. For example, suppose the city planning department has funds to add a further 2 x 10^6 m^2 of retail floorspace to the stock of shops in our example city. One suggested plan is to allocate all the additional floorspace to zone 2 because the amount of retail floorspace in this zone, W_2 = 3 x 10^6 m^2, is relatively small in comparison with the zone's weekly

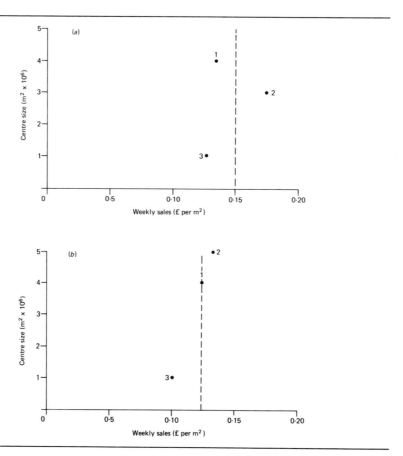

Figure 5.6.   Impact analysis: shopping expenditure in the three-region city. (a) Shopping centre sales prior to building. (b) Shopping centre sales after building. Broken lines represent ideal sales lines for all centres.

shopping expenditure, $O_2 = £6 \times 10^5$. To test the impact of this plan the model predictions are calculated using the data listed in figure 5.4, with the exception that W_j is given the new values

$$j \longrightarrow n$$

	1	2	3
$W =$	4	5	1

The predicted pattern of shopping expenditure corresponding to this planned change in the distribution of retail floorspace is listed in table 5.2. The projected sales per square metre in each of the three centres are plotted against centre size in figure 5.6(*b*). Notice that the addition of 2×10^6 m^2 of retail floorspace has reduced the average sales per square metre in the city to £0·12 per m^2. The general effect of the plan would be to reduce marginally the competitiveness of centre 3 at the expense of centres 1 and 2. Indeed, if the planners felt it was desirable for each centre to capture a proportionately equal share of the market, then the model could be tested with different allocations of floorspace, until a particular allocation was found which gave each centre the same sales per square metre. From figure 5.6(*b*) it can be seen that the equal sales per square metre plan would occur when the majority of the additional floorspace was assigned to centre 2 and the remainder to centre 3.

Although urban planners are usually most interested in the shopping model's ability to assess the impact of changes in shopping centre size, the model can also be used to assess the impact of future changes in the value of the other independent variables in the model equation. For instance, the effects of forecasted changes in consumer expenditure on centre sales can be assessed by calculating the model using the projected values of O_i — the zonal expenditure. Similarly, the impact of transportation improvements on shopping centre sales is measured by calculating predictions based on the reduced values of c_{ij} — the cost of travelling between zones.

The validity of impact analysis depends upon a close correspondence between the initial prediction and the observed pattern of shopping expenditure flows. In practice, a reasonable fit can usually be achieved by finding the value of the cost exponent which minimizes the differences between the observed and predicted flows. It is possible to improve the degree of fit further by adding an exponent, λ, to the shopping

Table 5.2. Impact analysis: flows of shopping expenditure in the three-region city with $W_2 = 5 \times 10^6$ m^2.

		$j \longrightarrow n$			
		1	2	3	O_i
	i 1	2·29	1·43	0·28	4
$T_{ij} =$	2	1·60	4·00	0·40	6
	n 3	0·73	0·91	0·36	2
$T._j$		4·62	6·34	1·04	

centre size variable, W_j. The shopping model equations now become

$$T_{ij} = A_i O_i W_j^\lambda c_{ij}^{-\alpha} \; , \qquad (5.16)$$

where

$$A_i = 1 \bigg/ \left[\sum_j^n W_j^\lambda c_{ij}^{-\alpha} \right] . \qquad (5.17)$$

The value of λ which gives the best fit is related to the scale economies associated with the size of shopping centre: For example, if λ is large, W_j^λ increases more rapidly than W_j (size), and therefore large shopping centres are assumed to be relatively more attractive than small centres. The scale economies measured by λ will be related to the large centre's ability to provide a wider range of goods and services at lower prices than smaller centres. Finally, the value obtained for the cost exponent α will be related to the average trip cost in the study area. As α becomes larger, $c_{ij}^{-\alpha}$ decreases more rapidly than c_{ij} increases. Therefore, the larger the value of α, the lower the average trip cost within the study area. Once the best fitting values of α and λ have been found, these values are held constant during the subsequent impact analysis.

THE ATTRACTION-CONSTRAINED GRAVITY MODEL

In the production-constrained gravity model the trip origins were known, and therefore constrained, and the attention of the research was focused on the model's allocation of flows to destinations. The reverse is true of the attraction-constrained model. Here the trip destinations are known and the allocation of flows to origins is the central interest. The most common form of the attraction-constrained gravity model is a simple residential allocation model which assigns workers in zone j to residences in zone i.

The model predicts the matrix $\{T_{ij}\}$ for n zones where each element T_{ij} is the number of workers living in zone i who work in zone j. The model is derived from the usual gravity assumptions:

$$T_{ij} \propto c_{ij}^{-\alpha}, \qquad (5.18)$$

where c_{ij} is the average cost of the journey to work between zone i and zone j,

$$T_{ij} \propto W_i, \qquad (5.19)$$

where W_i is some measure of the attractiveness of zone i as a residential location,

$$T_{ij} \propto D_j, \qquad (5.20)$$

where D_j is the number of jobs available in zone j. It is assumed that the total number of jobs in each zone is known prior to prediciton and therefore the predicted trip matrix of journey-to-work flows must satisfy the constraint

$$\sum_i^n T_{ij} = D_j. \qquad (5.21)$$

This constraint asserts that the total number of journey-to-work trips arriving in zone j

must be equal to the number of jobs in that zone, that is

$$T_{11} + T_{21} + , \ldots, + T_{n1} = D_1$$
$$T_{12} = T_{22} + , \ldots, + T_{n2} = D_2, \text{etc.}$$

In arithmetic terms the constraint states that the sum over the flows in any column j in the predicted trip matrix must equal the known column total D_j.

The predicted journey-to-work matrix is obtained from the equation

$$T_{ij} = B_j D_j W_i c_{ij}^{-\alpha}. \tag{5.22}$$

The only undefined term in this equation is B_j, which is a scaling constant calculated for each destination zone to ensure that the constraint (equation 5.21) is satisfied. For any zone j, B_j is calculated as the ratio between the known number of jobs D_j, and the sum of the unscaled predicted journey-to-work flows arriving in zone j from each zone i, that is,

$$B_j - D_j \bigg/ \left[\sum_i^n D_j W_i c_{ij}^{-\alpha} \right].$$

Again, D_j cancels out to give the simplified expression

$$B_j = 1 \bigg/ \left[\sum_i^n W_i c_{ij}^{-\alpha} \right]. \tag{5.23}$$

An arithmetic example of this model has not been given because the equations are simply mirror images of the production-constrained equations. However, the format of the output of the equations is illustrated in table 5.3. An $n \times n$ journey-to-work matrix is

Table 5.3. Attraction-constrained gravity model output for an n-zone city.

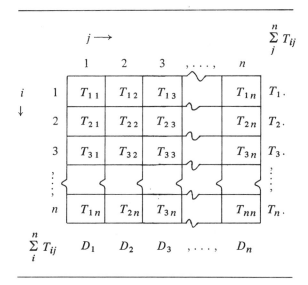

produced which sums over each column to give the zonal employment D_j. Like the production-constrained model, the most important prediction is not the individual flows T_{ij}, but the sum over the flows in each row i, that is,

$$T_{i.} = \sum_{j}^{n} T_{ij}. \tag{5.24}$$

Each value $T_{i.}$ is the predicted number of workers living in zone i. The attraction-constrained gravity model is a residential location model in the sense that it uses knowledge of the distribution of jobs, D_j, the residential attractiveness of each zone, W_i, and journey-to-work costs, c_{ij}, to assign workers to households in each zone. If we assume no unemployment and one-worker families then $T_{i.}$ is a prediction for the number of residences in each zone i.

The attraction-constrained gravity model is a highly simplified representation of the process by which individuals choose household locations. The purpose of introducing the topic of residential locations is to illustrate the idea of constraining trip destinations to make predictions about the location of trip origins. A good deal of research has been devoted to the topic of residential location modelling, and the interested reader is referred to the works of Alonso (1964), Herbert and Stevens (1960), Muth (1969) Senior and Wilson (1974), and Wilson (1974) for a thorough grounding in the subject. Typically, more advanced residential location models include variables such as house type, income levels, average house prices, average income available for housing expenditure, and size of the housing stock to make the model equations more realistic.

THE PRODUCTION–ATTRACTION-CONSTRAINED GRAVITY MODEL

The model equations

We have seen that constraining either trip origins or trip destinations allows the gravity model to predict the total number of flows either ending (production-constrained) or originating (attraction-constrained) in each zone. Thus a singly constrained gravity model is essentially a location model. However, if both trip origins and trip destinations are constrained in the model design, our attention again returns to predicting the size of the individual flows T_{ij}. Such a production–attraction-constrained gravity model is most commonly applied to the prediction of journey-to-work flows within an urban area.

The journey-to-work model assumes that both the number of workers living in each zone, O_i, and the number of jobs in each zone, D_j, are known prior to predicting the journey-to-work matrix $\{T_{ij}\}$. Therefore, the predicted flows must satisfy two constraints: firstly,

$$\sum_{j}^{n} T_{ij} = O_i, \tag{5.11}$$

which as usual states that the total number of predicted flows originating in each zone i must be equal to the number of workers living in zone i; and secondly,

$$\sum_{i}^{n} T_{ij} = D_j, \tag{5.21}$$

which asserts that the total number of predicted trips ending in each zone j must be equal to the number of jobs in zone j. The doubly constrained gravity model takes the form

$$T_{ij} = A_i O_i B_j D_j c_{ij}^{-\alpha},\qquad(5.25)$$

where A_i are the scalars which ensure the origin constraint is satisfied, and B_j are the scalars for the destination constraint. As usual, A_i is the ratio between O_i and the sum of the unscaled predictions for all journey-to-work trips leaving zone i defined by the equation

$$A_i = O_i \bigg/ \left[\sum_j^n O_i B_j D_j c_{ij}^{-\alpha}\right].$$

which simplifies to

$$A_i = 1 \bigg/ \left[\sum_j^n B_j D_j c_{ij}^{-\alpha}\right].\qquad(5.26)$$

Similarly, B_j is the ratio between D_j and the sum of the unscaled predictions for all work trips arriving in zone j defined by the equation

$$B_j = D_j \bigg/ \left[\sum_i^n A_i O_i D_j c_{ij}^{-\alpha}\right],$$

which simplifies to

$$B_j = 1 \bigg/ \left[\sum_i^n A_i O_i c_{ij}^{-\alpha}\right].\qquad(5.27)$$

The scaling algorithm

To illustrate the calculations needed to solve equations (5.25–5.27) we will use the example data for the three-zone city (figure 5.7) and assume that the exponent of distance decay is known to be $\alpha = 1$. As usual we begin by obtaining the values of the, two scalar variables, A_i and B_j. However, inspection of equations (5.26) and (5.27) will show that we have to know the values of B_j in order to solve for A_i and, conversely, the values of A_i must be known before we can solve for B_j. There is no direct method of solving this problem. However, the values of A_i and B_j can be solved experimentally using an *algorithm*. Algorithms are simply sets of arithmetic rules which are applied initially to obtain an approximate solution to a problem. Repeated application of these rules causes the approximate solution to converge gradually on the true answer. Each repetition of the rules is termed an *iteration*.

The algorithm of solving A_i and B_j proceeds in the following way. First we set the values B_j equal to some arbitrary value. It is usual to set all $B_j = 1 \cdot 0$. These arbitrary values are substituted in equation (5.26) to obtain first estimates, $A_i^{(1)}$, of the origin

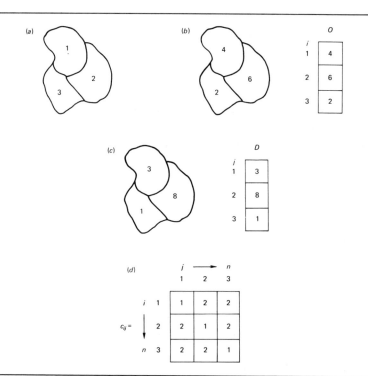

Figure 5.7.   Journey-to-work data for the three-region city. (*a*) Zone numbers, (*i, j*). (*b*) Distribution of workers (x 10^5). (*c*) Distribution of jobs (x 10^5). (*d*) Journey-to-work costs (£).

scalars; accordingly,

$$A_1^{(1)} = 1/[B_1 D_1 c_{11}^{-1} + B_2 D_2 c_{12}^{-1} + B_3 D_3 c_{13}^{-1}]$$
$$= 1/[(1 \times 3/1) + (1 \times 8/2) + (1 \times 1/2)]$$
$$= 1/7 \cdot 5 = 0 \cdot 1333,$$
$$A_2^{(1)} = 1/[B_1 D_1 c_{21}^{-1} + B_2 D_2 c_{22}^{-1} + B_3 D_3 c_{23}^{-1}]$$
$$= 1/[(1 \times 3/2) + (1 \times 8/1) + (1 \times 1/1)]$$
$$= 1/10 = 0 \cdot 0100,$$

and in the same way

$$A_3^{(1)} = 1/6 \cdot 5 = 0 \cdot 1538.$$

These first estimates of $A_i^{(1)}$ are now substituted in equation (5.27) to obtain first estimates of $B_j^{(1)}$ as

$$B_1^{(1)} = 1/[A_1^{(1)} O_1 c_{11}^{-1} + A_2^{(1)} O_2 c_{21} + A_3^{(1)} O_3 c_{31}^{-1}]$$
$$= 1/[(0 \cdot 1333 \times 4/1) + (0 \cdot 0100 \times 6/2) + (0 \cdot 1538 \times 2/2)]$$
$$= 1/0 \cdot 7171 = 1 \cdot 3945;$$

in the same way, $B_2^{(1)} = 2 \cdot 0812$ and $B_3^{(1)} = 1 \cdot 6545$. This calculation completes the first iteration.

A second estimate $A_i^{(2)}$ is now made using the values $B_j^{(1)}$ to give

$$A_1^{(2)} = 1/[B_1^{(1)}D_1 c_{11}^{-1} + B_2^{(1)}D_2 c_{12}^{-1} + B_3^{(1)}D_3 c_{13}^{-1}]$$
$$= 1/[(1 \cdot 3945 \times 3/1) + (2 \cdot 0812 \times 8/2) + (1 \cdot 6545 \times 1/2)]$$
$$= 0 \cdot 0750;$$

similarly, $A_2^{(2)} = 0 \cdot 0511$ and $A_3^{(2)} = 0 \cdot 0828$. These second estimates of $A_i^{(2)}$ are then used to give second estimates of $B_j^{(2)}$ and so complete the second iteration. The iterations continue by using the values $B_j^{(2)}$ to obtain revised estimates of $A_i^{(3)}$ which, in turn, provide estimates of $B_j^{(3)}$, and so on. The results obtained from the algorithm during the first five iterations are listed in table 5.4. Inspection of the record will show that the scalar values converge on a stable value by the fifth interation. For example, the value A_1 stays constant at $0 \cdot 0707$ between the fourth and fifth iterations. The algorithm is terminated when the scalar values stabilize at some pre-defined level of accuracy. In the present example, termination of the algorithm at the fifth interation gives scalar estimates which are accurate to a level of $0 \cdot 001$ (see B_3, table 5.4). If accuracy to five or six decimal places is required, it is usually necessary to compute seven or eight iterations before a satisfactory convergence is obtained. Although the mechanics of the algorithm are quite straightforward it will be apparent that numerous calculations are necessary before the final estimates are obtained. Such tedious tasks are best left to the electronic computer, and a program for solving A_i and B_j is listed in Baxter (1976).

The journey-to-work trip matrix is calculated by substituting the final estimates of A_i and B_j in equation (5.25). For the three-region city the following results are obtained

$$T_{11} = A_1 O_1 B_1 D_1 c_{11}^{-1}$$
$$= (0 \cdot 0707 \times 4 \times 1 \cdot 9055 \times 3)/1 = 1 \cdot 62$$
$$T_{12} = A_1 O_1 B_2 D_2 c_{12}^{-1}$$
$$= (0 \cdot 0707 \times 4 \times 1 \cdot 8369 \times 8)/2 = 2 \cdot 08, \text{ etc.}$$

Table 5.4. Record of iterations for A_i and B_j.

Scalar	Iteration number				
	1	2	3	4	5
A_1	0·1333	0·0750	0·0710	0·0707	0·0707
A_2	0·0100	0·0511	0·0535	0·0537	0·0537
A_3	0·1538	0·0828	0·0810	0·0809	0·0809
B_1	1·3945	1·8653	1·9026	1·9055	1·9055
B_2	2·0812	1·8539	1·8382	1·8369	1·8369
B_3	1·6545	2·1322	2·1524	2·1539	2·1540

Table 5.5. Predicted journey-to-work flows
(persons × 10^5) in the three-region city.

		1	2	3	O_i
	i 1	1·62	2·08	0·30	4
$T_{ij} =$	2	0·92	4·73	0·35	6
	n 3	0·46	1·19	0·35	2
	D_j	3	8	1	

The full set of predictions are listed in table 5.5 and it will be noticed that the introduction of two sets of scalars has ensured that elements in each row sum to give the number of workers in each zone, O_i, and the sum of the elements in each column is equal to the number of jobs, D_j, in each zone.

The doubly constrained journey-to-work model is used in urban planning to assess how future changes in the values of O_i and D_j will affect the pattern of work trips. For instance, if it was planned to build a new housing estate in a particular region, the model is calculated using the projected value of O_i, and a prediction is obtained for the new work trip pattern. Similarly, the impact of a new industrial estate on work trip patterns can be obtained by using the projected zonal employment, D_j, in the model equations. Such forecasts help the urban planner to assess whether the present capacity of the existing transportation system could cope with the projected patterns of journey-to-work.

Criticism of gravity models

The gravity model has been presented as a law capable of predicting flows, and it is not surprising that such an attempt to model human behaviour has been subjected to some severe criticisms. Broadly, these criticisms can be divided into technical problems which occur when the model is operationalized and, more seriously, criticism of the assumptions of the model.

Technical problems arise in relation to the measurement of independent variables and the delimitation of the system of n regions. Gravity models are constructed using measures of production, attraction, and distance and often these can be represented in a number of different ways. The most problematic variable is distance, d_{ij}. Ideally, d_{ij} should measure the average cost of a trip from region i to region j for the population of zone i. However, data are rarely available to compute the desired cost measure and often the shortest road distance between zone centres is used as a crude approximation for average travel costs. Furthermore, intra-zonal distances are usually estimated by some approximate convention which can sometimes lead to inaccuracies in the prediction of the large number of flows which take place within zonal boundaries.

Gravity model predictions also vary with the way in which the study area is divided into a system of regions. The number, size, and shape of areal units all determine the values of the independent variables in the model equations. A number of attempts have been made to define optimum regional systems for gravity model studies. For instance, Broadbent (1970) has suggested that, if the prediction of individual flows, T_{ij}, is the main research interest, then a regional system should be delimited where at least 85 per cent of the flows occur between the regions and less than 15 per cent of all flows take place within regions. When an observed interaction matrix is available for a large number of regions within the study area, it is possible to use computer routines (see Masser and Brown 1975) to group these basic regions together to obtain a more aggregated observed matrix where the patterns of flows correspond to Broadbent's optimality criterion. A final technical problem arises with flows that either begin or end outside the boundaries of the study area. Such flows have not been accounted for in the models we have described, yet the magnitude of these flows is often important, particularly so in impact analysis. For instance, prediction of sales per square metre for a peripheral shopping centre will most likely be under-estimated if flows of cash to the centre from outside the study area are ignored. However, a number of methods have been devised for estimating external flows and the reader is referred to Wilson (1974) for an introduction to this topic.

Technical difficulties can usually be overcome and are not serious so long as the research worker is fully aware of the possible sources of error. However, theoretical criticisms of the gravity model cannot be explained away so easily. These objections all arise because the basic gravity propositions of attraction, production and distance decay are derived from trends in flow data and not from a theory of individual trip-making behaviour. Thus, although the gravity model equations are deduced from a set of propositions, the propositions themselves are merely descriptions of the average behaviour of many thousands of individuals. The gravity model does not explain why a particular individual makes a particular journey, it simply makes use of observed regularities in group behaviour to make aggregate predictions. This weakness of the gravity model has implications for impact analysis. There is no reason to assume that the distance exponent which best fits the observed pattern of interaction will remain constant in the future because the factors which influence its value are not specified in the model assumptions. Since the value of the distance exponent has a major influence on the predictions it is clearly unwise to use the gravity model to assess the impact of long-term planning decisions.

ENTROPY-MAXIMIZING MODELS

The most recent theoretical developments in spatial interaction modelling have managed to overcome many of the problems posed by the gravity propositions. These developments are due largely to Wilson (1967, 1970) who derived a journey-to-work model which mixed deterministic assumptions, such as origin and destination constraints, with the probabilistic idea of entropy-maximizing. Many of the mathematical ideas in Wilson's model are adapted from a theoretical paper by Jaynes (1957). A rigorous mathematical treatment of these topics is well beyond the scope of this book. Instead, this section gives a mathematical appreciation of the Wilson model using the nomenclature of state descriptions.

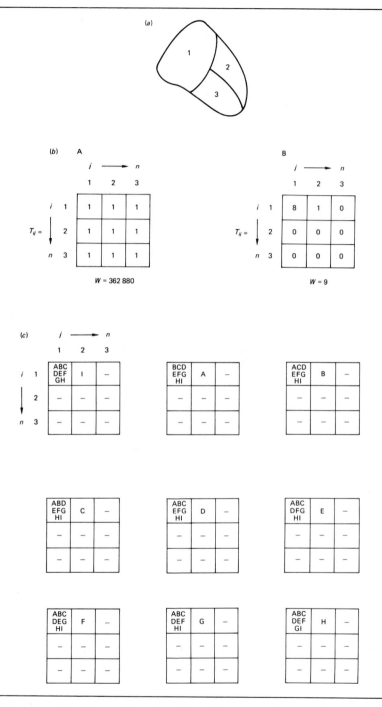

Figure 5.8.   Feasible trip matrices in a city with $n = 3$ regions and $T = 9$ trip-makers. (*a*) Regions and identity numbers, (i,j). (*b*) Two feasible, meso-state trip matrices. (*c*) The $W = 9$ micro-states associated with meso-state B.

Entropy and state descriptions

Suppose we have divided a city into $n = 3$ regions (zones) and the only information we know about the journey-to-work is that, each day, $T = 9$ people travel to work (figure 5.8.). We shall term the information $T = 9$ and $n = 3$, which is known *prior* to making a prediction, *a macro-state description* of the journey-to-work system. As usual, our intention is to predict a trip matrix, $\{T_{ij}\}$, whose elements count the number of people using each of the $n^2 = 9$ journey-to-work routes. Because we know $T = 9$, any trip matrix we predict must satisfy the condition, or constraint, that

$$\sum_i^n \sum_j^n T_{ij} = T. \tag{5.28}$$

This constraint asserts that the sum over the elements, T_{ij}, in any predicted trip matrix, must be equal to the total number of trip-makers, T, living in the city. For the example problem, the constraint is evaluated as

$$T_{11} + T_{12} + T_{13} + T_{21} + T_{22} + , \ldots , + T_{33} = 9.$$

Clearly, a large number of different trip matrices can be formed whose individual elements sum to T, and we will call each of those feasible trip matrices a *meso-state description* of the journey-to-work system. Two of the millions of meso-states which sum to $T = 9$ are shown in figure 5.8(b). To proceed further we need to find some criterion which allows us to select *one* of the feasible meso-states as our predicted trip matrix. To help us make this selection we will introduce the probabilistic notion of the *entropy*, or likelihood, of a meso-state trip matrix. The entropy, W, of any feasible meso-state $\{T_{ij}\}$ is measured by the coefficient

$$W(\{T_{ij}\}) = \frac{T!}{\prod_i^n \prod_j^n T_{ij}!} \tag{5.29}$$

that is, the factorial number[5] for the total number of trips, $T!$, divided by the product, Π, of the factorial numbers for each value T_{ij}. For example, the meso-state trip matrix B (figure 5.8b) has an entropy of

$$W(\{T_{ij}\}) = T!/[T_{11}! \times T_{12}! \times T_{13}! \times T_{21} \times , \ldots , \times T_{33}!]$$
$$= 9!/[8! \times 1! \times 0! \times 0!, \ldots , \times 0!]$$
$$= 362880/[40320 \times 1 \times 1 \times 1 \times , \ldots , \times 1]$$
$$= 362880/40320 = 9.$$

Similarly, the reader may wish to check that the entropy of the trip matrix A (figure 5.8b) is evaluated as $W = 362880$.

To explain the meaning of these entropy values we must introduce the idea of a *micro-state description*. If we identify each of the $T = 9$ trip-makers individually by the letters A to I, then each micro-state description is defined as a different, distinguishable arrangement of the T individuals among the cells of the trip matrix (figure 5.8c). The entropy formula (5.29) simply counts the number of micro-states that can be formed from a particular set of meso-state trip numbers. For example, the $W = 9$ micro-states, or arrangements of individuals, that can be formed from the trip matrix B are shown in figure 5.8(c).

As its name suggests, the principle of entropy-maximization asserts that we choose the predicted trip matrix to be the particular meso-state description for which W takes on a maximum value. Therefore, the entropy-maximizing trip matrix is that set of trip numbers, $\{T_{ij}\}$, where the T trip-makers are able to rearrange themselves among the n^2 journey-to-work routes in the greatest number of ways. When we construct entropy-maximizing models it is assumed that we will never bother to find out which micro-state actually occurred in reality, that is, we never find out to which route each of the T individuals actually assigned themselves. Given this assumption, the entropy-maximizing criterion has a behavioural meaning because we select the solution which maximizes the T individuals freedom to choose between the available journey-to-work routes. For this reason, the entropy-maximizing solution is said to be the *most likely* trip matrix.

It turns out that, when T and n are the only macro-state constraints on W, the entropy-maximizing trip matrix is the meso-state where all the values of T_{ij} are equal; that is, where all

$$T_{ij} = T/n^2. \tag{5.30}$$

For our example, this relationship states that the entropy-maximizing trip matrix is the one where all $T_{ij} = 9/3^2 = 1$, which is the matrix A shown in figure 5.8(b). However, the rule defined by formula (5.30) only holds when formula (5.28) is the only constraint on W. When additional constraints are added to the problem, finding the entropy-maximizing solution is a more complex procedure.

Adding constraints and reducing the entropy

So far our treatment of journey-to-work in the example city has been extremely simple, because our entropy-maximizing prediction has ignored a great deal of macro-state information that is usually available to the research worker *before* he makes his prediction. Two obvious macro-states in journey-to-work modelling are: the set of values $\{O_i\}$ which count the number of workers living in each of the n regions, and the values $\{D_j\}$ which count the number of jobs in each region. Figure 5.9(a) lists the values of O_i and D_j which we will use to illustrate our example problem.

We now consider the effects of incorporating these pieces of macro-state information as constraints on the entropy-maximizing trip matrix. The distribution of workers is termed the origin constraint and, as usual, the constraint asserts that the number of trips which *begin* (originate) in zone i *must* be equal to the number of workers living in that zone, O_i. Recall, we write this statement mathematically as

$$\sum_{j}^{n} T_{ij} = O_i. \tag{5.11}$$

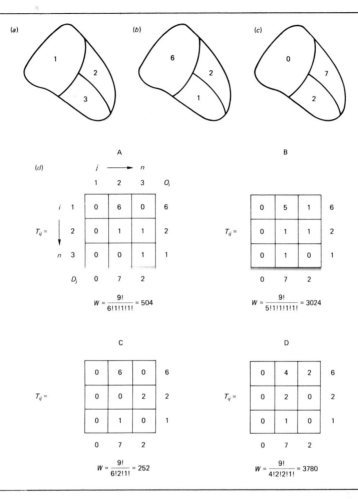

Figure 5.9. Feasible trip matrices with origin and destination constraints in the three-region city with nine trip-makers. (*a*) Identity numbers *i*,*j*. (*b*) Distribution of workers, O_i. (*c*) Distribution of jobs, D_j. (*d*) Feasible trip matrices.

For the example problem (figure 5.9) the constraint (equation 5.11) is evaluated as

$$T_{11} + T_{12} + T_{13} = O_1 = 6$$
$$T_{21} + T_{22} + T_{23} = O_2 = 2$$
$$T_{31} + T_{32} + T_{33} = O_3 = 1.$$

Similarly, the distribution of jobs (the destination constraint) asserts that the number of trips which *end* in zone *j* must be equal to the number of jobs in that zone, D_j. This constraint is written as

$$\sum_i^n T_{ij} = D_j. \tag{5.21}$$

Again, for the example problem the constraint (equation 5.21) is evaluated as

$$T_{11} + T_{21} + T_{31} = D_1 = 0$$
$$T_{12} + T_{22} + T_{32} = D_2 = 7$$
$$T_{13} + T_{23} + T_{33} = D_3 = 2.$$

The incorporation of these two constraints as a macro-state description dramatically changes the entropy-maximizing solution. By applying the new constraints, only four meso-state trip matrices exist which satisfy the distributions of jobs and workers (figure 5.9). Previously (figure 5.8), when $T = 9$ and $n = 3$ were the only constraints, many thousands of meso-state trip matrices[6] each satisfied the macro-state descriptions. Thus by adding the origin and destination constraints to our problem we have, in effect, restricted the freedom of the values T_{ij} to vary and thereby reduced the number of feasible meso-scales. For this expanded problem, the entropy-maximizing trip matrix can be obtained arithmetically by evaluating the entropy (formula 5.29) for each of the four feasible trip matrices. Inspection of the four values of W (figure 5.9d) will reveal that matrix D is the entropy-maximizing solution with $W = 3780$.

The reader may have noticed that the inclusion of the origin and destination constraints has reduced the entropy of the most likely trip matrix by a factor of 96. This result is obtained by dividing the entropy of the initial maximal solution (figure 5.8) by the entropy of the constrained maximal solution (figure 5.9, matrix D), that is, 362 880/3780 = 96. This reduction indicates that the origin and destination constraints restrict the freedom of the $T = 9$ trip-makers to choose between routes. Thus the use of additional macro-state information has reduced the number of behavioural choices (micro-states) we assume for the T trip-makers.

The cost constraint

By far the most novel feature of Wilson's model is his incorporation of a cost constraint into the macro-state description. For most cities, two final pieces of information are usually available and may be included in the macro-state descriptions. First, a matrix of values $\{c_{ij}\}$ each measuring the cost of a work trip beginning in region i and ending in region j. These trip costs for the example city are shown in figure 5.10(c). It can be seen that the cost of travelling *within* any of the $n = 3$ zones, $c_{i=j}$, is £0, while the cost of travel *between* any pair of zones is £1. The second bit of available information is an estimate of a single value, C, which measures the amount of money available to all T trip-makers for expenditure on the daily journey-to-work. For our example city, we will assume that $C = £7$ is available to the nine trip-makers for expenditure on the journey-to-work.

The cost constraint asserts that if each value T_{ij} is *multiplied* by the cost, c_{ij}, of travelling from region i to region j, then the *sum* of these products *must* be equal to the total available expenditure on journey-to-work trips, C. This constraint is written symbolically as

$$\sum_{i}^{n} \sum_{j}^{n} T_{ij} c_{ij} = C. \tag{5.31}$$

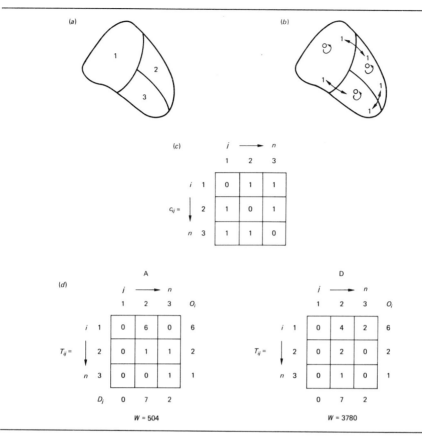

Figure 5.10. Feasible trip matrices with a cost constraint (C = £7.00), origin and destination constraints, in the three-region city with nine trip-makers. (*a*) Identity numbers (i,j). (*b*) Trip costs, c_{ij} (£). (*c*) The cost matrix, $\{c_{ij}\}$ (£). (*d*) Feasible trip matrices (reduced from figure 5.9).

For our example city, this constraint is interpreted arithmetically as (see figure 5.10(*c*) for the values c_{ij})

$$[(T_{11} \times c_{11}) + (T_{12} \times c_{12}) + (T_{13} \times c_{13}) + (T_{21} \times c_{21}) + , \dots , + (T_{33} \times c_{33})] = C$$

$$[(T_{11} \times 0) + (T_{12} \times 1) + (T_{13} \times 1) + (T_{21} \times 1) + , \dots , + (T_{33} \times 0)] = 7.$$

The addition of the cost constraint, C = £7, to the macro-state description has the effect of leaving only two trip matrices (A and D in figure 5.10*d*) as feasible solutions to the problem[7]. Again, the reader may wish to check that the trip matrices B(C = £8) and C(C = £9) do not satisfy the cost constraint of C = £7. A comparison of the feasible trip matrices in figure 5.9(*d*) with those in figure 5.10(*d*) will reveal that the entropy-maximizing trip matrix, D, remains unaltered by the addition of the cost constraint[8]. There is, however, an extremely important difference between the two-constraint solution (figure 5.9*d*) and the three-constraint solution (figure 5.10*d*). This difference concerns our confidence about the *actual* occurrence of the trip patterns,

Table 5.6. The likelihood of the entropy-maximizing trip matrix in the example city.

	$L[W(\{T_{ij}\})]$				Total number of
	A	B	C	D	micro-states
(i) Origin and destination constraints only (figure 5.9*d*)	6·7	40·0	3·3	50·0	7560
(ii) Origin, destination and cost constraints (figure 5.10*d*)	11·8	—	—	88·2	4284

$\{T_{ij}\}$, contained in this entropy-maximizing solution. This confidence may be measured by defining the *likelihood* (max L) of the entropy-maximizing trip matrix [max $W(\{T_{ij}\})$]. Verbally, maximum likelihood is defined as

$$\text{max } L[\text{max } W(\{T_{ij}\})] = \frac{\begin{array}{c}\text{Number of micro-states associated with}\\ \text{the entropy-maximizing trip matrix}\end{array}}{\begin{array}{c}\text{Total number of micro-states associated}\\ \text{with all feasible trip matrices}\end{array}} \times 100.$$

From this descriptive formula it can be seen that our confidence in this prediction is expressed as the *percentage* of all possible micro-states that are created solely from the entropy-maximizing trip matrix. For example, the maximum likelihood for the three-constraint problem in the case of the three-region city is calculated from figure 5.10(*d*) as

$$\text{max } L[W_D(\{T_{ij}\})] = \frac{W_D(\{T_{ij}\}) \times 100}{W_A(\{T_{ij}\}) + W_D(\{T_{ij}\})}$$

$$= \frac{3780 \times 100}{504 + 3780}$$

$$= 88\cdot2 \text{ per cent.}$$

This result shows that, because the entropy-maximizing trip matrix accounts for 88·2 per cent of all micro-states, we can be 88·2 per cent confident in the actual occurrence of this trip matrix. When the same analysis (see table 5.6) is undertaken for the two-constraint problem (figure 5.9*d*), we are only 50 per cent confident about the occurrence of matrix D because the other feasible matrices, A, B, and C, together account for 50 per cent of all possible micro-states.

What have we learned altogether from the journey-to-work in the example city? Firstly, by adding macro-state constraints we automatically reduced the entropy of the maximum entropy trip matrix (meso-state). This finding suggests that our entropy model becomes more realistic as we constrain the meso-state with known, measurable information (data)[9]. Secondly, the addition of constraints increases the likelihood that, in reality, the entropy-maximizing trip matrix actually occurs. This second result may lead the reader to suspect that the entropy-maximizing solution is so likely that it is hardly worth testing the model. Real cities, however, have a nasty habit of playing tricks on the theorist and, in all model-building exercises, it is always wise to test the assumptions of the model. For an evaluation of the performance of the entropy-maximizing model as a predictor of journey-to-work flows in Merseyside, the reader is referred to a paper by Thomas (1977).

The model equations[10]

When we model the journey-to-work in real cities the numbers never work out so simply as our example city, and so a set of equations is needed to predict the entropy-maximizing trip-matrix, $\{T_{ij}\}$. The general structure of the journey-to-work model is illustrated in figure 5.11, and we can state the problem succinctly using the equations discussed in the previous section. The journey-to-work model requires us to maximize the value of the entropy

$$W(\{T_{ij}\}) = \frac{T!}{\prod_i^n \prod_j^n T_{ij}!} . \tag{5.29}$$

Subject to the origin constraint

$$\sum_j^n T_{ij} = O_i \tag{5.11}$$

the destination constraint

$$\sum_i^n T_{ij} = D_j \tag{5.21}$$

and the cost constraint

$$\sum_i^n \sum_j^n T_{ij} c_{ij} = C. \tag{5.31}$$

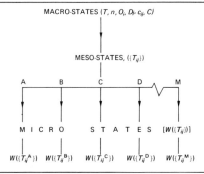

Figure 5.11. The journey-to-work model: state descriptions.

Finding the equations, which predict the set of values $\{T_{ij}\}$ that maximize the value of W, is a mathematical problem involving the method of Lagrangian multipliers. This complicated topic is well beyond the scope of this book, and the interested reader is referred to Wilson (1970) and Wilson and Kirkby (1975) for a formal discussion of the method. Here, all that is necessary is to record that, for any city divided into n zones, each value T_{ij} in the entropy-maximizing trip matrix is predicted by the equation,

$$T_{ij} = O_i A_i D_j B_j e^{-\beta c_{ij}}. \tag{5.32}$$

where

$$A_i = 1 \bigg/ \left[\sum_{j}^{n} B_j D_j e^{-\beta c_{ij}} \right], \tag{5.33}$$

and

$$B_j = 1 \bigg/ \left[\sum_{i}^{n} A_i O_i e^{-\beta c_{ij}} \right]. \tag{5.34}$$

The solution of these equations, for any known values of the constraints (equations 5.11, 5.21, and 5.31), is quite complicated and the procedure is outlined in some detail in the following section. Here, we will only discuss the general role of the terms A_i, B_j, and β in the model equations.

The parameter β is the most important of these terms, because the solution to the model equations (5.23–5.34) ensures that the value given to β both maximizes the entropy (formula 5.29) and satisfies the known values of the constraints (equations 5.11, 5.21, 5.31). A rough estimate of β may be obtained from the formula

$$\hat{\beta} \simeq 1/\bar{c},$$

where

$$\bar{c} = C/T.$$

Therefore, β is approximately equal to the reciprocal of the average cost of a work trip, $\bar{c}$, within the study area. The values of A_i and B_j are both dependent on the value obtained for β. In the solution of the model equations the values of A_i are scalars which ensure that the origin constraint, O_i, is met, while the scalars B_j ensures that the destination constraint, D_j, is satisfied. Indeed the sets of values $\{A_i\}$ and $\{B_j\}$ play exactly the same role in the entropy-maximizing journey-to-work model as they did in the doubly constrained gravity model (formulae 5.25–5.27)[11].

Calibrating the journey-to-work model

To calculate the entropy-maximizing trip matrix, $\{T_{ij}\}$ equations 5.32–5.34 must be solved. For any practical application of this model the values of n (the number of zones) and T (the number of trip-makers) will be large and the equations have to be solved using a computer program. In this section we will outline the mathematical procedures embodied in a computer program published in Baxter (1976) which prints the entropy-maximizing trip matrix and uses mathematical results presented in Hyman (1969).

Recall that, in the model equations, we know the value of the origin constraint, O_i, the destination constraint D_j, the zonal travel costs, c_{ij}, and the total journey-to-work expenditure, C. The problem is to find values for A_i, B_j, and β which both satisfy the three constraints and maximize the value of entropy, W. Baxter's program solves the unknown parameters in the model equations using an algorithm which requires $m = 1$, $2, \ldots, 10$ iterations. The program begins by giving $\hat{\beta}$ a starting value of

$$\hat{\beta}_{m=1} = 1/\bar{c}, \tag{5.35}$$

where $\bar{c}$ is the known average trip cost (see p162). Using $\hat{\beta}_1$, the program then calculates the values of A_i and B_j which satisfy the origin and destination constraints for this starting value of β. The procedures for solving A_i and B_j in the entropy model equation are identical to those used in the doubly constrained gravity model (see pp148–149) except $e^{-\beta c_{ij}}$ replaces $c_{ij}^{-\alpha}$ in the calculations.

It is unlikely that our starting value of $\hat{\beta}_1$ will satisfy the cost constraint, $\bar{c}$. The logic of the Baxter program is to modify systematically $\hat{\beta}_m$ until the value of $\bar{c}_m$ *implied* by one of the trial values of $\hat{\beta}_m$ is more or less equal to the *known* value of $\bar{c}$. Therefore, having just calculated starting values ($m = 1$) for $\hat{\beta}_1$, A_i, and B_j, we substitute these initial values in the model equation (5.32) to obtain an initial trip matrix $\{T_{ij}^{m=1}\}$. This matrix is then used to calculate an implied value of $C_{m=1}$ from the rearranged cost constraint equation (formula 5.31) defined by

$$C_{m=1} = \sum_i^n \sum_j^n T_{ij}^{m=1} c_{ij}. \tag{5.36}$$

The implied average trip cost, $\bar{c}_{m=1}$, is obtained by dividing the implied total journey-to-work expenditure by T; that is

$$\bar{c}_{m=1} = C_{m=1}/T. \tag{5.37}$$

Obtaining a value for $\bar{c}_1$ completes the first ($m = 1$) iteration in the Baxter algorithm.

The starting value of $\hat{\beta}_1$, defined by formula (5.35) is chosen because the implied value of $\bar{c}_1$ it generates is usually *larger* than the known value of $\bar{c}$. The second ($m = 2$) iteration begins by setting β to an assumed value which is known to generate an implied value of $\bar{c}_2$, which is usually *less* than $\bar{c}$. This second value $\hat{\beta}_2$ is defined by the formula

$$\hat{\beta}_2 = \bar{c}_1 \hat{\beta}_1/\bar{c}. \tag{5.38}$$

The same routine we used to obtain the implied value $\bar{c}_1$ is now repeated to obtain a second implied value of $\bar{c}_2$.

From the $m = 3$ iteration onwards, the Baxter algorithm uses a standard method of mathematical optimization known as *linear interpolation*. This technique is illustrated graphically in figure 5.12. Remember that we are systematically modifying $\hat{\beta}_m$ until we find its true value. This true value will then generate an implied value of $\bar{c}_m$ that is more or less equal to the known value of $\bar{c}$. The first two iterations have yielded results for $\bar{c}_1$ and $\bar{c}_2$ which correspond with the respective values of $\hat{\beta}_1$ and $\hat{\beta}_2$. These four values are plotted on figure 5.12(a) and are identified by the points numbered 1 and 2.

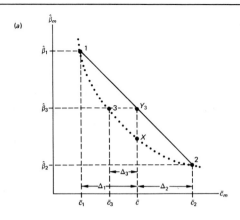

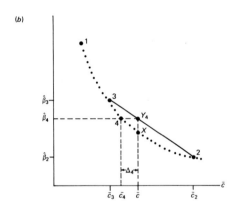

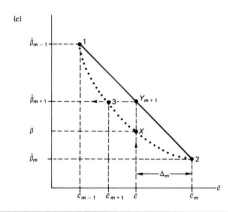

Figure 5.12.   Calibrating the journey-to-work model. (*a*) Iterations $m = 1$ to $m = 3$.
(*b*) Iteration $m = 4$. (*c*) Linear interpolation: general structure.

The dotted curve on the graph represents the *unknown values* of the functional relationship between β and $\bar{c}$. Linear interpolation is a method designed to help us find the true value of β which corresponds with the point X (figure 5.12a) in as few iterations as possible. X is our objective because its position on the unknown functional relationship is located by the known value of $\bar{c}$. Linear interpolation proceeds by assuming that the unknown, nonlinear relationship between the parameter to be estimated, $\hat{\beta}$, and the known variable, $\bar{c}$, is described by a simple linear function. This *assumed* linear relationship is depicted by the line connecting points 1 and 2 on figure 5.12(a). To improve our estimate of β we simply use the known value of $\bar{c}$ to interpolate between our fixed points 1 and 2. Therefore, on the $m = 3$ iteration of the general Baxter algorithm, we read off the value $\hat{\beta}_3$ which corresponds with the point $Y_{m=3}$ on figure 5.12(a). Using $\hat{\beta}_3$ in the model equations we can now calculate a third implied value $\bar{c}_3$. Notice that the linear interpolation has greatly *improved* our estimate of β. For example, if we define the difference (the error—Δ_m) between any implied value $\bar{c}_m$ and the true value $\bar{c}$ by the formula

$$\Delta_m = |\bar{c}_m - \bar{c}| \tag{5.39}$$

then it can be seen from figure 5.12(a) that the value of $\Delta_3 = |\bar{c}_3 - \bar{c}|$ will be much smaller than either $\Delta_1 = |\bar{c}_1 - \bar{c}|$, or $\Delta_2 = |\bar{c}_2 - \bar{c}|$.

The $m = 4$ iteration (figure 5.12b) proceeds in the same way as the third iteration with the exception that an improved linear relationship is assumed by connecting the point marked 3, which we have just located, with the point 2. The resulting implied value $\bar{c}_4$ is an improvement on $\bar{c}_3$ because the error $\Delta_4 = |\bar{c}_4 - \bar{c}|$ is smaller than Δ_3. Subsequent iterations ($m = 5, 6, \ldots$) use the point Y_{m+1} (figure 5.12c) to improve the assumed linear relationship and so reduce the error Δ_m still further. In practice, by the iteration $m = 10$, the error Δ_{10} is usually so close to zero that $\hat{\beta}_{10}$ can be safely taken as representing the true value of β. The Baxter program concludes by substituting $\hat{\beta}_{m=10}$ into the model equation and calculating the entropy-maximizing trip matrix, $\{T_{ij}\}$, together with the final scalar values of $\{A_i\}$ and $\{B_j\}$.

Obviously, the computer program does not solve linear interpolation graphically and, at the conclusion of any iteration $m = 2, 3, \ldots, 10, \ldots$, the improved estimate of $\hat{\beta}_{m+1}$ (figure 5.12c) is obtained using the relationship

$$\hat{\beta}_{m+1} = \hat{\beta}_m + \frac{(\bar{c} - \bar{c}_m)(\hat{\beta}_m - \hat{\beta}_{m-1})}{(\bar{c}_m - \bar{c}_{m-1})}. \tag{5.40}$$

It can be seen from this interpolation formula that values of all the terms on the right-hand side are known before the $m + 1$ th iteration begins.

Linear interpolation is the simplest of the many techniques that form the mathematical subject of parameter optimization. Computer algorithms such as the Baxter program may be improved greatly by assuming *nonlinear* relationships between the known data points ($m - 1$, m). Such assumptions usually need fewer interations to obtain a good parameter estimate and therefore save vast amounts of computer time. However, optimization is a complex subject best left to trained mathematicians, but the interested reader is referred to an excellent text by Burley (1974) which gives a thorough introduction to the subject.

CONCLUDING REMARKS

Spatial interaction modelling provides a neat example of a system of interest where geographers have made considerable progress by successively devising more realistic predictive equations. The initial gravity assumptions, which were simply general-izations of observed migration trends, have gradually been replaced by models which combine probabilistic explanations of aggregate human behaviour with precise functional relationships between the structural variables. The choice of examples to illustrate this progress may have led the reader to believe that the journey-to-work is the only type of movement to have benefited from these developments. However, the theory of human migration has been considerably refined by mixing probabilistic and deterministic assumptions and the reader is referred to a paper by Hyman and Gleave (1978) for a discussion of recent developments in this field.

Theories of human behaviour can never be completely rational and, not surpris-ingly, the recent developments in spatial interaction modelling have not escaped the scrutiny of the critics. For instance, Sayer (1976) has criticized the entropy model assumptions for failing to represent adequately the behavioural process which leads an individual to select a particular journey-to-work. He argues that the mathematical procedures for solving the origin and destination constraints imply that an individual selects the home-place and work-place locations simultaneously and, by this assump-tion, the model ignores the individual space-searching that must accompany both these decisions. In essence, Sayer is criticizing the model for stressing the importance of group behaviour at the expense of individual behaviour. In a similar vein, Curry (1972) has argued that we usually possess a vast amount of *qualitative* information about movements which obviously cannot be quantified and incorporated in the analysis as constraints on the entropy function. For example, the journey-to-work model ignores the social and economic characteristics of the zones which form the study area.

Notes

1   Inspection of the initial trip matrix will show that each value T_{ij} above the principal diagonal is equal to the corresponding value T_{ji}, below the diagonal. For example, $T_{12} = T_{21} = 5 \cdot 6$, which means that the predicted flows from region 1 to region 2 are the same as those from region 2 to region 1, etc. This symmetry occurs because regional population has been used to measure both W_i and W_j. Clearly, if different variables are used to measure trip generation and trip attraction, the symmetry will disappear.

2   In the shopping expenditure model, O_i is a composite variable made up of the population of zone i multiplied by the average expenditure per resident of zone i on shopping goods in a unit time period.

3   In urban planning studies it is usual to replace distance with some measure of the generalized cost of travelling between zones.

4   In the example city, each region contains one shopping centre of size W_j.

5 *Factorial numbers.* These numbers frequently turn up in probability modelling and the expression $T!$ is evaluated as the product of all the numbers from one up to the number T. For

example, if $T = 4$, then

$$T! = T(T - 1)(T - 2)(T - 3)$$
$$4! = 4(4 - 1)(4 - 2)(4 - 3)$$
$$= 4 \times 3 \times 2 \times 1 = 24.$$

Furthermore, both the numbers $0!$ and $1!$ have defined values of one.

6   These additional trip matrices (feasible meso-states) are not shown in figure 5.8 for obvious reasons of space.

7   The worked example is not typical because, in real journey-to-work studies, the number of zones, n, is usually large, and in such cases the cost constraint is likely to eliminate a large proportion of the trip matrices that were feasible when only the origin and destination constraints were applied.

8   This result is again untypical of real journey-to-work studies. More often than not the entropy-maximizing trip matrix will change with the addition of the cost constraint.

9   A fourth constraint is often added to the journey-to-work model. This constraint makes use of information about modal split (the different costs of public and private transport) to reduce still further the entropy of the maximum solution. This constraint, however, is quite complicated mathematically, and the reader is referred to Wilson (1974) for a full discussion of its properties.

10   From here to the end of this section the discussion becomes quite technical and is intended for the more mathematically minded geography student.

11   The reader may have noticed that similarity between the entropy model (equations 5.32–5.34) and the doubly constrained gravity model (equations 5.25–5.27). The only difference between the two sets of equations is that in the entropy model we find that the term $e^{-\beta c_{ij}}$ replaces the distance term $c_{ij}^{-\alpha}$ in the gravity model, where α is the exponent of distance decay. This difference occurs because the two models are derived from different assumptions. The gravity model is deterministic and assumes that the correct value of α can be specified from our prior knowledge about the problem. The entropy model *does not* assume a distance decay effect, and the term $e^{-\beta c_{ij}}$ arises

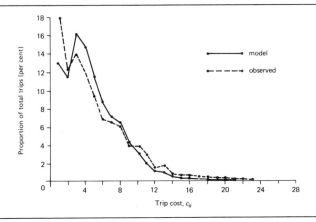

Plots against cost of observed and predicted trip distributions. Source: R W Thomas (1977) p 827.

algebraically as a result of maximizing the entropy subject to the constraint equations. Therefore, the entropy model is more realistic than the gravity model because the weak distance decay assumption is replaced by a mathematical method which allows us to find the most likely trip matrix subject, especially, to the constraint imposed by C, the known total expenditure on journey-to-work.

It is worth noting that, because the solution to the entropy model contains a negative exponential term, the model usually predicts a trip cost frequency distribution characterized by large number of short-distance (low-cost) trips and relatively few long-distance (high-cost) trips. This effect is illustrated by the graph which compares the observed trip cost distribution on Merseyside (1966), with the distribution predicted by the entropy model. However, this effect was produced *not* by assuming distance decay, but because a declining trip cost distribution turns out to be the most likely way in which a large number of workers will allocate a finite resource (total journey-to-work cost − C) amongst themselves.

CHAPTER 6

Spatial Allocation Models

The subject of this chapter is, again, the flows of goods and people across the Earth's surface. However, the approach taken here differs from the spatial interaction modelling style in a number of ways. Spatial interaction models attempt to predict exactly observed patterns of movement. In contrast, spatial allocation models attempt to answer the question: 'What is the best, or *optimum*, pattern of flow?' In this sense spatial allocation models are said to be *normative* because their concern is 'what ought to be', rather than 'what is'. The linchpin of spatial allocation analysis is a model called the *transportation problem*. The solutions to this model are trip matrices which are optimal according to some pre-defined criterion. For instance, the simpler forms of the transportation problem predict trip matrices which minimize either the total cost of transportation or the total distance travelled. The value of such solutions, which identify the 'best' pattern of flow, is that their predictions suggest ways in which the observed pattern of flow can be modified to make the trading system operate more efficiently. Furthermore, because spatial allocation models are concerned with optimality, their mathematical structures provide considerable theoretical insights about the trading mechanisms operating in idealized spatial economies.

THE TRANSPORTATION PROBLEM AND LINEAR PROGRAMMING

The transportation problem falls within the scope of a branch of mathematics known as *linear programming*. The originator of linear programming was the American mathematician Dantzig who, in 1947, published an algorithm known as the simplex method. The importance of this procedure was that it provided solutions to problems which had previously been thought to be unrelated. Linear programming problems are typified by the fact that some objective is to be obtained, such as maximum profit, minimum cost, or minimum production time. One of the earliest linear programs solved by Dantzig was the diet problem which involved finding a combination of foods that provided an acceptable nutritional diet for the least possible cost. This description illustrates a second feature of linear programming problems which is that the objective is subject to constraints. The minimum-cost diet is constrained by the fact that an acceptable nutritional level must be attained. In effect, the constraint prevents us making the trivial solution to the diet problem of producing a zero-cost diet by starving people.

Stated in more mathematical terms, a linear programming problem is set up by specifying a linear objective function containing several variables whose value we wish to minimize or maximize. The constraints on the variables in the objective function must also be specified as a set of linear equations. To illustrate these statements we can consider how journey-to-work patterns may be modelled as a transportation problem. The objective is to find the trip matrix $\{T_{ij}\}$ which minimizes the total expenditure on work trips. The objective function is a linear equation relating numbers of work trips T_{ij}, to work trip costs, c_{ij}. The constraints on the minimization of journey-to-work costs are the same as the constraints in the production–attraction-constrained gravity model; that is, the values in each row of the trip matrix must sum to the number of workers living in each zone and values in each column must sum to give the number of jobs in each zone.

A method of solving the transportation problem was given by Hitchcock (1941) prior to the general development of linear programming methods. Fortunately, the transportation problem contains several simplifying features which make the solutions relatively easy to compute. This simplicity enables us to avoid explaining the complexities of Dantzig's simplex algorithm, although the interested reader is referred to the textbook by Dorfman *et al* (1968) for an introduction to this important general method.

The structure of the transportation problem

In essence, the transportation problem describes a simple spatial economy. The description of this economy requires knowledge of the following information. Firstly, a set of $i = 1, 2, \ldots, n$ locations each producing O_i units of some good or commodity; secondly, a set of $j = 1, 2, \ldots, m$ locations each demanding D_j units of the commodity[1] and thirdly, a matrix of costs $\{c_{ij}\}$ each measuring the cost per unit of production for transportation of the commodity from each origin i to each destination j. In reality, the variables O_i and D_j are written in units such as tonnes and movements costs, c_{ij}, in units such as £/tonne. Clearly, in this economy the commodity will move from points of production to points of demand and, as usual, we designate the magnitude of each possible flow by the variable T_{ij}. Notice that the total transportation cost incurred by any flow T_{ij} will be the product $T_{ij}.c_{ij}$. The primary objective of the transportation problem is to find the particular set of values $\{T_{ij}\}$ which minimizes the total cost of all commodity flows. Symbolically, we write this objective as minimizing the value of the function

$$Z = \sum_i^n \sum_j^m T_{ij} . c_{ij}. \tag{6.1}$$

Notice that Z is simply the sum of the cost incurred by each individual flow.

The values of T_{ij} which minimize our objective function must satisfy a set of constraints. The first constraint is written as

$$\sum_{j=1}^m T_{ij} \leqslant O_i. \tag{6.2}$$

and asserts that the total shipments of goods out of a production point i must always be less than or equal to the production capacity, O_i, of the source. The second constraint is written as

$$\sum_{i}^{n} T_{ij} = D_j,$$ (6.3)

and asserts that the sum of the shipments into the j destination must be exactly equal to the total demand for the good, D_j, at that destination. In the simplest form of the transportation problem the production constraint is written as an equality which assumes that all production is consumed and no surplus production capacity exists in the economy. Notice that when the transportation constraints are stated as equalities they are identical in structure to the constraints in the production—attraction-constrained gravity model. A final side condition in the transportation problem is that all

$$T_{ij} \geqslant 0,$$ (6.4)

for, in reality, it is clearly impossible to represent the magnitude of a flow by a negative number. Taken together, equations (6.1–6.4) constitute a linear program because all the algebraic relations contained within each expression are linear in form.

THE TRANSPORTATION ALGORITHM

The solution to the transportation problem is provided by an iterative procedure known as the transportation algorithm. In common with most linear programming routines the transportation algorithm begins by finding a feasible set of values $\{T_{ij}\}$, which obviously satisfy the constraints (equations 6.2–6.4), but do not necessarily minimize the value of the objective function (equation 6.1). The algorithm proceeds by successively modifying the initial feasible solution so that the values of $\{T_{ij}\}$ gradually converge on the final optimum solution where the total cost of transportation is indeed at its minimum value. The description of the transportation algorithm given here is adapted from the presentation given by Scott (1971a).

The initial feasible trip matrix

The operations involved in the transportation algorithm will be explained with reference to the example problem listed in figure 6.1. The problem consists of $n = 3$ factories each producing O_i units of the commodity and $m = 3$ markets (destination points) each demanding D_j units. The costs, c_{ij}, of transporting a unit of the commodity from each factory i to each market j are also given in figure 6.1. The problem is to find the flow matrix $\{T_{ij}\}$ which minimizes the total cost of transporting the 100 units of the commodity from the factories to the markets. The first stage of the transportation algorithm is to find an initial feasible trip matrix $\{T_{ij}^{(1)}\}$ using a procedure known as the *northwest corner rule*. This rule is the simplest of a number of methods that are available for finding a flow matrix which satisfies the origin and destination constraints. The rule is to begin assigning commodity shipments in the cell$_{11}$ (the northwest corner of the flow matrix) by comparing the values of the origin

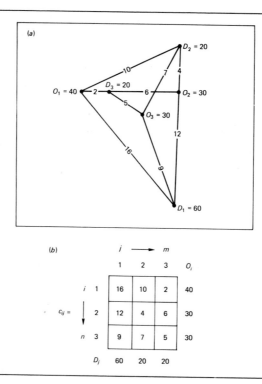

Figure 6.1.   Data for the example transportation problem. (*a*) Map of supply points, O_i, demand points, D_j, and unit transport costs, c_{ij}. (*b*) Matrix respresentation.

and destination constraints, and then work towards the southeast corner (the cell$_{nm}$) of the flow matrix at each stage making the largest possible assignment permitted by the constraints,

The northwest corner solution to any problem may be found by repetition of the following steps.

(1) Set up a blank matrix along with the values of the origin and destination constraints.

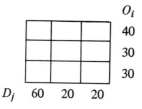

(2) Let T_{11} (the northwest corner flow) equal the smaller of the first origin constraints ($O_1 = 40$) and the first destination constraint ($D_1 = 60$). The smaller of these constraint values, 40, is the maximum possible assignment because we cannot

make an assignment which exceeds either constraint value. Reduce both constraint values by the new value of T_{11}.

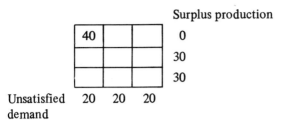

Surplus production

40 | | | 0
| | | 30
| | | 30

Unsatisfied demand   20   20   20

(3) Application of step (2) will either exhaust the production capacity of the first production point, or satisfy all the demand at the first market. Repeat step (2) using the northwest element of the matrix obtained by deleting the row or column whose constraint has been fully satisfied. In this instance, row one is ignored and T_{21} is the new northwest element.

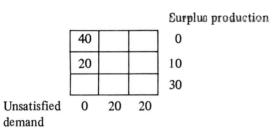

Surplus production

40 | | | 0
20 | | | 10
| | | 30

Unsatisfied demand   0   20   20

(4) Repeat step (2) until all production, O_i, is used and all demand is satisfied.

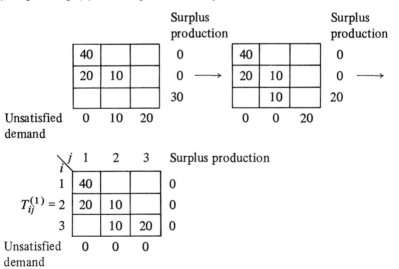

Surplus production

40 | | | 0
20 | 10 | | 0
| | | 30

Unsatisfied demand   0   10   20

⟶

Surplus production

40 | | | 0
20 | 10 | | 0
| | 10 | 10

0   0   20

⟶

$i \backslash j$	1	2	3	Surplus production
1	40			0
$T_{ij}^{(1)} =$ 2	20	10		0
3		10	20	0

Unsatisfied demand   0   0   0

The northwest corner rule yields a feasible matrix $\{T_{ij}^{(1)}\}$ which satisfies the constraints and makes the smallest possible number of positive shipments to the cells of the flow matrix.

The transportation cost, $Z^{(1)}$, of the initial feasible solution is obtained from the objective function (equation 6.1) as

$$Z^{(1)} = [(T_{11}^{(1)} \cdot c_{11}) + (T_{12}^{(1)} \cdot c_{12}) + (T_{13}^{(1)} \cdot c_{13}) + (T_{21}^{(1)} \cdot c_{21})$$
$$+ , \ldots , + (T_{33}^{(1)} \cdot c_{33})]$$
$$= [(40 \times 15) + (0 \times 10) + (0 \times 2) + (20 \times 12) + , \ldots , + (20 \times 5)]$$
$$= 1090.$$

The costs, c_{ij}, used in the calculation are listed in figure 6.1(b). The northwest corner rule gives a feasible solution but the solution is unlikely to be optimal because transport costs are not considered in the process of assigning flows to cells.

Shadow prices and opportunity costs

The remainder of the transportation algorithm involves repetition of an iterative procedure which enables the initial flow matrix to be successively modified until the optimum cost-minimizing solution is found. Each iteration is designed to find an unused cell in the feasible flow matrix, where the assignment of a positive shipment to that cell would reduce the total transportation cost. The maximum flow possible is then assigned to this cell. The iterations continue until no cell is found where an improvement to costs can be made.

To identify such cells the transportation algorithm makes use of computational devices termed shadow prices and opportunity costs. For the moment, we shall define a shadow price as the *relative* price of the commodity at each factory and each market. The factory shadow prices are denoted by the symbol u_i and the market shadow prices by v_j. Because shadow prices are relative measures we can arbitrarily set the shadow price at any one location equal to zero and measure the remaining prices

Table 6.1. Shadow prices for the initial flow matrix.

i	u_i \ v_j	1 / 16	3 / 8	3 / 6
1	0	16 / 40	10 / −	2 / −
2	4	12 / 20	4 / 10	6 / −
3	1	9 / −	7 / 10	5 / 20

Note: In each cell the value above the diagonal is the transportation cost c_{ij} and values below the diagonal are the shipments $T_{ij}^{(1)}$.

relative to this numerical reference point. It is a convention to set the price at $u_1 = 0$. The shadow prices for the initial trip matrix are illustrated in table 6.1. The following argument is used to obtain the remaining shadow prices. Because there is a positive flow of commodity from factory 1 to market 1, the shadow price v_1 must be equal to the relative price at factory 1, u_1, *plus* the unit transportation cost, c_{11}, that is $v_1 = u_1 + c_{11} = 0 + 16 = 16$. The generalization of this argument is that the algebraic condition

$$v_j = u_i + c_{ij} \tag{6.5}$$

must hold true for any market j which has received a positive shipment from factory i. Rearranging this expression we obtain the condition

$$u_i = v_j - c_{ij} \tag{6.6}$$

which must be satisfied for any factory i which made a positive shipment to market j. For example, we already know the shadow price v_1, and that a positive shipment $T_{21}^{(1)}$ was made, therefore, the shadow price at factory 2 is obtained from (6.6) as $u_2 = v_1 - c_{21} = 16 - 12 = 4$. Knowledge of u_2 enables us to calculate v_2 from condition (6.5) as $v_2 = u_2 + c_{22} = 4 + 4 = 8$. It is important to note that the conditions (6.5 and 6.6) may only be used to compute shadow prices when a *positive* shipment has been made between the factory i and the market j. The reader may wish to calculate the remaining shadow prices listed in table 6.1.

The shadow prices are used to calculate *opportunity costs*, $\bar{c}_{ij}$, for each cell assigned a *zero* shipment in the initial matrix $\{T_{ij}^{(1)}\}$. Opportunity costs are defined by the relationship

$$\bar{c}_{ij} = v_j - u_i. \tag{6.7}$$

For example, the opportunity cost $\bar{c}_{12}$ is obtained from the shadow prices in table 6.2 as

$$\bar{c}_{12} = v_2 - u_1 = 8 - 0 = 8.$$

The opportunity costs for the remaining unoccupied cells are listed in table 6.2. An opportunity cost is interpreted as the cost to the spatial economy of leaving a cell unoccupied. It follows that, if the opportunity cost, $\bar{c}_{ij}$, for any cell is greater than the corresponding transportation cost, c_{ij}, then the assignment of a commodity flow to that cell will reduce the total transportation cost, $Z^{(1)}$, of the initial solution.

Table 6.2. Opportunity costs
for the initial trip matrix.

$\bar{c}_{ij} =$ / v_j \ u_i	16	8	6
0	—	8	6
4	—	—	2
1	15	—	—

Table 6.3. Differences between
opportunity costs and
transportation costs for the
initial trip matrix.

j	1	2	3
1	–	−2	+4
$(\bar{c}_{ij} - c_{ij}) =$ 2	–	–	−4
3	+6	–	–

Adjusting the initial trip matrix

The initial solution is improved by finding the unoccupied cell with the greatest
positive difference between the opportunity cost and transportation cost, and then
making the largest possible assignment to that cell. From table 6.3 it can be seen that
for the initial solution the greatest difference (+6) occurs in cell$_{31}$. This cell now acts
as a pivot around which we adjust the flows in the initial solution. The adjustments are
made by adding flows to the pivot cell and then adjusting flows in the occupied cells
to satisfy the origin and destination constraints. Table 6.4 illustrates the method of
making adjustments. The procedure begins by entering a plus sign in the pivot cell$_{31}$
to indicate that a shipment is to be added to this cell. An implication of this addition
is that factory 3 will now not be able to supply the needs of markets 2 and 3 to the
same extent as the initial solution. Therefore, the magnitude of either T_{32} or T_{33}
must be reduced if the pivot flow is to be feasible. It is decided to reduce the flow
T_{32}, and this reduction is indicated by the minus sign in the change cycle. However, if
the flow T_{32} is reduced there will be a deficit at market 2. The only other source of
supply to market 2 is factory 2, and therefore we enter a plus sign in cell T_{22} to
indicate that this flow must be increased. In turn, this addition implies that factory 2
must reduce its supplies to market 1 in order to stay within its production capacity.
Accordingly, we place a minus sign in cell$_{21}$ to indicate the necessary reduction. In
this instance, although the reduction of T_{21} implies a deficit at market 1, this deficit
has already been met by the original addition of flows to the pivot cell. The balancing
of signs in column 1 indicates that a feasible pattern of adjustment has been found.
The general principle for designing adjustment cycles is to find a path from the pivot
cell through the occupied cells with the property that, in each row and column, a
negative adjustment is complemented by a positive adjustment[2].

It remains to determine the increment of flow, Δ_T, which is to be added or
subtracted from the cells in the adjustment cycle. The idea is to find the value of Δ_T
which results in the maximum possible assignment of flows to the pivot cell. The
maximum permissible value of Δ_T is equal to the smallest value of $T_{ij}^{(1)}$ from amongst
the cells assigned minus signs in the adjustment cycle. If Δ_T were to exceed this
defined value, negative flows would occur in the adjusted flow matrix. Inspection of
tables 6.4(*a*) and (*b*) will show that, in this example, $\Delta_T = T_{32}^{(1)} = 10$, and the necessary adjustments to the initial solution are given in the matrix, $\{T_{ij}^{(2)}\}$ (table 6.4*c*). The

Table 6.4. Improving the initial solution.

(*a*) Initial solution.

$T_{ij}^{(1)} =$	$i \backslash j$	1	2	3	O_i	
	1	40			40	
	2	20	10		30	
	3	*	10	20	30	*Pivot cell
	D_j	60	20	20		

(*b*) Change cycle ($\Delta_T = 10$).

	$i \backslash j$	1	2	3	O_i
	1				40
	2	−			30
	3	+	−		30
	D_j	60	20	20	

(*c*) Second improved solution.

$T_{ij}^{(2)} =$	$i \backslash j$	1	2	3	O_i
	1	40			40
	2	10	20		30
	3	10		20	30
	D_j	60	20	20	

total transportation cost of the shipments $\{T_{ij}^{(2)}\}$ is $Z^{(2)} = 1030$, which is an improvement on the cost of the initial solution $Z^{(1)} = 1090$.

The second iteration of the transportation algorithm is simply a repetition of the same procedures to the adjusted matrix $\{T_{ij}^{(2)}\}$. The shadow prices and opportunity costs for $\{T_{ij}^{(2)}\}$ are listed in table 6.5(*a*). The fact that the opportunity cost $\bar{c}_{13} = 12$ exceeds the transportation cost, $c_{13} = 2$ indicates that it is possible to improve upon the assignments in $\{T_{ij}^{(2)}\}$ and further reduce the total transportation cost. The cycle of adjustments to the matrix $\{T_{ij}^{(2)}\}$ are given in table 6.5(*b*) and the resulting matrix $\{T_{ij}^{(3)}\}$ with a total transportation cost of $Z^{(3)} = 830$ is listed in table 6.5(*c*). The shadow prices and opportunity costs for the matrix $\{T_{ij}^{(3)}\}$ are listed in table 6.6. In this instance, the opportunity cost for each unoccupied cell is less than the respective transportation cost, which indicates that $T_{ij}^{(3)}$ is the optimum assignment of flows[3]. Therefore, $Z^{(3)} = 830$ is the minimum transportation cost permitted by the constraints O_i and D_j and the algorithm is terminated at this point.

Table 6.5. Improvements to the second solution.

(a) Shadow prices and opportunity costs for the matrix $\{T_{ij}^{(2)}\}$.

	j	1	2	3
	v_j	16	8	12
i	u_i			
1	0	●	10 / 8	2 / 12
2	4	●	●	6 / 8
3	7	●	7 / 1	●

●Cells where a positive shipment is made in the matrix $\{T_{ij}^{(2)}\}$.
Values above the diagonal in the unoccupied cells are the transportation
costs, c_{ij}. Values below the diagonal are the opportunity costs, $\bar{c}_{ij}$.

(b) Change cycle.

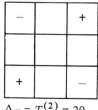

$\Delta_T = T_{33}^{(2)} = 20$

(c) Third improved solution.

	j	1	2	3	O_i
i					
	1	20		20	40
$T_{ij}^{(3)} =$	2	10	20		30
	3	30			30
D_j		60	20	20	

$Z^{(3)} = 830$

Table 6.6. Shadow prices and opportunity costs for the
optimal matrix, $T_{ij}^{(3)}$.

---

	j	1	2	3
	v_j	16	8	2
i	u_i			
1	0	●	10 / 8	●
2	4	●	●	6 / −2
3	7	●	7 / −3	5 / −5

c_{ij} /
/ $\bar{c}_{ij}$

The cost-ranking starting method

It is often possible to reduce the number of iterations needed to solve the transportation algorithm by using a cost-ranking method to obtain the first feasible flow matrix instead of the northwest corner rule. This alternative method is based on ranking the unit transportation costs from lowest, rank 1, to highest, rank $n \times m$ (table 6.7a). The first feasible matrix is obtained by assigning the maximum possible shipment permitted by the origin and destination constraints to the cell ranked 1, and then sequentially making the maximum possible assignment to the second and third ranked cells, and so on, until the constraints have been fully satisfied. During this procedure it is possible for cells with relatively low transportation costs to be given zero assignments because an earlier assignment has exhausted one of the constraints. For example, cell$_{33}$ (rank 3) is not assigned a commodity shipment because the full demand at market 3 ($D_3 = 20$) has already been met by the assignment $T_{13} = 20$. In this instance, the transportation cost-ranking method produces an initial feasible matrix which is the optimum solution (compare table 6.7b with table 6.5c).

Table 6.7. An alternative method of finding the first feasible matrix.

(*a*) Table of ranked costs.

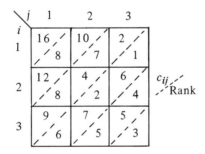

(*b*) Initial feasible solution.

	j 1	2	3	O_i
i 1	20	0	20	40
$T_{ij}^{(1)} =$ 2	10	20	0	30
3	30	0	0	30
D_j	60	20	20	

Degenerate solutions

The reader may wish to check that throughout the computation of the example problem the number of occupied cells in the feasible trip matrix remained constant at 5. This result illustrates a fundamental theorem of linear programming which asserts that the number of occupied cells in the optimal solution can never exceed $n + m - 1$ ($3 + 3 - 1 = 5$). However, during the computations for the transportation algorithm the constraint values may be such that a feasible matrix with *less* than $n + m - 1$ cells is produced. When this phenomenon occurs the program is said to be *degenerate* and the normal computational procedures break down. For example, application of the northwest corner rule to the transportation problem listed in table 6.8(*a*) produces an initial trip matrix containing only three occupied cells, instead of the usual $n + m - 1 = 2 + 3 - 1 = 4$ occupied cells for a problem of this size. When we try to calculate shadow prices over the occupied cells of the matrix listed in table 6.8(*b*) the procedure breaks down at u_2 because cell$_{2\,1}$ is unoccupied. To overcome this degeneracy the cell responsible for the breakdown is designated the θ cell (table 6.8*c*) and is treated as though it were occupied in the degenerate matrix. The computations then proceed in the usual manner. In the example, the pivot is cell$_{1\,3}$ and the θ cell is assigned a positive assignment in the adjustment cycle (table 6.8*d*). When the adjustments are made to the initial flow matrix both the θ cell and the pivot cell receive positive assignments and therefore the degeneracy disappears because the matrix $\{T_{ij}^{(2)}\}$ now contains the usual four occupied cells (table 6.8*e*).

Table 6.8. A degenerate transportation problem.

(*a*) Costs and constraints.

$$
c_{ij} = \quad \begin{array}{c} j \longrightarrow m \end{array}
$$

i	1	2	3	O_i
↓ 1	3	4	1	20
n 2	2	1	2	40
D_j	20	15	25	

(*b*) Northwest corner rule.

$$
T_{ij}^{(1)} = \quad \begin{array}{c} j \longrightarrow m \end{array}
$$

i	1	2	3
↓ 1	20		
n 2		15	25

(*c*) Shadow prices and opportunity costs.

$$
\bar{c}_{ij} =
$$

u_i \ v_j	3	2	3
0	•	2	3
1	θ	•	•

(*d*) Change cycle.

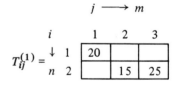

−		+
+		−

$\Delta_T = T_{11}^{(1)} = 20$

(*e*) Non-degenerate second solution.

$$
T_{ij}^{(2)} =
$$

		20
20	15	5

In other cases of degeneracy it may occur that the θ cell receives a *negative* shipment in the adjustment cycle. In such circumstances readjustment of the degenerate matrix is impossible because the maximum possible readjustment, Δ_T, is the zero shipment of the θ cell. This problem is overcome by redesignating the θ cell as the pivot cell of the failed adjustment cycle. This redesignation will cause the θ cell to receive a positive adjustment in the redesigned change cycle and so eliminate the degeneracy.

The reader will no doubt appreciate that in real applications of the transportation problem the values of n and m will be relatively large and therefore a large number of tedious calculations will be made before the optimal solution solution is found. For this reason, the transportation algorithm is usually solved on the computer.

THE DUAL PROBLEM

The dual equations

The transportation algorithm used shadow prices and opportunity costs as computational devices in the search for the optimal flow matrix. In this section the mathematical and economic meaning of shadow prices are examined in more detail in relation to the idea of the *dual* problem in linear programming.

Previously it was stated that a linear programming problem involved finding the minimum or maximum value of an objective function which is subject to a set of linear constraint equations. The transportation problem was to find the minimum total transportation cost subject to the origin and destination constraints. This is termed the *primal* problem because it defines the objective of the research. However, for every primal problem there exists a *dual* problem which will involve the maximization or minimization of a linear relationship between the variables which constrain the primal problem. If the primal problem involves minimization of the objective function the dual will involve maximization and vice versa. A formal mathematical treatment of the relationship between primal and dual problems is beyond the scope of this book, but the topic is dealt with in most standard texts on linear programming (Dantzig 1963, Dorfman *et al* 1958).

In the transportation problem, when the value, Z, of the primal objective function (equation 6.1) is at a minimum, the value, Z', of the following dual function is at a maximum:

$$Z' = \sum_{j}^{m} D_j v_j - \sum_{i}^{n} O_i u_i,$$ (6.8)

subject to the constraint

$$v_j - u_i \leqslant c_{ij}, \qquad \begin{array}{l} (j = 1, 2, \ldots, m) \\ (i = 1, 2, \ldots, n) \end{array}$$ (6.9)

and the side conditions

$$v_j \geqslant 0 \qquad \text{and} \qquad u_i \geqslant 0.$$ (6.10)

Evaluation of the dual function (equation 6.8) for the shadow prices associated with the cost-minimizing matrix $\{T_{ij}^{(3)}\}$ (table 6.5c) gives the result

$$Z = [D_1 v_1 + D_2 v_2 + D_3 v_3] - [O_1 u_1 + O_2 u_2 + O_3 u_3]$$
$$= [(60 \times 16) + (20 \times 8) + (20 \times 2)] - [(40 \times 0) + (30 \times 4) + (30 \times 7)]$$
$$= 1160 - 330 = 830.$$

This result illustrates that in the optimal solution the minimum value of the primal objective function is equal to the maximum value of the dual function[4]. Recall that v_j was defined as the relative price of a unit of the good at market j, then $\Sigma_j^m D_j v_j$ is a measure of the total revenue obtained from the sale of goods in all m markets. Similarly, because u_i is the relative price of the good at each factory, then $\Sigma_i^n O_i u_i$ is a measure of the value of all goods before transportation. The dual is simply the difference between these two summations and therefore the problem can be stated as finding the set of shadow prices which maximize the value added to the goods in transportation. The dual problem is itself constrained by equation (6.9) which asserts that the difference between any market price v_j and factory price u_i must be less than or equal to the transportation cost between the two. Moreover, the constraint equation (6.9) is an explicit definition of the optimality criterion for solving the primal problem. Recall that the primal problem was solved when the opportunity costs $\bar{c}_{ij}$, for unoccupied cells were all less than or equal to the respective transportation cost, c_{ij}. Symbolically, we write this optimality criterion as $\bar{c}_{ij} \leqslant c_{ij}$. Since opportunity costs are defined by the relation $\bar{c}_{ij} = v_j - u_i$ it can be immediately be seen that the constraint equation (6.9) can be written as $\bar{c}_{ij} \leqslant c_{ij}$, which is the *optimality criterion*.

Spatial price equilibrium

A more detailed economic interpretation of the optimum shadow price has been given by Stevens (1961). He interprets u_i as the *location rent* accruing to each factory, and v_j as the *equilibrium market price* of the commodity at each destination. To understand the meaning of these definitions we shall consider a simple spatial economy consisting of three factories (sellers) and one market (buyer), as illustrated in figure 6.2. The following assumptions are made about this simple system: firstly, the demand, D_1, for the commodity at the market is constant quantity; secondly, that together the three factories can supply more than the market's demand; and finally, that the unit production costs at each factory are all equal. The last assumption implies that variations in the delivered price of the good at the market will be due solely to differences in the transportation cost, c_{ij}, from each factory. Therefore, to simplify the discussion we shall assume that production costs at each factory are zero and, without any loss of generality, can be ignored. As a consequence, the delivered price of a unit of the good at the market from each factory is equal to the transportation cost, c_{ij} (figure 6.2b).

To understand how a system of prices develops in this simple competitive system, consider how the buyer acquires the quantity of the good, D_1, demanded at the market. Naturally, he will turn to the cheapest source of supply. In the example, factory 1 is the cheapest source with a delivered price equal to c_{11}. However, from figure 6.2(b) it can be seen that the quantity, O_1, of the good produced at the first

(a)

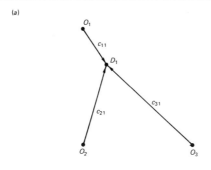

(b)

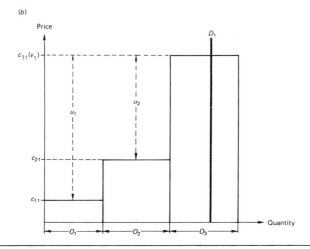

Figure 6.2. A transportation problem with three factories and one market. (a) Location map. (b) Graph of the pricing system.

factory does not fully satisfy the demand, D_1. The buyer is forced to turn to the next cheapest source of supply, factory 2, to make up the deficit in his demand. Again, figure 6.2(b) illustrates that the supplies O_2 only satisfy a part of this deficit, and the buyer has to make up the outstanding demand by purchasing a part of the supply at factory 3 with its delivered price c_{31}.

So far we have only considered prices from the point of view of the buyer. However, in a competitive spatial economy the sellers at each factory will attempt to obtain the highest price possible for their production and, therefore, it is not necessarily true that the buyer will be able to purchase his supplies at the minimum prices c_{11}, c_{21}, and c_{31}. For example, consider the pricing policy open to the seller at factory 1 whose minimum delivered price is c_{11}. It will be apparent to this seller that he can dispose of all the production capacity O_1 at any price that is equal to or below c_{31}. The seller cannot raise his delivered price above c_{31} because the buyer then has a cheaper alternative source of supply from the surplus capacity available at factory 3. By the same reasoning, the seller at factory 2 can raise his delivered price from c_{21} to

c_{31} without fear of the buyer switching his supply to the surplus at factory 3. The seller at factory 3 is the *farthest* from the market and has no competitive advantage over the other two factories. He is termed the *marginal* seller and is forced to sell at his minimum delivered price of c_{31}. Therefore, because the buyer's demand is fixed at the quantity D_1, the outcome of the competition between factories to supply this demand is that all goods are bought in the market at the price c_{31}. The equilibrium market price (the shadow price, v_j) is equal to the minimum delivered price of the marginal seller which, in this instance, is the transportation cost, c_{31}. The market is said to be in equilibrium when v_1 prevails because no seller can raise his price without fear of being undercut by surplus capacity and, conversely, no cheaper alternative source of supply is available to the buyer.

A feature of the equilibrium solution is that all factories, with the exception of the marginal factory, make profits in excess of their minimum delivered prices. For example, the excess profit on each unit of production at factory 1 is equal to the difference between the equilibrium price, $v_1 = c_{31}$, and the minimum delivered price, c_{11}. Similarly, excess profits at factory 2 are given by $v_1 - c_{21}$. Notice that excess profits are defined by the same relationship as the shadow prices u_i (see equation 6.5). In this example the full set of shadow prices are given by the relationships

$$v_1 = c_{31},$$

$$u_1 = v_1 - c_{11},$$

$$u_2 = v_1 - c_{21},$$

$$u_3 = v_1 - c_{31} = 0.$$

Notice too, that the marginal factory 3 makes a zero excess profit because the minimum delivered price, c_{31}, is equal to the market equilibrium price, v_1. The excess profits earned by factories 1 and 2 are due entirely to more favourable transportation costs in comparison with the marginal factory. Thus the excess profits, or rents, are due to geographical advantages and for this reason the shadow prices u_i are termed *location rents*.

This interpretation of shadow prices in the optimal solution can be extended from the case of $n = 3$ sellers and $m = 1$ buyer to the general case of n sellers and m buyers. With more than one market the m buyers compete for the cheapest source of supply and the outcome is a set of v_j equilibrium prices such as those listed in table 6.6. The values of the equilibrium prices reflect the competitiveness of the various markets with respect to the sources of supply. In the example problem, the most competitive market is 3 with $v_3 = 2$, and the least competitive is market 1 with $v_1 = 16$. The competiveness of a particular market is dependent on the existence of relatively low transportation costs between the market and marginal factory with zero location rents. For this reason, Stevens (1961) also terms v_j location rents because their value is determined by the geographical position of the market with respect to the various sources of supply. The optimal set of shadow prices describes a market equilibrium where no seller can increase his location rent by selling to an alternative market because he would be undercut by a more competitive seller; similarly, no buyer can reduce the equilibrium price by switching supplies to an alternative production point because he would be undercut by a more competitive market.

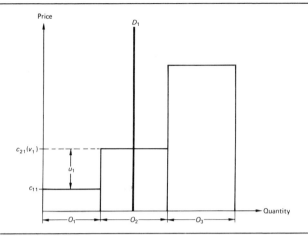

Figure 6.3. The effect of reduced demand, D_1, on the pricing system for the transportation problem with three factories and one market.

From our discussion of equilibrium shadow prices for the case of three sellers and one market (figure 6.2) it was apparent that the optimal values of the shadow prices were determined by the transportation cost c_{31} incurred by the marginal seller who made the location rent $u = 0$. Similarly, in any system of n sellers and m markets the equilibrium shadow prices are determined by transportation costs of a seller who is marginal to the entire system. For our example problem the marginal seller[5] is seller 1 (table 6.6) with a location rent $u_1 = 0$.

So far we have argued that equilibrium shadow prices are dependent on transportation costs. However, this assertion is only true if it is assumed that production capacities, O_i, and market demands, D_j, are fixed quantities. To illustrate how shadow prices are dependent on these constraint values consider the equilibrium conditions that prevail in the problem of three sellers and one market (figure 6.2) when demand at the market is reduced to the level illustrated in figure 6.3. The reduction in demand forces seller 3 out of business and seller 2 takes on the role of marginal seller with $u_2 = 0$. The fall in demand has created a new market equilibrium in which both the equilibrium price, v_1, and the location rent u_1 are reduced in value.

To complete this discussion of the dual problem two points are worth stressing. First of all, the transportation algorithm uses properties of the dual problem to obtain the solution to the primal problem. For instance, recall that the transportation algorithm was terminated when all opportunity costs in unoccupied cells were less than the respective transportation costs. This rule is derived from the constraint (equation 6.9) on the dual function which states that the difference between the shadow price v_j and the shadow price u_i must always be less than or equal to the transportation cost c_{ij}. Secondly, the dual variables describe how the system of rents and prices in the spatial economy reach a condition of market equilibrium. Therefore, in addition to the least-cost pattern of flow, the solution to the transportation problem also provides sets of shadow prices measuring the locational competitiveness of both production points and markets.

EXTENSIONS TO THE TRANSPORTATION PROBLEM

Variations in production costs

So far, the assumptions we have made about the spatial economy have been extremely simplistic. However, the basic transportation problem can be modified in a number of ways to make the spatial economy more realistic. For instance, it has been assumed that producers at each of the n locations each face the same unit production costs. However, it is a simple matter to incorporate variations in production costs into the primal objective function. Let p_i be the unit production cost of the ith factory, then the minimum delivered cost at market j of a good from the ith factory is $p_i + c_{ij}$. Similarly, the cost of delivering T_{ij} units of the good is the product $T_{ij}(p_i + c_{ij})$. Each buyer will now attempt to meet the market demand by purchasing supplied from the factory where the joint cost of production and transportation is at a minimum. This modification allows buyers to purchase supplies from relatively distant factories provided the high transportation costs are offset by low production costs. The prime objective is to find the set of flows $\{T_{ij}\}$ which minimize the joint costs of production and transportation and this objective function is written symbolically as minimizing the value of

$$Z = \sum_i^n \sum_j^m T_{ij}(p_i + c_{ij}). \tag{6.11}$$

The minimization of Z is subject to the usual origin, destination, and non-negativity constraints.

The modified transportation problem can be solved using the transportation algorithm. The only change to the iterative procedure described previously is that the unit production costs $\{p_i\}$ are added to the unit transportation cost matrix $\{c_{ij}\}$ before the algorithm begins (table 6.9). Therefore, the calculations for shadow prices and opportunity costs to solve the algorithm use joint production and transportation costs instead of just transportation costs.

Systems with slack capacity

A second modification to the transportation problem is to relax the assumption that the total supply of the good is equal to the total demand for the good; that is, the

Table 6.9. The matrix of unit production and unit transportation costs.

		$j \longrightarrow$			m
		1	2	$, \ldots,$	m
i	1	$p_1 + c_{11}$	$p_1 + c_{12}$		$p_1 + c_{1m}$
$\downarrow$	2	$p_2 + c_{21}$	$p_2 + c_{22}$		$p_2 + c_{2m}$
$(p_i + c_{ij}) =$	$\vdots$				
	n	$p_n + c_{n1}$	$p_n + c_{n2}$		$p_n + c_{nm}$

Table 6.10. An example transportation problem with
slack production capacity.

		$j \longrightarrow m$				
		1	2	3	slack	O_i
i	1	16	10	2	0	50
$c_{ij} = \downarrow$ 2		12	4	6	0	40
n	3	9	7	5	0	40
	D_j	60	20	20	30	

assumption that

$$\sum_i^n O_i = \sum_j^m D_j.$$

This modification to the assumption makes it necessary to devise a method for solving
the transportation problem for cases where either more of the good is produced at the
n factories than is demanded at the m markets or vice versa.

The solution to such problems is obtained by including *slack* variables in the $n \times m$
flow matrix. To illustrate this idea we will modify the example problem shown in
figure 6.1. Suppose that, in this problem, transportation costs, c_{ij}, and market
demands, D_j, remain unchanged, but the production capacity at each of the $n = 3$
factories is increased by ten units. The variable O_i now takes on the values $O_1 = 50$,
$O_2 = 40$, $O_3 = 40$, and a full matrix representation of the problem is shown in table
6.10. To cope with the excess of 30 units of production capacity a slack column is
included in the transportation matrix. Values of T_{ij} in the slack column represent the
amount of unused production capacity at each factory i. Moreover, because this
unused capacity will never be transported to market, the transportation costs in the
slack column are all set equal to zero. Finally, it may be noted that the slack column is
given a constraint value of $D_{j=\text{slack}} = 30$, indicating that the entire excess production
capacity in the system must be assigned to an artificial slack market. In this way, the
total demand in the system is artificially made equal to the total production capacity.

Table 6.11. Optimal flow matrix for the transportation
problem with slack capacity.

		$j \longrightarrow m$					
		1	2	3	slack		Used
	v_j	12	4	2	0	O_i	capacity
	u_i						
i	1 0			20	30	50	20
$T_{ij} = \downarrow$ 2 0		20	20			40	40
n	3 3	40				40	40
	D_j	60	20	20	30		
	$Z = 720, Z' = 720$						

The minimum transportation cost flow matrix is obtained by applying the normal procedures of the transportation algorithm to the data listed in table 6.10.

If the northwest corner rule is used to obtain the first feasible flow matrix, four iterations of the transportation algorithm are necessary to obtain the optimal flow matrix (table 6.11) for the slack production capacity example[6]. A map of the optimal pattern of flow is shown in figure 6.4(b). Notice that in the optimum solution the entire excess capacity of 30 units is assigned to factory 1. Thus only 50 − 30 = 20 units of the production capacity is sold and the 30 units of slack remains unused.

It is instructive to compare the optimal solution for our original example problem (tables 6.5(c) and 6.6) with the slack capacity solution (table 6.11) in the light of the

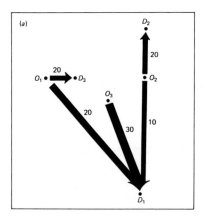

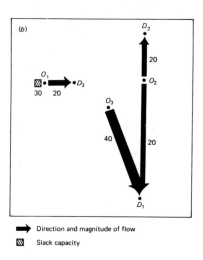

Figure 6.4. Maps of optimal flow patterns for the example problem. (*a*) Flows prior to excess production capacity (see table 6.5(c)). (*b*) Flows with excess production capacity (see table 6.11).

economic interpretation given to the dual problem. Recall that in the original example the minimum total cost of transportation was 830 cost units and factory 1 was the marginal seller making zero location rent ($u_1 = 0$). When excess capacity is included in the problem the minimum transportation cost falls to 720 units. This improvement occurs because market 1 is now able to meet demands ($D_1 = 60$) by obtaining all its supplies from factories 1 and 2 with respective transportation costs $c_{21} = 12$ and $c_{31} = 9$, whereas previously 20 units of market 1's demand was supplied from factory 1 with a transportation cost $c_{11} = 16$. In this way the marginal position of factory 1 in the original problem is exposed by the increased production capacity at factories 2 and 3, and the result is unused capacity at factory 1. A second effect of the increased production capacity is an improvement in the overall ability of the markets to compete with factories for sources of supply. This improvement is reflected by the fact that the equilibrium market prices, v_j, in the slack capacity problem are lower than in the original problem. These results reinforce the point that shadow prices are only totally dependent on transportation costs when the constraint values remain constant. Any change in the constraint values create a new market equilibrium which is reflected in changed rents and prices. Finally, notice that factory 2 is also marginal in the slack capacity problem which indicates that further increases to this factory's production capacity will remain unused unless market demands also change.

The transportation problem with network constraints

So far it has been implicitly assumed that the transportation route connecting each production point i to each market j is capable of handling all goods it is assigned in the optimum solution. However, in reality, network carrying capacities are finite and, for any link in the network, an upper limit to the carrying capacity can often be identified. In this section the way in which network capacity constraints can be incorporated into the basic structure of the transportation problem will be examined.

Figures 6.5(a) and (b) present the structure of our original transportation problem example with the addition of a maximum carrying capacity for each link in the network. The upper limit for the quantity of goods that can be transported on each route is denoted by the variable N_{ij}. As usual, the primary objective of this *capacitated transportation problem* is to find the matrix T_{ij} which minimizes the total cost of transportation. The linear program for this problem is written symbolically as

$$\text{minimize } Z = \sum_i^n \sum_j^m T_{ij} c_{ij}, \tag{6.12}$$

subject to the constraints

$$\sum_j^m T_{ij} = O_i, \tag{6.13}$$

$$\sum_i^n T_{ij} = D_j, \tag{6.14}$$

$$T_{ij} \geqslant 0, \tag{6.15}$$

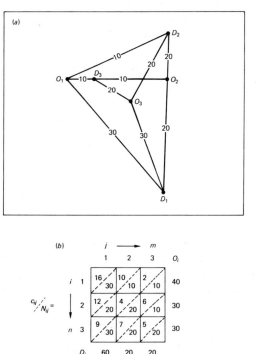

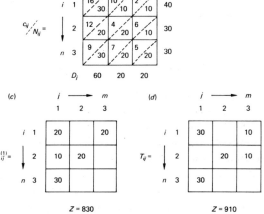

Figure 6.5. Example transportation problem with network capacity constraints, N_{ij}. (a) Map of network capacities, N_{ij}. (b) Matrix representation of the problem. (c) Initial optimal solution, uncapacitated network. (d) Optimal solution with network capacities.

and

$$T_{ij} \leq N_{ij}. \tag{6.16}$$

The only difference between this problem and the basic transportation problem is the inclusion of the constraint (equation 6.16) which asserts that the magnitude of each shipment T_{ij} must not exceed the route carrying capacity N_{ij}.

The general solution to the capacitated transportation problem is beyond the scope of this book, although the interested reader may consult Wagner (1959) for a discussion of the method. However, the main components of the method can be outlined with reference to the problem listed in figure 6.5. The first stage is to use the transportation algorithm to obtain an optimal solution which ignores the capacity constraint (equation 6.16)[7]. Thus, for the example problem, this initial solution (figure 6.5c) is the same as the solution listed in table 6.5(c). The true optimal solution is now obtained by making the necessary adjustments to the first solution to comply with the requirements of the capacity constraint. For example, a comparison of the flows in figure 6.5(c) with the network capacities in figure 6.5(b) will reveal that the flow $T_{13}^{(1)} = 20$ exceeds the route capacity of $N_{13} = 10$. Accordingly, a cycle of adjustment is applied to the initial matrix in order to re-assign the excess capacity of 10 units to under-utilized routes. The result of this procedure is the final optimal matrix listed in figure 6.5(d).

The general effect of introducing network capacity constraints is to decrease the efficiency of the optimal solution. For example, in our problem the total transportation cost of the uncapacitated solution was 830 units, compared with minimum transportation cost of 910 units in the capacitated solution. A corollary to the inherent inefficiency of solutions to the capacitated transportation problem is the question of improving efficiency by investment in additional network capacity. It turns out that network investment problems may also be solved within the general structure of the transportation problem and the reader is referred to Garrison and Marble (1958), Quandt (1960), Ridley (1969), and MacKinnon and Barber (1977) for a discussion of many of the complex issues involved in operationalizing such programs. However, to give an impression of the mathematical structure of network investment programs we shall consider a relatively simple problem described by Scott (1971). Suppose an entrepreneur has a fixed budget, B, to spend on increasing network capacity. His problem is to allocate this budget among the various routes in such a way that the maximum possible reduction in total transportation costs is made. The first part of the entrepreneur's problem is described by equations (6.12–6.14), and is that he wishes to find the matrix of flows which minimizes total transportation costs subject to the usual origin and destination constraints. However, the entrepreneur's fixed budget imposes a number of additional constraints on the objective function (equation 6.12). To state these constraints the following variables need to be defined: y_{ij} is the increased capacity of each route from i to j in the optimal solution, and b_{ij} is the cost of increasing the capacity of each route from i to j by one unit. The additional constraints now take the form

$$T_{ij} - y_{ij} \leqslant N_{ij}, \tag{6.17}$$

and

$$\sum_{i}^{n} \sum_{j}^{m} b_{ij} y_{ij} \leqslant B. \tag{6.18}$$

Equation (6.17) is a revision of the network capacity constraint (equation 6.16), and asserts that the size of any flow T_{ij} minus the increased capacity y_{ij} cannot exceed the original route capacity N_{ij}. In equation (6.18) the individual terms $b_{ij} y_{ij}$ measure the cost of increasing the capacity on a route from i to j by y_{ij} units. Therefore the constraint (equation 6.18) asserts that the total cost of all the increased capacity cannot exceed the fixed budget B.

GEOGRAPHICAL APPLICATIONS

The efficiency of commodity flows

The most obvious application of the transportation problem is to compare minimum transportation cost flow patterns with observed patterns and then suggest improvements to the existing trading system. However, there are pitfalls in this approach. For many goods transportation costs form a small proportion of total operating costs and therefore the optimum solution will be of little monetary consequence. Most geographical applications of the transportation problem have avoided this difficulty by analysing flows of foodstuffs and raw materials where transportation costs are an important element in total operating costs. In one of the earliest applications, Henderson (1958) analysed the efficiency of the coal trade in the USA, while Cox (1965) has compared inter-state flow patterns of aluminium bars with the optimal flow pattern. Examples of foodstuff are provided by Morrill and Garrison (1960) who analysed patterns of trade in wheat flour across the United States and, more recently, Osayimwese (1974) has calculated the optimum flow of the groundnut export crop in Nigeria from production areas to export ports.

To give an indication of the type of results obtained from such studies Henderson's (1958) study of coal movements in the USA will be described in more detail. He divided the United States into 14 regions and predicted optimum flow patterns for the years 1947, 1949, and 1951. The following discussion refers only to the 1947 results. The constraint D_j was measured as the 1947 coal consumption in each of the 14 regions. Eleven of the regions were coal producers, and each of these regions was assigned two production capacities, O_i, namely a shaft (underground) mining capacity and an opencast mining capacity. Therefore, the optimum pattern of flow was presented in an $n = 22 \times m = 14$ matrix. Furthermore, because production capacity exceeded total demand by 20 per cent an additional column was added to assign slack opencast and shaft capacity to each of the 11 production regions. Data were not available for the actual pattern of flow; however, the observed levels of opencast and shaft production in each region were known which make it possible to compare the observed pattern of slack capacity with the optimum pattern. The main difference between these two production patterns was that in the optimum solution all the slack capacity was assigned to shaft mining and the relatively cheap opencast capacity was fully utilized whereas, in reality, much opencast capacity remained unused at the expense of shaft capacity. The sub-optimal use of inefficient shaft capacity was particularly apparent in the traditional mining regions of West Virginia and Pennsylvania, and in total Henderson estimated that actual production and movements costs were almost 9 per cent above these costs in the optimal solution. Finally, Henderson also calculated the dual variables for the optimum solution. It turned out that both location rents and shadow prices were relatively low in the competitve eastern regions where both levels of production and demand were high, while in the less competitive and relatively isolated western regions, both location rents and equilibrium prices were high.

Applications of the transportation problem are not necessarily restricted to goods which are distributed in a competitive space economy. Often the supply and distribution of goods operate under monopoly conditions. In centrally planned economies, such as the USSR, the monopolist is the State, whereas in market economies the monopolist might well be a government corporation controlling the affairs of a nationalized industry. Where total control exists, the monopolist can use the transportation problem to plan the optimum patterns of distribution.

Hinterland delimitation

Although the transportation problem was originally designed to predict least-cost commodity flows, the model has been adapted by geographers to design spatially efficient administrative systems. A problem that often occurs in planning administrative regions is how to design a system of catchment areas around a set of public facilities such that the total distance travelled by consumers to facilities is at a minimum. For example, education authorities often have to draw catchment areas around schools, and it makes sense to delimit these regions so that the total distance (or cost) travelled by pupils is at a minimum. A similar problem is faced by health authorities who have to allocate patients to hospitals and assign populations to general practitioners. Such problems may be solved within the framework of the transportation problem where the optimum trip matrix is used to construct a map of catchment areas around each public facility. In this application of the transportation problem it is important to remember that the locations of the public facilities are given and therefore we are not trying to find optimum facility locations within the study area.

Table 6.12 shows how the transportation problem is adapted to allocate consumers to public facilities and we will explain its structure in the context of designing school catchment areas. The first task is to map the residential location of each pupil within the study area. A grid of n regions is then superimposed on the residential location map and the number of pupils in each region is counted. The number of pupils in each region forms the value of each term O_i in the origin constraint. Notice that the number of regions, n, is chosen arbitrarily by the research worker. Clearly the subsequent analysis will be more accurate with large numbers of regions; however, the size of n has

Table 6.12. The allocation of consumers to facilities as a transportation problem.

		Facilities, j					
Consumer regions, i		1	2	3	,....,	m	O_i – Consumer populations
	1	c_{11}	c_{12}	c_{13}		c_{1m}	O_1
	2	c_{21}	c_{22}	c_{23}		c_{2m}	O_2
	3	c_{31}	c_{32}	c_{33}		c_{3m}	O_3
	n	c_{n1}	c_{n2}	c_{n3}		c_{nm}	O_n
D_j – Facility capacities		D_1	D_2	D_3	,...,	D_m	

to be balanced against the amount of computer storage available to perform the numerous calculations needed to solve the transportation algorithm[8].

The destinations for trips are the locations of the m public facilities distributed across the study area and the values of the destination constraint. The D_j's are the service capacity of each facility. In this instance, D_j is the number of places in the jth school. The total number of places will equal the total number of pupils and thus slack capacity does not enter this problem. The remaining information required to solve the linear program is the matrix of distances or cost, $\{c_{ij}\}$, representing the average cost of travel from each of the n regions to each of the m schools. Given these definitions, the optimum transportation cost flow matrix $\{T_{ij}\}$ can be obtained by applying the usual procedures of the transportation algorithm.

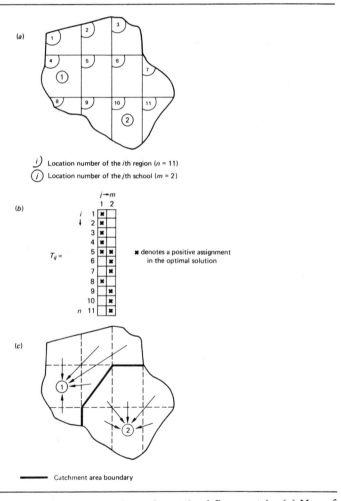

Figure 6.6.   Constructing catchment areas from the optimal flow matrix. (*a*) Map of region and school locations. (*b*) Structure of the optimum flow matrix. (*c*) Construction of school catchment areas.

The final stage of the analysis is to convert the optimal flow matrix into a system of catchment areas around each school. Figure 6.6 illustrates the problem for a study area which has been divided into $n = 11$ origin regions and contains $j = 2$ schools. From our previous work we know that the optimum flow matrix will contain no more than $n + m - 1 = 12$ positive flows of pupils from origin regions to schools (figure 6.6b). An origin region is assigned to the catchment area of a school if a positive shipment of pupils between the region and school has been made in the optimum flow matrix (figure 6.6c). Inspection of the optimum flow matrix shows that with the exception of region 5 all the pupils in each region are assigned to a single school. In the case of region 5 the pupils have been divided between both schools, and, therefore, the catchment area boundary partitions region 5 such that the required number of pupils are assigned to each school. In this example it is only necessary to partition one region. This result is typical of consumer allocation problems and makes the construction of catchment areas relatively straightforward. Few partitions are necessary because the number of origin regions, n, will normally be much larger than the number of facilities, m. Moreover, because no more than $n + m - 1$ assignments are made in the optimum flow matrix, only $m - 1$ origin regions will have their consumer population assigned to more than one facility.

One of the earliest applications of linear programming to the construction of school catchment areas was carried out by Yeates (1963). He used the method we have just outlined to design catchment areas around the 13 high schools in Grant County, Wisconsin, which minimized the total distance travelled to school. The objective of this study was to compare the actual school catchment areas as they existed in 1961 (figure 6.7a) with the optimum distance-minimizing regions (figure 6.7b). The comparison showed (figure 6.7c) that, in terms of travel distances, approximately 18 per cent of pupils were assigned to the wrong school. Yeates estimated that this spatial inefficiency cost the county school system approximately $4000 per annum.

A more complex school-districting problem has been analysed by Maxfield (1972) for the 13 primary schools in Athens, Georgia. In addition to the usual objective of minimizing total travel distances, Maxfield also wished to ensure that each school was racially balanced. Approximately 66 per cent of primary school pupils in Athens were white and 34 per cent were black, and the analysis was designed so that the capacity of each school was filled in accordance with this 66/34 ratio. To achieve this objective two transportation problems were solved: the first used the white pupil population distribution as the origin constraint and 66 per cent of each school's total capacity as the destination constraint, while the second used the black pupil population distribution as the origin constraint and 34 per cent of each school's capacity as the destination constraint. The result of this procedure is that each school is assigned two catchment areas, one for black pupils and one for white pupils. An interesting feature of the racially balanced solution is that the value of the total distance travelled is greater than the corresponding value in a single solution which ignores the balancing constraint and assigns pupils to the nearest school irrespective of race. This result occurs because the ratio of white to black students in each origin region does not conform to the 66/34 average for the whole study area. Thus the availability of different linear programming solutions raises the delicate issue of weighing the social objective of racial balance against the economic objective of cost minimization.

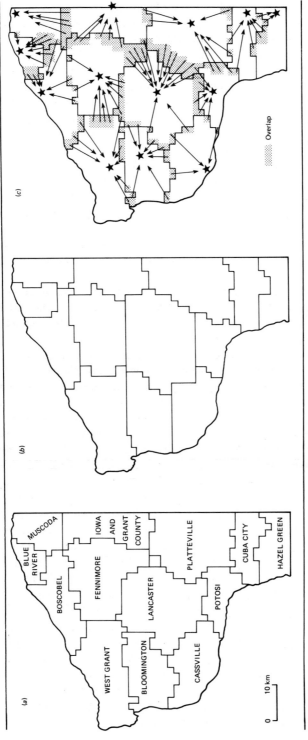

Figure 6.7. School districting for Grant County, Wisconsin. (*a*) Actual school districts. (*b*) Linear programming solution. (*c*) Spatial inefficiency of actual districts. Reprinted with permission from M H Yeates 1974, *Introduction to Quantitative Analysis in Human Geography*, McGraw-Hill, New York, figure 10.4.

CONCLUDING REMARKS

Two themes have emerged from our discussion of the transportation problem. If the transportation problem is interpreted as a theoretical model of a simplified spatial economy, then the dual variables provide an interesting explanation of how a trading system evolves to reach a condition of market equilibrium. Alternatively, the primal problem can be viewed as a purely pragmatic planning device which enables geographers to design marketing systems where the effects of the uneven distribution of population and resources are minimized. To conclude the discussion we will outline some of the theoretical limitations of the model and point to some of its more ambitious practical applications.

The most serious defect of the transportation problem is the assumption that economic relationships are all linear in form. Thus the model takes no account of the technical economies of scale which operate in most production and transportation processes. In reality, large quantities of goods can be produced and transported at relatively lower costs than small quantities. A second problem is that the allocation mechanisms of the transportation algorithm assume that quantities of goods can be divided into an infinite number of smaller quantities. Yet actual trading systems operate with fixed batches of goods which often preclude the attainment of an optimal pattern of flow. However, some of these problems can be overcome by using more advanced methods such as quadratic programming and integer programming which fall outside the field of linear programming. The relevance of these methods for geographers and planners has been reviewed by Scott (1971b) and Macmillan (1978).

Finally, it is worth drawing attention to a complex set of practical location problems which become apparent if we relax some of the constraints in the transportation problem. Recall that to solve the hinterland delimitation problem it was necessary to assume that both the capacities and locations of the facilities were known prior to the analysis. However, an important question in locational analysis is how to locate new facilities in a study area so that the known distribution of consumers is served in the most efficient way. This problem is highly complex because, in addition to finding the optimum pattern of flow, it is necessary to find optimal locations and capacities for the new facilities. Indeed, to date, no method has been devised to identify an exact optimum solution to the problem. Instead, a number of heuristic (trial and error) programming methods have been devised which cannot be guaranteed to provide optimal solutions, but nevertheless are known to give near optimum solutions the majority of the times they are applied. The reader is referred to Taylor (1977) for a clear introduction to this complex topic.

APPENDIX

The relationship between the entropy model and the transportation problem

It is appropriate that we conclude our discussion of deterministic modelling by presenting some theoretical results which integrate the transportation problem with the entropy-maximizing journey-to-work problem described in the previous chapter.

Recall that the entropy model entailed finding the trip-matrix, $\{T_{ij}\}$, which maximized the value of the entropy function given by

$$W(\{T_{ij}\}) = \frac{T!}{\prod_{i}^{n} \prod_{j}^{n} T_{ij}!}, \tag{6.19}$$

subject to the origin constraint

$$\sum_{j}^{n} T_{ij} = O_i, \tag{6.20}$$

the destination constraint

$$\sum_{i}^{n} T_{ij} = D_j, \tag{6.21}$$

and the cost constraint

$$\sum_{i}^{n} \sum_{j}^{n} T_{ij} c_{ij} = C. \tag{6.22}$$

The solution to this problem is given by the equation

$$T_{ij} = A_i O_i B_j D_j e^{-\beta c_{ij}}, \tag{6.23}$$

where A_i and B_j are scalars and β is a parameter which is calculated to maximize the entropy (equation 6.19) subject to the *known* distribution of workers (equation 6.20), distribution of jobs (equation 6.21), and the total expenditure on journey-to-work (equation 6.22) (see Chapter 5, pp161–162). In this way we maximize the behavioural choices of the T trip-makers subject to known information about the journey-to-work system.

The following discussion elaborates upon some theoretical properties of the entropy model obtained by S P Evans (1973). The entropy model assumes that we *know*, prior to prediction, the value of either the total journey-to-work expenditure, C, or the average trip cost, $\bar{c}$, given by

$$\bar{c} = C/T. \tag{6.24}$$

This knowledge enables us to calculate and approximate value of the β parameter (see Chapter 5, pp162–163) from the relationship

$$\beta = 1/\bar{c}. \tag{6.25}$$

Now, suppose $\bar{c}$ is *unknown*. What happens to our entropy-maximizing problem if the value of β is *not* constrained by equation (6.22)? Rearranging equation (6.25) we obtain the approximate relationship

$$\bar{c} = 1/\beta, \tag{6.26}$$

which shows that the average trip cost, $\bar{c}$, *implied* by an arbitrary β value, will become less as β increases in value. Furthermore, as we let β become increasingly large, and reduce still further the implied average trip cost, we must eventually reach a limit of the *minimum possible* average trip cost or journey-to-work expeditions. This limit is imposed by the origin and destination constraints (equations 6.20–6.21) which ensure that all workers must be assigned a job.

The reader should recognize this limiting case of the entropy model as a form of the transportation problem. If we set up the journey-to-work system as a spatial allocation problem, then the *prime objective* is to *minimize* total journey-to-work expenditure subject to the origin and destination constraints. Symbolically, we write this problem as minimizing

$$Z = \sum_{i}^{n} \sum_{j}^{n} T_{ij} c_{ij} = \min C, \tag{6.27}$$

subject to the constraints

$$\sum_{j}^{n} T_{ij} = O_i, \tag{6.20}$$

and

$$\sum_{i}^{n} T_{ij} = D_j. \tag{6.21}$$

Therefore, the objective of this problem is to find the allocation of work trips, $\{T_{ij}\}$, which minimizes the expenditure on the journey-to-work. Mathematically, we can express the relationship between the two models in the following way; if we let $\beta \rightarrow$ (tend to) infinity, the trip matrix predicted by equation (6.23) will converge on the solution linear programming problem described equations (6.20, 6.21 and 6.27).

Figure 6.8 shows the relationship between β and $\bar{c}$ for the journey-to-work distribution on Merseyside in 1966 (see Thomas 1977). The entropy-maximizing problem is solved with a value of $\beta = 0.354$, which predicts exactly the known average trip cost of $\bar{c} = 4.21$ cost units. The graph shows what happens to the value of $\bar{c}$ if β is given values different from the entropy-maximizing solution. Notice, that as β becomes increasingly large (approximately 20 to 30), so the average travel stabilizes at the linear programming minimum value of min $\bar{c} = 0.10$. Figure 6.8 also illustrates the effect of a

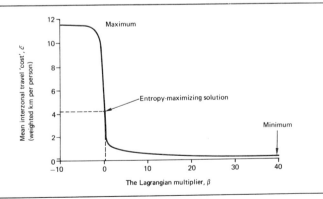

Figure 6.8. The relationship between β and $\bar{c}$ for journey-to-work on Mersyside, 1966. (From R W Thomas 1977, p824.)

second boundary condition on $\bar{c}$ noted by A W Evans (1971). When β becomes increasingly large in a negative direction ($\rightarrow -\infty$) then, its value must eventually imply an average travel cost that is the *maximum* cost permitted by the origin and destination constraints. For Merseyside, a value of $\beta = -3 \cdot 000$ (figure 6.8) gives the maximum average travel cost of $\bar{c} = 11 \cdot 60$. In relation to these boundary conditions, the entropy-maximizing solution ($\bar{c} = 4 \cdot 21$, $\beta = 0 \cdot 3544$) is much nearer to the cost-minimizing linear programming solution ($\beta = 20 \cdot 000$, $\bar{c} = 0 \cdot 10$) than the cost-maximizing solution ($\beta = -3 \cdot 000$, $\bar{c} = 11 \cdot 60$), which suggests that the minimization or work trip costs is an important part of the 'average' Merseysider's journey-to-work decision.

The relationship between entropy-maximizing and linear programming models also extends to the dual problem and the reader is referred to a paper by Wilson and Senior (1974) for a discussion of this rather complicated relationship.

Notes

1   We have modified the notation used to identify origin and destination points in this section. In the transportation problem we have assumed that the $i = 1, 2, \ldots, n$ production or origin points are spatially distinct from the $j = 1, 2, \ldots, m$ destination or demand points. However, it is quite permissible to set up transportation problems using the gravity model format where the $ij = 1, 2, \ldots, n$ regions are both potential origins and destinations for trips. Indeed, the journey-to-work transportation problem sketched on p170 would be defined according to the second scheme.

2   There is only one feasible adjustment cycle for any pivot cell. Therefore, quite often the search for a feasible cycle will lead to an impasse. For instance, in table 6.4(*b*) suppose that, instead of making the negative adjustment to the flow T_{32}, we make the negative adjustment to the flow T_{33}. The construction of the cycle breaks down at this point because there is no other positive shipment in column 3. This impasse would then force us to construct the cycle by making T_{32} the deficit flow.

3   In other numerical examples it may occur that in the optimal solution one or more of the opportunity costs is equal to (but not greater than) the respective transportation costs. The occurrence of these equalities indicate that another matrix of flows exists which equally minimizes the total cost of transportation. Clearly, a preference for one of these optimal solutions cannot be made in terms of cost minimization, although in reality there may be other criteria, not included in the transportation problem equations, which enable the choice to be made.

4   The reader may wish to check that the dual takes on a maximum value in the optimal solution by calculating the value of the dual function using the sub-optimal shadow prices listed in either table 6.1 or 6.5

5   The reader may recall that in the transportation algorithm the shadow price u_1 was arbitrarily set equal to zero to provide a relative reference point for measuring the remaining shadow prices. In the example problem, factory 1 coincidentally turned out to be the marginal seller. However, in other problems where factory 1 is not the marginal seller, then one or more of the remaining values of u_i will be negative. In such circumstances, the largest negative value of u_i in the optimum solution will identify the marginal seller.

6   Notice that the optimum matrix, table 6.11, is degenerate because only five cells are occupied instead of the usual $n + m - 1 = 6$ cells for problems of this size.

7   This statement only applies if there is a direct transportation route between each source and each destination, which is the case in figure 6.5(*a*). In problems where indirect routes (that is, routes from *i* to *j* passing through one or more intermediate points) connect pairs of places; the cells corresponding to these routes are kept unoccupied during the computation of the transportation algorithm.

8   When *n* is large, it is likely that some regions will contain no pupils, giving a value of $O_i = 0$. By definition, no trips will begin in a zero region and, therefore, the size of the trip matrix can be reduced by omitting these regions from the analysis.

PART III

Probability Models

CHAPTER 7

The Elements of Probability Theory

So far this book has considered models based on deterministic assumptions about the processes which mould the landscape. To test these assumptions we measure observed values of the independent variables in the model equation. By using these observations to evaluate or calibrate the model equation, we obtain exact predictions about reality. In this way deterministic models either ignore, or average out, the effects of *chance* which can often cause substantial differences between prediction and reality. In Part III of this book we examine models which allow for the effects of chance to be built explicitly into our assumptions about geographical processes. This style of analysis will focus our attention, not on an exact set of predictions, but on a variety of different outcomes which each have a different probability of occurring in reality. To study the effects of chance in geography necessitates a reasonable working knowledge of the mathematics of chance, namely, probability theory. Accordingly, this chapter introduces the reader to some of the simpler mathematical components of probability theory. The treatment follows the usual sequential structure of a mathematical theory: the definition of terms and their measurement, the rules (axioms) for manipulating these terms, and the deduction of theorems which allow us to make predictions about reality. A subsidiary intention is to illustrate some of the simpler geographical applications of probability theory, although substantial geographical applications are treated in Chapters 8 and 9.

THE MEASUREMENT OF PROBABILITIES

Events and experiments

Problems which involve an element of chance or uncertainty are usually termed *experiments*. Games of chance such as coin-tossing, dice-rolling, and card-playing are experiments that will be familiar to the reader. Similarly, many geographical problems are inherently chance-like. We may speak of the experiment of a river flooding in a given year, or the experiment of a farmer deciding which crop to plant in the face of uncertain weather conditions. Because such problems are chance-like they must have more than one outcome; the coin will either show heads or tails, the river either floods or stays within its banks. We refer to each of these individual outcomes as an *event*, and all the possible outcomes as the *sample space* of the experiment. For example, the sample space of a dice-throwing experiment is the list of the six events represented by the numbers on the six faces of the dice:

Sample space: 1, 2, 3, 4, 5, 6.

The individual events in this sample space may be written symbolically as

E_1 : the outcome is   1
E_2 : the outcome is   2
etc, to E_6 : the outcome is   6.

In more general terms, we speak of E_i, the *i*th individual outcome in a sample space formed of *n* events.

A question that arises immediately from this definition is how to measure the probability that some event occurs as the outcome to the experiment at some time in the future? In other words, how do we measure expressions of the form $p(E_i)$ when p is the probability of the occurrence of the event *i*? The process of measuring probabilities has intrigued mathematicians since the theory's original development in the gaming rooms of the Paris court during the seventeenth century. Indeed, we are less sure today about the best method of measuring probabilities because mathematicians have devised so many alternative procedures. It is important for geographers to be aware of these measurement procedures because they each lead to different styles of probability modelling. What follows is by necessity selective, but we will present the three measurement procedures which have had the greatest influence upon the geographical application of probability theory, namely, the classical view, the relative frequency view, and the subjective view of probability.

The classical view of probability

It is generally agreed that the classical view of probability theory developed from a correspondence between the French mathematicians Pascal and Fermat. This correspondence was about the fairness of various bets in gambling. Before we discuss their ideas, it is worth indicating the one aspect of probability measurement about which there is most mathematical agreement, this being the scale of measurement. The probability of some event occurring is measured on a scale between 0 and 1 such that $0 \leqslant p(E) \leqslant 1$. Values of $p(E)$ are zero if an event is impossible, or yet to be observed, and are given the value one (unity) when an event is certain to occur[1]. Between these limits we find the interesting, uncertain events of a probabilistic nature.

The classical view of probability measurement rests on the assumption that all the events in the sample space are equally likely to occur in any one trial on an experiment. From this assumption it follows that the probability of any event is the ratio of 1 to *n* (the total number of events in the sample space), which is

$$p(E) = 1/n. \tag{7.1}$$

Therefore, $p(E_1)$, the probability of our dice showing a 1, is immediately seen to be 1/6.

Experiments such as tossing a coin and throwing a dice are extremely simple because they both contain a small number of discrete events in their sample space. The classical measurement of probability was founded on the symmetry (or geometry) of such probability experiments. This last statement means that because the sides of our dice are of equal area, it is assumed that each side is equally likely to show up in a single throw (trial). Thus the classical approach to measuring probabilities stresses the importance of the physical structure of the object of the probability experiment (our dice or our coin) as a determinant of the object's future behaviour. The idea of symmetry may easily be extended to deal with probability experiments where the number of elementary events in the sample space is infinite. For a geographical example of this mathematical idea, imagine that the square depicted in figure 7.1 is a hypothetical square city. Nothing very surprising ever happens in this city because it is inhabited by people who slavishly obey the rules of chance. A shop is to be located within the boundaries of this city and, for the purposes of illustration, we will assume that our shop has an equal chance of being located anywhere within the city boundary. Notice that this city is a more complicated object of a probability experiment than a dice or a coin because, in the city, there are an infinite number of points (events) within the boundary which are possible locations for the shop. The dice is simpler because it possesses six *discrete* sides. To overcome the problem posed by the continuity of the city we again make use of its geometry. Given our assumptions, it is easy to see that the probability of our shop being located in a specific discrete region of the city (figure 7.1) is given by

$$p(E) = \frac{\text{area of specific region}}{\text{area of city}}.$$

Because the city is a square of side length 4 (figure 7.1) the required probability, $p(E)$, is obtained as

$$p(E) = 1^2/4^2 = 1/16.$$

Many real geographical entities are readily representable as lines (rivers, roads) and areas (regions) and are amenable to classical measurement by reference to their geometrical characteristics.

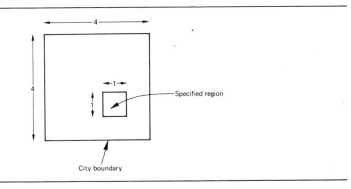

Figure 7.1. The dimensions of an imaginary city.

The major mathematical criticism of classical measurement is that we often possess observational information about an experiment which contradicts the assumption of equally likely elementary events. For example, if the population of our city were distributed unevenly within its boundaries we would be unwise to assume equally likely shop locations because we know geographically that shop locations tend to be market-orientated. Therefore, a more sensible strategy on our part would be to incorporate our knowledge of the underlying population distribution into the measurement of probabilities. To circumvent this problem, classical probability theorists invoke what has been called the *principle of insufficient reason*, which asserts that elementary events can only be assigned equal probabilities in the absence of any real evidence to suggest a more appropriate assignment.

The relative frequency view of probability

Whereas classicists adopted a restricted, but logical proposition, to evaluate probabilities, the relative frequency view is founded on the assumption that probabilities can only be measured from observational evidence. The frequency of an event, $freq(E_i)$, is defined simply as the number of times the event E_i is observed as the outcome in r trials on an experiment. This *absolute frequency* is converted into a *probability or relative frequency* by dividing its value by r. Therefore, if 'heads' were observed 48 times in 100 throws of a coin then the possibility of the coin showing heads, $p(E_H)$, is given the value

$$p(E_H) = freq(E_H)/r$$
$$= 48/100 \qquad\qquad (7.2)$$
$$= 0.48.$$

The relative frequency view asserts that as r tends to infinity (that is, becomes extremely large) the relative frequency will tend to the true probability. Symbolically, we can write this statement as

$$\frac{freq(E_i)}{r} \longrightarrow p(E_i) \text{ as } r \longrightarrow \infty, \qquad\qquad (7.3)$$

where $\longrightarrow$ is read as 'tends to'.

Table 7.1 and figure 7.2 are designed to illustrate the implications of the relative frequency view of probability measurement. Table 7.1 lists the relative frequency

Table 7.1. Relative frequences in a coin-tossing experiment.

E_i	H	T	T	T	H	H	T	H	H	T
r	1	2	3	4	5	6	7	8	9	10
$freq(E_H)/r$	1.00	0.50	0.33	0.25	0.40	0.50	0.43	0.50	0.56	0.50

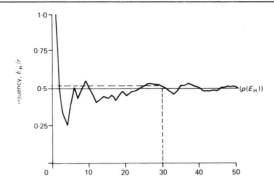

Figure 7.2. Relative frequencies during 50 tosses of an unbiased coin.

estimate of the probability of what we will assume is an unbiased coin showing heads at each stage in an observed sequence of ten tosses of the coin. The graph (figure 7.2) extends the results of this experiment to 50 tosses. The important feature of the graph is that, although in the first few tosses the relative frequency estimate varies greatly about the true value of 0·5, after about 30 tosses the estimate stabilizes in a progressively narrowing band around the true value. In other words, the error in the relative frequency estimate decreases rapidly as the number of trials, r, increases. This statement implies that the relative frequency obtained from quite a small number of observations can often provide quite accurate estimates of unknown (true) probabilities.

Also implicit in the relative frequency view is the idea of a *random sample*. Because the coin-tossing experiment listed in table 7.1 was conducted so that the result of any one toss did not influence the outcome of any other toss (that is, the sequence of events were *independent* of one another), the results obtained in table 7.1 are a random sample taken from the infinite number of tosses that could be made with our unbiased coin. Therefore, a random sample is an unbiased group of observations selected to represent some larger group of possible outcomes. For a geographical example of random sampling, assume we wish to estimate the number of people over 65 years of age living in the imaginary city, whose total population is known to be exactly 100 000. To save interviewing the whole population we could select a random sample of 100 people to represent the 100 000 people with the city boundaries. Here, *at random* means that each time we select a person to include in our sample each member of the total population has an equally probability ($p = 1/100\,000$) of being selected for our random sample. If we found that 19 people in our sample were over the age of 65, then the absolute frequency estimate of the total number of people aged over 65 would be 19 000. We express this results as a relative frequency estimate of 0·19 (19/100), which we should interpret as the probability of a single individual, selected at random from the total population, being over the age of 65.

The relative frequency view is important because it is the measure of probability which forms the basis of conventional statistical reasoning expounded in such texts as

Cochran (1953), Fisher (1956), and Lindgren (1975). These mainstream statistical methods and techniques have been widely applied in geography during the past 25 years and a number of introductory geography texts have been published on the subject, examples being Gregory (1968), King (1969), and Yeates (1974). Conventional statistical reasoning is founded on the mathematical properties of a random sample. It has been proved that, as long as the *estimated* value of a relative frequency probability is obtained from a random sample of reasonable size, the *errors* inherent in the probability estimate will obey known probability laws. We will examine this idea in more detail when we come to consider the central limit theorem which is the theoretical basis of random sampling. However, we may note that the probability theory used for assessing random sampling errors permits us to make statements concerning our confidence in the unknown, true value of some probability occurring within a range of sampling error around the relative frequency estimate of the true value. Such procedures are known as *statistical inference*, because we infer characteristics of true values on the basis of estimated probabilities obtained from random samples.

Because relative frequency measurement is founded on randomly collected observations the results obtained from this method must of necessity lack generality. For example, the information about the coin-tossing experiment listed in figure 7.2 refers only to the coin used in the experiment, not to all coins. Similarly, our geographical example of estimating the number of old-aged people living in the city refers only to the imaginary city, not all cities. Therefore, even to gain only a limited understanding of some geographical object (such as, cities, drainage basins and weather stations) a geographer needs large numbers of random samples measured on different objects of the same type. It follows that, for relative frequency measurement to be of any use to the geographer, the relative frequency method demands large amounts of data. Furthermore, because relative frequency measurement can only be used in conjunction with random sampling, the method precludes the direct use of any other information we possess about the problem in hand. By other information we mean the qualitiative and quantitative knowledge that exists at any point in time, but which is *not available* in the form of random samples.

Subjective probability

Neither the classical nor the relative frequency view of probabilities require the experimenter to take his own view into account when assigning values to probabilities. So long as the rules for these two methods are obeyed, different experimenters will give the same value to probabilities in the same experiment. These measured probabilities are then fed into the relevant probability theory to obtain the results of the experiment. In this sense, both classical probabilities and relative frequency probabilities *obey* the axioms of probability theory.

The more modern subjective view of probability measurement is very different from classical and relative frequency measurement. An essential part of the subjective view is that the experimenter (be he a geographer, statistician, or a decision-maker) is expected to weigh all the evidence he thinks is relevant to the problem at hand and then make an informal measure of probability that is *consistent* with his *present beliefs.* Naturally, different experimenters will come to different numerical conclusions about the same problem, hence the term subjective probability. The subjective method is particularly appropriate for experiments (situations) which can never be repeated exactly. For example, betting on a horse race is a good example of such an experiment. The bookmaker considers the horses' recent form, the state of the ground, the weight of the jockeys, and any other information he feels will influence the outcome of the race, and then offers odds on each horse winning. The odds are really subjective probabilities − 3-to-1 represents the bookmaker's view that the probability of the horse winning is 0·25. The race will never be repeated again under the same conditions and with the same horses, so the next time the horse runs it is unlikely that the bookmaker will offer the same odds of 3-to-1. The punter may also be regarded as an experimenter in the horse-racing experiment. He also considers the form, including the bookmaker's odds, and reaches his own subjective assessment of the outcome of the race, which he then translates into a bet on the most favourable outcome (one of the horses). Horse racing may at first sight appear far removed from geography. However, the case of the punter is similar in structure to the case of a farmer trying to decide which crop combinations will bring him the best returns in the face of uncertain weather conditions. The crop combinations are equivalent to the horses because the farmer knows from past experience which returns different combinations are likely to produce under different weather conditions. Long-term weather forecasts provide the farmer with expert evidence to judge the best crop combination in the same way that the bookmaker's odds represent an expert evaluation of the most likely outcome of the race. In both these examples the experimenter (punter or farmer) receives information in the form of subjective probabilities which he can use as the basis for making decisions about any future outcome of the experiment. Therefore, in the same way as relative frequency measurement leads to geographical applications founded on the probability theory of random samples so the subjective view leads, quite naturally, to geographical applications concerned with decision-making in the face of an uncertain environment. Indeed, in recent years decision theory has developed as a specialized branch of probability theory and this subject is discussed both in introductory texts such as Chernoff and Moses (1959) and in advanced mathematical treatments of the theory such as Lindley (1965).

Subjective probabilities are, by their very nature, difficult to define symbolically; however, we gain some understanding of the formal structure of subjective experiments by elaboration of a scheme presented by Good (1962), which is illustrated in modified form in figure 7.3. This scheme begins with the experimenter weighing the evidence concerning an experiment and then expressing his judgment as a set of prior, subjective probabilities. These prior probabilities are then fed into the formal mathematics of probability theory. Certain theorems in the theory allow us to predict outcomes to the experiment in the light of our present evidence. These predictions are termed discernments and are expressed as posterior probabilities. Thus discernments follow logically from the judgments. The experimenter now decides whether or not the discernments are reasonable. If his decision is favourable, the discernments are absorbed into the judgments (prior probabilities) and the whole process begins again. The purpose of this structure is to expand and improve the reliability of the prior probabilities. The most difficult link to understand in this scheme is the manner in

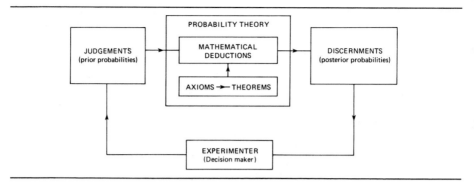

Figure 7.3.   The structure of subjective probability experiments. (Adapted from
I J Good 1962, p321.)

which the theory can transform a set of prior probabilities (judgements) into a set of
posterior probabilities. The mathematical reasoning that allows for this transformation
will become clearer when we consider Bayes' theorem later in this chapter.

SOME AXIOMS IN PROBABILITY THEORY

The building materials from which all mathematical theories are deduced are termed
axioms or *primitive postulates*. Most people are familiar with the axioms of arithmetic
although they do not know them by this name. Arithmetic axioms are simply the rules
for adding, subtracting, multiplying, and dividing numbers. We hope you will recognize
immediately that the statement $1 + 2 = 3$ satisfies an axiom of simple arithmetic while
the statement $12 + 1 = 1$ does not. However, the second statement does satisfy an
alternative arithmetic founded on a different set of axioms. The result would be
perfectly true if we were counting on a 12-hour clock.

Popper (1959) is among many philosophers of science who have speculated about
the properties a set of axioms belonging to some mathematical theory should possess.
He argues that for a set of axioms to be acceptable they must be free from
contradiction, independent, sufficient, and necessary. *Free from contradiction* means
the axioms must be stated in an unambiguous symbolic language. *Independence*
requires that no one axiom should be deducible from any other axiom. Sufficient and
necessary are paired conditions. *Sufficient* means that the entire theory must be
deducible from the axioms. *Necessary* means there must be no surplus axioms such
that the entire theory can be deduced without them. Although Popper's definition of
an axiomatized theoretical system is concise, mathematicians have often found that it
is possible to deduce the same theory from different sets of axioms[2]. But such subtle-
ties need not concern us here because what follows is a presentation of the most
commonly used axioms for the deduction of probability theory. Our treatment is
similar to that presented in Gray's (1967) text on probability theory. Axioms are
defined as the basic rules for manipulating probabilities whose values have already
been established by one of the three probability measurement methods.

The addition axiom for mutually exclusive[3] events

This axiom states: 'if E_1 and E_2 are two mutually exclusive events, the probability that *either* E_1 or E_2 happens is the sum of their individual probabilities.' Symbolically, we write this statement as

$$p(E_1 \cup E_2) = p(E_1) + p(E_2), \tag{7.4}$$

where $\cup$ is termed union and is read as either/or. For example, the probability that our dice shows either a 5 or a 1 ($E_5 \cup E_1$) in a single throw is obtained from equation (7.4) as

$$p(E_5 \cup E_1) = p(E_5) + p(E_1) = \tfrac{1}{6} + \tfrac{1}{6} = \tfrac{1}{3}.$$

Examination of the axiom shows that the symbol for the grouping of mutually exclusive events, $\cup$, is synonymous with the addition of probabilities.

Two other simple results follow from the addition axiom. Firstly, in a single trial on an experiment, the probability that *any one of all the* n *events* in the sample space occurs is equal to one. We write this statement as

$$p(E_1 \cup E_2 \cup E_3, \dots, E_n) = \sum_{i=1}^{n} p(E_i) = 1. \tag{7.5}$$

The statement simply asserts that in a single trial one of the elementary events must occur as the outcome. Indeed, in any probability experiment the sum of the probabilities for all elementary events must equal one (unity).

Secondly, the addition axiom can be used to define the idea of a complementary event. A specified event will either occur or not as the outcome to a single trial on an experiment. When the specified event, E_i, does not occur, we say that the *complementary event* has occurred and denote it by $\bar{E}_i$. Thus the complementary event is the union of all events other than the specified event. For instance, in a particular year (trial) a river may flood, E_F, or not flood, $\bar{E}_F$, and, because these two events are mutually exclusive, we can use the addition axiom to write

$$p(E_F \cup \bar{E}_F) = p(E_F) + p(\bar{E}_F).$$

Furthermore, because the union of flooding with not flooding, $E_F \cup \bar{E}_F$, is certain to occur, it follows that the sum of the individual probabilities must be one; that is,

$$p(E_F) + p(\bar{E}_F) = 1.$$

This expression can be rearranged to give us a definition of the complementary event as

$$p(\bar{E}_F) = 1 - p(E_F),$$

which may be written more generally for the ith event as

$$p(\bar{E}_i) = 1 - p(E_i). \tag{7.6}$$

Returning to our unbiased dice, we can now obtain the probability of side 5 not occurring in a single throw as

$$p(\bar{E}_5) = 1 - p(E_5) = 1 - \tfrac{1}{6} = \tfrac{5}{6}.$$

The multiplication axiom for independent events

The multiplication axiom for independent events is used to evaluate the probability that a *sequence* of specified events occurs as the result of r trials on an experiment. This axiom holds for independent events, where we assume that the occurrence of any one event does not influence the occurrence of the other events forming the sequence.

The axiom states that, 'if E_1 and E_2 are two independent events, the probability that *both* E_1 and E_2 happen, is the product of their individual probabilities.' Symbolically, we write this statement as

$$p(E_1 \cap E_2) = p(E_1).p(E_2), \tag{7.7}$$

where $\cap$ is termed intersection and is read and/both. For example, the probability of throwing three heads in succession with an unbiased coin is obtained from equation (7.7) as

$$p(E_H \cap E_H \cap E_H) = p(E_H).p(E_H).p(E_H) = \tfrac{1}{2} \times \tfrac{1}{2} \times \tfrac{1}{2} = \tfrac{1}{8}.$$

More generally, for a specified sequence of length r the probability of occurrence is given by

$$p(E_1 \cap E_2, \ldots, \cap E_r) = \prod_{i=1}^{r} p(E_i). \tag{7.8}$$

Because evaluating the probability of a specified sequence occurring involves a multiplicative relationship, the probabilities for specified, independent sequences are usually exceptionally small in value even for quite short sequences. We can illustrate this tendency with reference to the random shop location problem in the example city (figure 7.1). Remembering that we obtained the probability, $p(E)$, of a single shop being randomly located within the specified region as $1/16$, then if four shops are to be located randomly within the city boundaries then the probability that they are all located in the specified region is

$$p(E \cap E \cap E \cap E) = p(E)^4 = (1/16)^4 = 1/75536,$$

which means it would be an extremely unlikely outcome to the experiment.

The multiplication axiom for dependent or general events

In reality, geographical events are rarely mutually exclusive or independent. More often than not, events occur in conjunction with one another or, alternatively, the occurrence of one event in a sequence influences the occurrence of subsequent events. When such relationships exist we term the events as *general* or *dependent*. For example, in the previous paragraph we were being unrealistic when we assumed that the four shops were located independently of one another. Suppose the four shops sell jewellery. It is well known in urban geography (see Sibley 1972) that jewellers tend to cluster together at the centre of cities because shoppers for jewellery compare different articles before deciding on a particular purchase. It is thought that such comparison shopping benefits all the shops in the cluster by increasing overall sales. This information is essentially subjective evidence which leads us to believe that the assumption of independent jeweller location in our city is unwarranted. If four jewellers are to be located in our city it is more than likely that the location of the first shop will *influence* subsequent locations by causing them to cluster in the same area of the city. These arguments lead us to conclude that our probability of 1/73556 for the four jewellers being located in the specifed region (figure 7.1) is unnecessarily low. Furthermore, this hypothetical discussion is an illustration of Good's scheme for measuring subjective probabilities (figure 7.3). We have used geographical evidence to refute discernments which were obtained from applying the multiplication axiom for independent events.

Formally, two events, E_1 and E_2, are said to be *dependent* if the probability that one occurs *varies* according to whether or not the other event happens. Dependence is simple to recognize because it is the antithesis of independence. Dependence arises when the multiplication axiom for independent events is seen to be incorrect, that is, when

$$p(E_1 \cap E_2) \neq p(E_1) \, p(E_2).$$

We can illustrate this idea with reference to some data on office decentralization from London listed in Rhodes and Kan (1971) (table 7.2). During the 1960s the British government, as an act of regional policy, offered a variety of economic inducements for offices located in central London to be relocated in less prosperous regions. Table 7.2(*a*) records information about both the type and distance moved of a random sample of 519 offices which were relocated during the period 1963–69. Partial moves refer to situations where only a subsection of the office was relocated, while complete moves refer to the relocation of all a firm's office activities. We can see immediately from the row and column totals (table 7.2*a*) that more complete moves were made than partial moves, and more short-distance (<20 miles) than long-distance moves (>40 miles).

The sample space of this experiment contains *two* classes of elementary events, that is, distance and type of move, which are *not mutually exclusive* of one another. If one of the 519 offices is selected at random it can be of both a certain distance and a certain type. In table 7.2(*a*) distance is represented as three elementary events and type by two. The individual cells of the table illustrate the six ways in which these elementary events can occur in conjunction with one another. We must now ascertain whether distance and type occur independently of one another, or alternatively, whether these two variables are in some way related and are therefore dependent.

If E_1 denotes the selection of a firm moving less than 20 miles and E_2 denotes the selection of a complete move, then if one firm is selected at random from the table, the probability of that firm moving less than 20 miles is obtained as

$$p(E_1) = 305/519 = 0{\cdot}59, \qquad \text{(see table 7.2}b\text{)}$$

and the probability of a complete move is

$$p(E_2) = 338/519 = 0{\cdot}65. \qquad \text{(see table 7.2}b\text{)}$$

Table 7.2(*b*) also gives us the probability of the relocation being *both* less than 20 miles *and* a complete move as

$$p(E_1 \cap E_2) = 217/519 = 0{\cdot}42.$$

Table 7.2. Office relocation from Central London: 1963–69 (Source: Location of Offices Bureau records, adapted from Rhodes and Kan 1971, p 19).

(a) *Observed Frequencies* (random sample of 519 office relocations).

Type of move	$j \rightarrow$ 1 <20	2 20–39	3 >40	Totals
Complete $i\downarrow$ 1	217	64	57	338
Partial 2	88	44	49	181
Totals	305	108	106	519

(b) *Relative frequency probabilities.*

$i\downarrow$	$j\rightarrow$ 1	2	3	p_i
1	0·42	0·12	0·11	0·65
2	0·17	0·09	0·09	0·35
p_j	0·59	0·21	0·20	1·00

(c) *Independent joint probabilities* (p_{ij}).

	1	2	3	p_i
1	0·38	0·14	0·13	0·65
2	0·21	0·07	0·07	0·35
p_j	0·59	0·21	0·20	1·00

(d) *Differences* (b–c).

	1	2	3
1	0·04	−0·02	−0·02
2	−0·04	+0·02	+0·02

Further, the multiplication axiom for independent events gives this last probability as

$$p(E_1 \cap E_2) = p(E_1) \cdot p(E_2) = 0·59 \times 0·65 = 0·38.$$

Clearly, in this case

$$p(E_1 \cap E_2) \neq p(E_1) p(E_2)$$

applies (see p215), and therefore type and distance are not independent of one another, but are in some way related. Obviously, we require a further multiplication axiom for dependent events which enables us to obtain the observed answer for $p(E_1 \cap E_2)$ as 0·42.

Before we can state this axiom we need to define the idea of a *conditional probability*. Conditional probabilities are denoted by symbols of the form $p(E_1 \mid E_2)$ or $p(E_2 \mid E_1)$. The first of these symbols is read as the probability of E_1 occurring when it is *assumed that E_2 has happened*, and in the same way the second conditional probability is read as the probability of E_2 occurring when it is assumed that E_1 has already happened. For the office relocation events the conditional probability $p(E_1 \mid E_2)$ can be identified as

$$p(E_1 \mid E_2) = 217/338 = 0·64,$$

which measures the probability of a less than 20 mile relocation, E_1, amongst the complete move, E_2, sub-group of offices. Similarly,

$$p(E_2 \mid E_1') = 217/305 - 0·71,$$

which indicates the probability of obtaining a complete move among the established sub-group of 305 relocations of less than 20 miles. Using these two conditional probabilities we can obtain the true probability $p(E_1 \cap E_2)$ from either of the following expressions

$$p(E_1 \cap E_2) = p(E_1)p(E_2 \mid E_1) \qquad (7.9)$$
$$= 0·59 \times 0·71 = 0·42$$
$$p(E_2 \cap E_1) = p(E_2)p(E_1 \mid E_2) \qquad (7.10)$$
$$= 0·65 \times 0·64 = 0·42.$$

These equations are the two alternative forms of the multiplication axiom for dependent events. Stated in English, these equations read, 'if E_1 and E_2 are *dependent* events, the probability that they *both* happen is the product of the probability of E_1 and the conditional probability of E_2 when E_1 happens, or vice versa (equation 7.10).

Contingency tables

Conditional probability, together with the multiplication axiom for dependent events, illustrates one of the ways in which probability theory can be used to examine relationships between two or more geographical variables. The data listed in table 7.3(a) contain frequency information about two variables: distance moved and type of move, which have been measured on the nominal (classified) scale. Such a data set provides simple information about geographical relationships, and is commonly termed a *contingency table*. Contingency tables are especially important because many geographical variables can only be measured in nominal form. Variables belonging to regional classification schemes (for example, vegetation type, soil type, landform type, and social area) do not lie along a neat measurable scale. Nevertheless we can record the frequency of occurrence of a geographical object among the classes of the overall geographical classification. For instance, in our decentralization problem the two types of relocation, complete and partial, cannot be distinguished on a numerical scale, yet we can measure the frequency of offices (objects) belonging to these two classes.

When we collect random samples for a contingency table we can use the multiplication axiom for *independent* events to gain a quite detailed understanding of any relationship that may exist between the two classifications. The independence axiom is used to obtain the *expected* probabilities that would occur if the two variables acted independently of one another; that is, if the variables were unrelated. We have previously obtained the independent probability of a single randomly selected office relocation being both less than 20 miles, E_1, and a complete move, E_2, as 0·38. If, instead of $p(E_1)$ and $p(E_2)$, we use the more general notation of p_{ij}, p_i and p_j given in table 7.3(c), we can calculate expected, independent, joint probabilities, p_{ij}, for each cell of the contingency table. These values p_{ij} are obtained from the multiplication axiom for independent events as

$$p_{ij} = p_i \cdot p_j. \tag{7.11}$$

In this revised notation the independent probability of randomly selecting a relocation which is both a complete move ($i = 1$) and less than 20 miles ($j = 1$) is obtained from formula (7.11) as

$$p_{11} = p_1 = p_1 = 0·65 \times 0·59 = 0·38.$$

Similarly, the same probability for a partial move ($i = 2$) of greater than 40 miles ($j = 3$) is obtained as

$$p_{23} = p_2 \cdot p_3 = 0·35 \times 0·20 = 0·07.$$

The independent probabilities for the remaining cells of the contingency table are illustrated in table 7.2(c). By subtracting the observed, dependent, relative frequency probabilities (table 7.2b) from the expected, independent probabilities (table 7.2c) we can see the form of the relationship between type of move and distance in table 7.2(d). Complete moves took place over shorter distances than one would expect, whereas partial moves were made over longer distances. This implies that, despite British governmental financial incentives, most major office relocations from London took place within London's immediate hinterland.

Contingency tables are a simple and general method for understanding geographical relationships. However, their use is not without its problems. Firstly, their interpretation is highly dependent on the classification of the variables. In the example, the qualitative distinction between partial and complete moves is unambiguous; however, the research worker is free to choose the distance classes in the table because distance is a ratio variable, which has been reduced to nominal form for the analysis of the contingency table data (table 7.2a). Therefore, by either increasing or decreasing the number of distance classes, we could easily alter our arithmetic, and subsequent interpretation of the data. Secondly, the relative frequency estimates (table 7.2b) are themselves subject to random sampling error which could be the major cause of the differences observed between the relative frequencies and the independent probabilities (table 7.2*d*). We will return to these problems when we discuss the chi-squared test later in this chapter.

SOME PROBABILITY THEOREMS AND THEIR GEOGRAPHICAL APPLICATIONS

The deductions which follow from the definitions and axioms of a mathematical theory are known as theorems, and these theorems are used to provide us with predictions about reality. A quick scan through any standard text on probability theory, such as Feller (1957), will reveal the enormous number of theorems that constitute the present theory. This section gives a minimal theoretical background to some of the theorems used in the present geographical applications of probability theory. The importance of some of these theorems has already been hinted at in the treatment of definitions and axioms; for example, Bayes' theorem, the central limit theorem (p231), and the chi-squared distribution (p236).

Bayes' theorem

This theorem was first proved by an English clergyman, Thomas Bayes, in 1763. The importance of the theorem lies in its ability to modify a set of prior probabilities in the light of some new experimental evidence and, as such, the theorem is one of the deductive mechanisms capable of transforming subjective prior probabilities into subjective posterior probabilities (figure 7.3). The theorem is derived from the multiplication axiom for dependent events and the idea of a conditional probability. To gain an impression of how the theorem works we will use an example adapted from Gray (1967, p38).

Each morning a man is equally likely to choose any one of three routes to the railway station. Referring to these routes as E_1, E_2, and E_3, then the *prior* probability of the man selecting any one of these three mutually exclusive events is

$$p(E_1) = p(E_2) = p(E_3) = \tfrac{1}{3}.$$

Suppose we observe the man's behaviour for a period of a year and calculate that the probabilities of his missing the train for each route are respectively $\tfrac{1}{16}$, $\tfrac{1}{8}$ and $\tfrac{1}{4}$. These probabilities are in fact the conditional probabilities of his missing the train after he has selected a particular route. These three conditional probabilities are

$$p(M \mid E_1) = \tfrac{1}{16}, \quad p(M \mid E_2 = \tfrac{1}{8}, \quad p(M \mid E_3) = \tfrac{1}{4},$$

where M is the event of missing the train. Suppose one day our man misses the train. What is the probability that he used route 3? Our prior guess would be $\tfrac{1}{3}$; however, Bayes' theorem gives a method for incorporating the observational evidence of the conditional probabilities into our estimations. The question requires the value of the conditional probability that route 3 was chosen when missing the train has been established, $p(E_3 \mid M)$. We know from the multiplication axiom for dependent events that the probability of his selecting any route E_i *and* missing the train is given by

$$p(E_i \cap M) = p(E_i) p(M \mid E_i). \tag{7.9}$$

Bayes' theorem states that the probability $p(E_i \mid M)$, for any event, E_i, conditional upon M, is the ratio between the joint probability $p(E_i \cap M)$ and the sum of this joint probability for all the events E_i, that is,

$$p(E_i \mid M) = \frac{p(E_i \cap M)}{\sum\limits_{j}^{n} p(E_j \cap M)} . \qquad (7.12)$$

By substituting the right-hand side of formula (7.9) into the right-hand side of formula (7.12) we obtain a more detailed expression of Bayes' theorem in the form

$$p(E_i \mid M) = \frac{p(E_i)p(M \mid E_i)}{\sum\limits_{j}^{n} p(E_j)p(M \mid E_j)} . \qquad (7.13)$$

For our example, Bayes' theorem gives the probability of his selecting route 3 when we know he misses the train as

$$p(E_3 \mid M) = \frac{p(E_3)p(M \mid E_3)}{p(E_1)p(M \mid E_1) + p(E_2)p(M \mid E_2) + p(E_3)p(M \mid E_3)}$$

$$= \frac{\frac{1}{3} \cdot \frac{1}{4}}{\frac{1}{3} \cdot \frac{1}{16} + \frac{1}{3} \cdot \frac{1}{8} + \frac{1}{3} \cdot \frac{1}{4}} = \frac{4}{7}.$$

By taking into account our knowledge that the train is most likely to be missed when route 3 is chosen, our *prior* estimate for route 3 has been revised from $\frac{1}{3}$ to a *posterior* estimate of $\frac{4}{7}$. Bayes' theorem also gives the two other posterior probabilities, that is, either selecting route 1 or route 2 when the train has been missed as

$$p(E_1 \mid M) = \tfrac{1}{7}, \qquad \text{and} \qquad p(E_2 \mid M) = \tfrac{2}{7}.$$

The reader may wish to check these last two results for himself. Notice that the three posterior probabilities sum to unity, and that the evidence of the conditional probabilities involves a substantial modification of the prior estimates.

Bayes' theorem is one of the most controversial results in probability theory. The objections to the theorem do not concern the proof of the theorem, which involves a simple rearrangement of the multiplication axiom for dependent events, but concern some of the applications that have been suggested for the theorem. When reliable quantitative estimates are available for both the priors and the conditional probabilities, no one questions the validity of the results obtained from the theorem, yet statisticians who hold the subjective view of probability argue that Bayes' theorem can be applied to experiments where the priors are estimated qualitatively. Subjectivists use the results of the theorem (the posteriors) to ascertain their degree of belief in a particular set of probabilities. The subjective probability theorist would approach our train-catching problem in the following way. He would know numerically only the conditional probabilities of the man missing the train after having selected a particular route $p(M \mid E_i)$. He would then pose the question, with what prior probability is the man *most likely* to select a particular route given the observational evidence of the conditionals and the posterior probabilities obtained from Bayes' theorem? In essence the subjective approach is concerned with an individual experimenter asking, 'which question gives me the most believable answer at the present time?' Objections to the subjective approach naturally centre on the fact that different experimenters will reach different conclusions about the same problem depending on their individual whims and preferences.

The binomial distribution

The binomial distribution is one of the oldest and most fundamental theorems of probability theory and is the theoretical mechanism which summarizes the long-term behaviour of sequences of independent events. We shall explain the properties of the binomial distribution with reference to the experiment of placing five balls independently of one another into a square grid consisting of nine equal-sized cells. This example problem has obvious geographical connotations because the cells may be thought of as equal-sized regions and the balls as points representing some geographical phenomenon.

The binomial distribution gives answers for the following class of questions: 'if r balls are placed independently, and one at a time, into a grid consisting of n equal-sized cells, what is the probability that a single, specified cell contains exactly x balls at the conclusion of the experiment?' We begin by restricting ourselves to finding the probability that a specified cell contains exactly one ball ($x = 1$) after five balls have been placed independently into the nine-cell grid. In figure 7.4 the specified cell is illustrated by broken lines. Clearly, from symmetry, if one ball is placed independently in the grid the probability of the ball landing in the specified cell, $p(E)$, is

$$p(E) = p = 1/n = 1/9.$$

Similarly, the probability of the complementary event occurring, $p(\bar{E})$, the ball landing in any one of the other cells, is

$$p(\bar{E}) = q = 1 - p = 8/9.$$

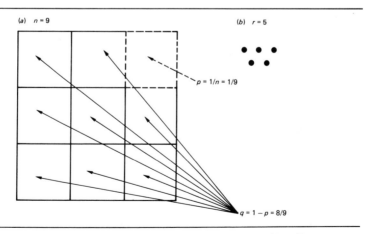

Figure 7.4. Elements of a binomial experiment. (*a*) Cells. (*b*) Balls.

Notice that these definitions hold irrespective of the particular cell we specify because all cells are equal-sized, and therefore each cell has an equal chance, p, of receiving a single ball.

A question that arises now is, 'in how many different ways (sequences) can balls be placed in the grid such that only *one* ball falls in the specified cell?' For example one favourable result is the sequence of events

$$p \cap q \cap q \cap q \cap q,$$

where the first ball landed in the specified cell and the remainder landed elsewhere. For experiments which contain only two elementary events, in this case $\{E, \bar{E}\}$, the total number of sequences where one of the events occurs exactly x times in r trials is counted using a number called a *binomial coefficient*. The formula[4] for such coefficients is

$$\binom{r}{x} = \frac{r!}{x! \, (r-x)!}. \tag{7.14}$$

For example, formula (7.14) counts the total number of different sequences in which $x = 1$ ball lands in the specified cell in $r = 5$ placements as

$$\binom{r}{x} = \binom{5}{1} = \frac{5!}{1! \, 4!} = \frac{1 \times 2 \times 3 \times 4 \times 5}{(1) \times (1 \times 2 \times 3 \times 4)} = 5.$$

These five sequences are in fact

$$p \cap q \cap q \cap q \cap q,$$
$$q \cap p \cap q \cap q \cap q,$$
$$q \cap q \cap p \cap q \cap q,$$
$$q \cap q \cap q \cap p \cap q,$$
$$q \cap q \cap q \cap q \cap p.$$

Because each of these sequences is composed of five independent events, we can obtain the probability of *any one* of these sequences occurring from the multiplication axiom for independent events, as

$$p \times q \times q \times q \times q = p^1 \cdot q^4 = \frac{1}{9} \left(\frac{8}{9}\right)^4 = 0 \cdot 069.$$

More generally, for sequences of length r, containing x occurrences of the specified event, we can write this joint probability as

$$p^x q^{r-x}. \tag{7.15}$$

The five sequences counted by the binomial coefficient each occur with the same probability (0·069); therefore, the probability, P_x, that the specified cell contains $x = 1$ ball is obtained from the product of the binomial coefficient (equation 7.14) and the joint probability (equation 7.15)

$$P_x = \binom{r}{x} p^x q^{r-x} \qquad x = 0, 1, 2, \dots, r \tag{7.16}$$

$$P_x = \binom{5}{1} \left(\frac{1}{9}\right)^1 \left(\frac{8}{9}\right)^4 = 0 \cdot 347.$$

The expression defined by formula (7.16) is the *binomial distribution*. Notice that the formula gives probabilities for all the values of x between 0 and r which represent all the possible outcomes to a binomial experiment. For example, the probability of of no balls (x = 0) landing in the specified cell at the conclusion of the experiment is obtained from formula (7.16) as

$$P_{x=0} = \binom{5}{0} \cdot \left(\frac{1}{9}\right)^0 \cdot \left(\frac{8}{9}\right)^5 = 0 \cdot 555.$$

Note that numbers of the form p^0 have defined values of 1 (here p = 1/9). Calculating P_x for all possible values of x (figure 7.5) gives a *probability distribution* which summarizes all the possible outcomes to this independent experiment. Because these binomial probabilities are exhaustive, that is, they describe all possible outcomes to the experiment, their values sum to unity. Symbolically, we write this property of a probability distribution

$$\sum_{x=0}^{r} P_x = 1. \tag{7.17}$$

The reader may wish to check formula (7.17) by summing the values in the binomial column in table 7.3 (p226).

It is easy to appreciate that, for binomial experiments where r is large, the binomial distribution will also contain a large number (r + 1) of individual probabilities. When

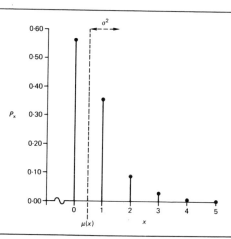

Figure 7.5. The binomial distribution for r = 5 and p = 1/9.

the number of individual probabilities, P_x, forming a distribution is large, clearly it is sensible to try to summarize these values with some number which represents their basic characteristics. Such summary measures of probability distributions are termed the *moments* or *parameters* of a probability distribution. The *first moment* of any probability distribution is variously termed the *mean*, $\mu(x)$, *average*, or *expected*, $Ex(x)$ values of the variable x. The mean is most easily thought of as the average value of x in a large number of repetitions of the same binomial experiment. In our ball-placing example we know from formula (7.16) that, in an infinite number of repetitions of placing five balls independently into the grid, we would observe a value of $x = 0$ in $P_{x=0} = 0.555$ of the experiments, a value of $x = 1$ in $P_{x=1} = 0.347$ of the experiments, and so forth, to a value of $x = 5$ in $P_{x=5} = 0.00002$ of the experiments (table 7.3). Therefore, the expected, or mean value of x will be the sum of the probability of x multiplied by the value of x, and we write this statement as

$$Ex(x) = \sum_{x=0}^{r} P_x . x = \mu(x). \tag{7.18}$$

For our example experiment the mean is calculated from formula (7.18) as

$$\mu(x) = 0.555 \times 0 + 0.347 \times 1 + 0.087 \times 2 + 0.011 \times 3 + 0.001 \times 4 + 0.00001 \times 5$$
$$= 0.556.$$

Fortunately, formula (7.18) can be proved to be equal to the probability of the ball landing in the specified cell in a single trial, p, multiplied by the number of trials, r, to give the mean of any binomial distribution more simply as

$$\mu(x) = rp = r/n \tag{7.19}$$
$$= 5 \times \tfrac{1}{9} = 0.556.$$

The *second moment* of the binomial distribution is usually defined as the expected, $Ex(x^2)$, or average value of x^2 in an infinite number of repetitions of the same binomial experiment and, symbolically, the second moment is written as

$$Ex(x^2) = \sum_{x=0}^{r} P_x . x^2 . \tag{7.20}$$

Further, if the *square of the first moment* is subtracted from the *second moment* (formula 7.20) we obtain the *variance* of a probability distribution, σ^2, as

$$\sigma^2 = Ex(x^2) - \{Ex(x)\}^2 . \tag{7.21}$$

The variance of any probability distribution is a measure of the *spread* or scatter of the values of x around the distribution's mean value, $\mu(x)$. Fortunately, the lengthy calculations implied by formula (7.21) are unnecessary, because it can be proved that, when P_x in formula (7.20) are binomial probabilities, formula (7.21) reduces to the equation

$$\sigma^2 = rpq, \tag{7.22}$$

which is the simplest way of calculating the variance of x in any binomial experiment. In fact, it is more usual to measure the spread of the values of x around the mean, $\mu(x)$, of any probability distribution by a parameter known as the standard deviation,

σ, which is simply the square root of the variance. Therefore, for binomial experiments, the standard deviation is obtained as

$$\sigma = \sqrt{rpq}. \tag{7.23}$$

The standard deviation is preferred to the variance as a measure of spread because its value is expressed in units of x, whereas the variance is not. For the example ball-placing problem, the values of the variance and standard deviation of this binomial experiment are obtained from formulae (7.22) and (7.23) as

$$\sigma^2 = 5 \times \tfrac{1}{9} \times \tfrac{8}{9} = 0.493 \qquad \sigma = \sqrt{0.493} = 0.703,$$

and the position of the mean and standard deviation in relation to the binomial distribution for the ball-placing experiment are illustrated in figure 7.5.

The Poisson distribution

The Poisson distribution is a special or limiting case of the binomial distribution. The reader will appreciate that the calculations for binomial coefficients become extremely lengthy for problems where r (the number of balls in our example) is large. In such circumstances, when certain conditions are satisfied, the Poisson distribution may be used to obtain quick approximation for the independent binomial probabilities (P_x in formula 7.16).

Imagine in figure 7.4 that, instead of being divided into nine equal-sized cells, the total area of the grid is divided into an extremely large number of equal-sized cells each with a minute area. Mathematically speaking, we are letting n (the number of cells) tend to infinity ($n \rightarrow \infty$). By definition, as $n \rightarrow \infty$ so p, the probability of a specified cell receiving a single independently placed ball, will tend to zero ($p \rightarrow 0$; see p221) to check this statement). It can be proved that as $p \rightarrow 0$, so the probabilities generated by the binomial distribution (formula 7.16) become more and more closely approximated by the Poisson formula given by

$$P_x = \frac{e^{-\lambda} . \lambda^x}{x!}, \qquad x = 0, 1, \ldots, \tag{7.24}$$

where $\lambda = r/n$ is the average number of balls per cell, and $e = 2.71828$ is the base of natural logarithms. In practice, the Poisson distribution is used only as an approximation for the binomial if the following conditions are satisfied:

$$n \geqslant 50 \qquad \text{and} \qquad \lambda < 5.$$

Although the ball-placing example does not meet these requirements because $n = 9$, the Poisson probabilities still give a remarkably close approximation to the binomial probabilities. The first three terms of the Poisson distribution are calculated as follows:

$$r = 5, n = 9, \text{ therefore } \lambda = 5/9 = 0.556$$

$$P_{x=0} = \frac{e^{-0.556} \times 0.556^0}{0!} = e^{-0.556} = 0.571.$$

Notice by definition $0! = 1$, and $2\cdot7183^{-0\cdot556} = 0\cdot571$ (see Appendix A)

$$P_{x=1} = \frac{e^{-0\cdot556} \times 0\cdot556^1}{1!} = 0\cdot5712 \times 0\cdot556 = 0\cdot318$$

$$P_{x=2} = \frac{e^{-0\cdot556} \times 0\cdot556^2}{2!} = \frac{0\cdot5712 \times 0\cdot556^2}{1 \times 2} = 0\cdot088.$$

The values of the remaining probabilities are given in table 7.3, together with the equivalent binomial probabilities. The closeness of the Poisson's approximation to the binomial can be seen from the small values in the difference column.

If the Poisson probabilities (P_x from formula 7.24) are substituted in formula (7.18) to obtain the mean value, μ, of the Poisson variable x, then it can be proved mathematically that the average or mean value of x in an infinite number of repetitions of the *same* Poisson experiment is given by

$$\mu(x) = rp = r/n = \lambda. \tag{7.25}$$

Therefore, for a Poisson experiment the mean is the average number of balls per cell, which is the same interpretation we gave to the mean of the binomial distribution. The variance of the Poisson distribution, σ^2, is obtained in the same way as we obtained the binomial variance (see p224), except that we substitute Poisson probabilities P_x in formula (7.20) instead of the binomial probabilities. Again it can be proved mathematically that the substitution for the Poisson variance reduces to the result

$$\sigma^2 = rp. \tag{7.26}$$

Comparing formula (7.25) with formula (7.26) reveals that a theoretical property of the Poisson distribution is that its mean equals its variance. In Chapter 8 we shall show this property of the Poisson distribution is used in the analysis of geographical quadrat sampling experiments.

Table 7.3. A Comparison of the binomial and
Poisson probabilities.

$n = 9, r = 5, \lambda = 0\cdot556$

x	(1) Binomial P_x	(2) Poisson P_x	Differences (1) − (2)
0	0·555	0·571	−0·016
1	0·347	0·318	+0·029
2	0·087	0·088	−0·001
3	0·011	0·016	−0·005
4	0·001	0·002	−0·001
5	0·00002	0·0003	−0·000

The normal distribution

The preceding discussion of the binomial and Poisson distributions made reference to the variable x obeying the assumptions of a probability distribution. In the worked example the variable x, the number of balls which were placed independently in the specified cell in a single experiment, took on discrete values between 0 and 5 in accordance with the binomial probabilities (P_x from equation 7.16). This statement can be interpreted in two ways. Firstly, if the experiment is repeated a large number of times, P_x represents the proportion of the time a specific value x will be observed. Alternatively, in a single experiment, P_x is the probability that the value x will be observed. In short, the binomial probabilities control the way in which x takes on discrete integer values between 0 and r. Any probability equation such as formula (7.16) which determines the way in which x takes on values is termed a *probability density function*. The probability density function for the binomial distribution formula (7.16)[5] was quite easily deduced from the ideas of a binomial coefficient and the multiplication axiom for independent events. However, probability theory also contains density functions which control the behaviour of continuous variables which can take on any value between $-\infty$ and $+\infty$. The mathematical methods which are used to deduce continuous probability distributions from first principles and quite complex. Therefore, to simplify explanation, the following discussion will assume that these continuous functions are given.

The most important continuous probability density function is known as the *standard normal distribution* which is defined by the following equation, where as usual $f(x)$ means x is a function of,

$$f(x) = \frac{1}{\sqrt{2\pi}} \cdot e^{-x^2/2} \qquad -\infty \leqslant x \leqslant +\infty. \qquad (7.27)$$

We can evaluate formula (7.27) for any value of x. For example, $x = 0$ gives the result

$$f(x = 0) = \frac{1}{\sqrt{2\pi}} \cdot e^{-0^2/2}.$$

Notice, by definition, that $0^2/2 = 0$

$$= \frac{1}{\sqrt{2 \times 3 \cdot 1416}} \cdot 2 \cdot 71828^{-0},$$

and since $e^{-0} = 1$

$$= \frac{1}{\sqrt{6 \cdot 2832}}$$

$$= 0 \cdot 399.$$

If we repeat this calculation for every possible value of x between $-\infty$ and $+\infty$, and then plot each value of x against $f(x)$, we would obtain the characteristic symmetrical, bell-shaped curve of the normal probability distribution shown in figure 7.6(a). Notice that outside the range $x = \pm 3$ the values of $f(x)$ obtained from formula (7.27) are too small to be depicted graphically. Therefore, although the potential range of value of x

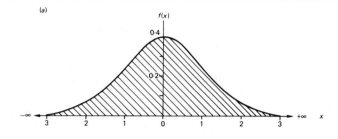

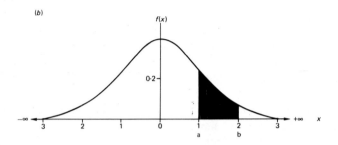

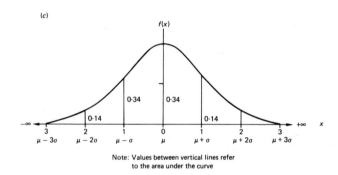

Note: Values between vertical lines refer
to the area under the curve

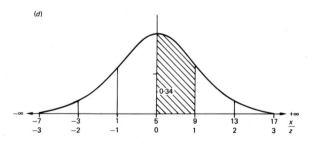

Figure 7.6.   (*a*)–(*c*) Properties of the standard normal distribution, and (*d*), the general normal distribution ($\mu = 5$, $\sigma = 4$).

is between ±∞, the standard normal distribution is only interesting between the range $x = ±3$.

It can be proved that the area under the normal curve (shaded in figure 7.6*a*) is equal to one (unity). Indeed, this statement is true for the curve generated by any continuous probability density function. The area under the curve is used to evaluate the probability of *x* taking on specific values. However, we cannot use continuous probability density functions to obtain the probability that *x* takes on an *exact* value because, for any exact value of *x*, the density function gives a value which is a line joining the point on the curve to the *x* axis. Instead, we measure the probability of *x* taking on a value in an interval under the curve (figure 7.6*b*). If a is the lower value of the interval and b is the upper value, then the probability that a single value of *x* lies in the interval [a,b] is defined as

$$p(a \leqslant x \leqslant b) = \frac{\text{Area under curve in } [a, b]}{\text{Total area under curve} = 1 \text{ (unity)}}. \tag{7.28}$$

For the interval under the curve [1, 2] shown in figure 7.6(*b*), this probability can be proved to be

$$p(1 \leqslant x \leqslant 2) = \frac{0 \cdot 14}{1 \cdot 0} = 0 \cdot 14.$$

This probability means that in a single experiment, where a variable *x* takes on values according to the standard normal distribution, the probability that the value of *x* we observe lies between $x = 1$ and $x = 2$ is 0·14. The mathematical procedures for calculating areas between intervals under the normal curve from formula (7.27) require the use of integral calculus. However, most statistical textbooks contain tables for the standard normal curve which enable you to read off quickly probabilities for all the intervals you are likely to encounter in practice. Such a table is included in Appendix C.

Moments (or parameters) of the standard normal distribution are obtained in the same way as the moments of the binomial and Poisson distributions. Without giving the proof, the mean, average, or expected value of a variable obeying the standard normal distribution is

$$\mu(x) = 0. \tag{7.29}$$

This may be interpreted as the average value of *x* we would observe in an infinite number of repetitions of the same experiment with a normal variable. The variance of the standard normal distribution can be proved to be

$$\sigma^2 = 1; \tag{7.30}$$

therefore, it follows that the standard deviation of the standard normal distribution, σ, is also one ($\sqrt{1} = 1$). The symmetry of the normal distribution enables us to give a more detailed interpretation of the normal standard deviation than was possible with the standard deviations of the binomial and Poisson distributions. Remembering that the standard deviation is expressed in units of *x*, it can be seen from figure 7.6(*c*) that 0·34 of the area under the curve is enclosed by the interval [0, + 1]. This is also the interval between the mean and the mean plus one standard deviation, $[\mu, \mu + \sigma]$. We can now give the following interpretations to the standard deviation of the normal distribution: firstly, if we observe the value of *x* in a large number of repetitions of a normal experiment a proportion of 0·34 of these values will be in the interval $[\mu, \mu + \sigma]$; and alternatively, in a single normal experiment, the probability that *x* takes on a value in the interval $[\mu, \mu + \sigma]$ is $p = 0·34$. Thus, the normal standard deviation can be attached to a normal *probability* which together describe the way in which the values of *x* will be scattered around the normal mean, formula (7.29). In fact, there are a variety of statements which usefully combine normal standard deviations and normal probabilities. It can be seen from figure 7.6(*c*) that there is a $p = 0·68$ (0·34 + 0·34) probability of *x* falling within the range $[\mu - \sigma, \mu + \sigma]$; similarly there is a 0·96 (0·68 + 0·14 + 0·14) probability of *x* falling in the range $[\mu - 2\sigma, \mu + 2\sigma]$.

So far our discussion has been concerned with the standard normal distribution whose variable *x* possesses a mean of zero and a standard deviation of one. However, the standard normal probability density function is a special case of the *general normal density function*[6], where the variable *x* may possess any mean and standard deviation between ±∞. For example, we can consider the case of the variable *x* which takes on values according to the general normal distribution and which possesses a mean, σ, of 5 and a standard deviation, σ, of 4 (figure 7.6*d*). From this diagram we see that the *x* scale has been transformed such that the normal probability, $p = 0·34$, of a single value of *x* being observed in the interval $[\mu, \mu + \sigma]$ now refers to the range of values [5, 9]. Similarly, the normal probability, $p = 0·96$, of *x* falling in the interval $[\mu - 2\sigma, \mu + 2\sigma]$ refers to the range [−3, 13]. Tables for the normal curve only give areas under the curve for standard normal density ($\mu = 0, \sigma = 1$); therefore, in cases where we require areas under the general normal curve, we transform the general values of *x* into standard values using the expression

$$z = \frac{x - \mu}{\sigma}, \tag{7.31}$$

where *z* = the standard value (score), μ = the general mean, σ = the general standard deviation, and, *x* = the general value of *x*. For our example, where $\mu = 5$ and $\sigma = 4$, a general value of *x* = 5 would be transformed to

$$z = \frac{5 - 5}{4} = 0.$$

Thus the general mean of 5 is converted to a standard mean of 0. Similarly, a general value of *x* = 17 is converted into

$$z = \frac{17 - 5}{4} = 3$$

which asserts that the general value *x* = 17 lies three standard deviations to the right of the mean. From these examples it should be clear that the *z* transformation converts general values of *x* onto a scale where the the general mean lies at zero and other general values of *x* are measured in units of one general standard deviation (figure 7.6*d*). The *z* transformation changes *general* values of *x* into a form suitable for use with *standard* normal probability tables (see Appendix C).

The central limit theorem

The variable x discussed in relation to the normal probability density function is an abstract idea, that is, we invent imaginary observations for x which take on values according to the assumptions of the normal distribution. In probability theory, such variables are termed *random variables* and they are simply artificial devices used to illustrate the *theoretical* properties of probability density functions. However, a probability density function is of no *practical* importance unless its random variable can be shown to behave in accordance with some real physical phenomenon. If a physical analogue can be found, then the assumptions and properties of the density function can be used as an explanation for the behaviour of the physical phenomenon. The normal distribution is important to the geographer because it can be proved that normal random variables approximate the behaviour of certain types of sampling error in statistics. Indeed, the normal distribution is the theoretical foundation ᴄf the procedure of statistical inference which allows us to make statements about the accuracy of data collected by random sampling. The formal connection between the normal distribution and random sampling theory is the *central limit theorem*. However, before we state this idea we need to understand some of the simpler terminology and procedures for sampling geographical data.

Table 7.4 presents farm size data taken from Theakstone and Harrison (1970) for the Scottish island of Rousay, which is one of the Orkney Islands. Under the column headed 'acreage' is listed the size, x_i, of all 44, N, farms on the island. These 44 values of the variable x_i constitute a simple geographical *population*. In this statistical sense a population is the set of all N observations that form some pre-defined variables. From the farm size data we have calculated the mean (μ) and standard deviation (σ, see table 7.4) and together, the values N, μ, and σ are referred to as the *parameters* of the population because they summarize the main numerical characteristics of the frequency distribution of all the observations forming the population. The position of the population mean and standard deviation in relation to the frequency distribution of farm sizes is shown as a histogram in figure 7.7(a). In this instance the population frequency distribution is positively skewed, there being many more small farms than large farms. Therefore, μ lies to the right of the main concentration of values x_i.

The majority of geographical populations contain many more observations than the $N = 44$ found in the farm size data. The populations of variables in urban geography often consist of observations which refer to the individuals who live in a town, and clearly such variables are likely to contain upwards of 50 000 individual values of x_i. Similarly, physical geographers often study populations which contain an infinite number of observations. For example, in any river valley there are an infinite number of slope angles that could be measured. Clearly, when N is large or infinite it is sensible to collect only a representative proportion, or sample of the total population of observations. To ensure that the sample observations are representative of the population, the sample is collected by some random procedure, that is, before collection each member of the total population has an equal chance, $1/N$, of being included in the sample of n observations to be used for analysis. For any single random sample of size n, we can calculate the sample mean, $\bar{x}$, and sample standard deviation, S_x and together $n, \bar{x}$, and S are termed the *statistics* of the sample. The values of $\bar{x}$ and

Table 7.4. Population and sample characteristics: farm size data, Rousay, 1966. (Source, Theakstone and Harrison 1970, pp101—2 used with permission.)

i Farm No.	x Total Acreage	Sample No.	i Farm No.	x Total Acreage	Sample No.
1	4	1, 2	23	100	
2	4	1	24	107	1, 2
3	4	1, 2	25	110	1, 2
4	5	1, 2	26	114	1, 2
5	6	1, 2	27	120	1, 2
6	8	1, 2	28	150	1, 2
7	13	2	29	158	1, 2
8	14	1, 2	30	160	2
9	20		31	171	2
10	25	1	32	187	2
11	40	1	33	198	1, 2
12	40	1, 2	34	212	
13	40		35	219	1, 2
14	44	1, 2	36	230	2
15	50	1	37	310	
16	51	1, 2	38	386	1
17	59	1, 2	39	391	
18	66	1, 2	40	435	1, 2
19	70	2	41	600	1, 2
20	72	2	42	960	1, 2
21	73		43	1900	1
22	76	1	44	3000	1, 2

Population Parameters

$N = 44$

$$\mu = \sum_{i}^{N} x_i/N = 250.$$

$$\sigma = \sqrt{\left[\sum_{i}^{N} (x_i - \mu)^2\right] / N} = 589.$$

$$\sigma_{\bar{x}} = \frac{\sigma}{\sqrt{n}} \cdot \sqrt{\frac{N-n}{N-1}}$$

$$= \frac{589}{\sqrt{30}} \cdot \sqrt{\frac{44-30}{44-1}} = 61.$$

Sample statistics

(1) $n_1 = 30$

$$\bar{x}_1 = \sum_{i}^{n} x_i/n_1 = 298.$$

$$S_1 = \sqrt{\left[\sum_{i}^{n} (x_i - \bar{x}_1)^2\right] / n_1} = 626.$$

$\hat{\sigma}_{\bar{x}_1} = 65.$

(2) $n_2 = 30.$

$\bar{x}_2 = 246.$

$S_2 = 548.$

Note. In this table the subscripts in n_1, n_2 or $\bar{x}_1, \bar{x}_2$, etc, are used to identify the results obtained from the two different random samples.

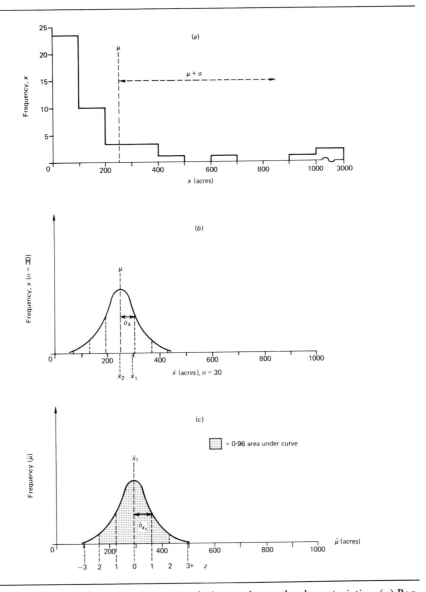

Figure 7.7. Rousay farm-size data: population and sample characteristics. (*a*) Population characteristics. (*b*) Distribution of sample means, $\bar{x}$, around population mean, μ. (*c*) Estimating the population mean: distribution of sample means around $\bar{x}_1$.

S for two different random samples from the farm size population of size $n = 30$ are given in table 7.4. The 30 farms in each sample were selected randomly by matching two sets of 30 random numbers (see Appendix B) with values of the farm numbers between 1 and 44 as listed in table 7.4.

It should be apparent to the reader that table 7.4 contains far more information than would be possessed in practice after sampling some large geographical population. In practice, we might know the value of N if the population was finite, but certainly would not know the values of the population mean, μ, the population standard deviation σ, and also the shape of the population frequency distribution (figure 7.7a). What we would possess is a single set of n randomly sampled observations, together with a single sample mean, $\bar{x}$, and the sample standard deviation, S_x. The question that now arises is, 'how representative is a single random sample mean of the unknown population mean?' The answer is provided by the central limit theorem which contains theoretical statements about the frequency distribution of sample means obtained by random sampling. The central limit theorem states[7] that if $\bar{x}_i$ are the values of a series of random sample means of size n taken from a population with a finite mean, μ, and standard deviation, σ, then the variable $\bar{x}_i$ obeys the general normal distribution with a mean μ and standard deviation given by $\sigma/\sqrt{n}$. This frequency distribution of all the possible random sample means of size n is termed the *sampling distribution of sample means* and is approximately normally distributed so long as n is greater than or equal to 30. The standard deviation of the sampling distribution ($\sigma/\sqrt{n}$) is generally referred to as the standard error and denoted by the symbol $\sigma_{\bar{x}}$. For populations where N is *infinite* the standard error is defined as above, but when N is finite the standard error is modified by a factor $\sqrt{(N-n)/(N-1)}$ to give

$$\sigma_{\bar{x}} = \frac{\sigma}{\sqrt{n}} \cdot \sqrt{\frac{N-n}{N-1}}. \tag{7.32}$$

The values of $\mu, \sigma, \sigma_{\bar{x}}$, and two sets of random sample statistics of size $n = 30$ for the farm size population are given in table 7.4, while the sampling distribution of sample means for this problem is illustrated in figure 7.7(b). Notice that both the sample means given in table 7.4 ($\bar{x}_1$ and $\bar{x}_2$) fall within the range $\mu = \pm 1\sigma_{\bar{x}}$.

To conclude our discussion of the central limit theorem we shall consider how the theorem is used in practice for *estimating* the unknown population mean. Returning to our farm size problem, imagine that the population parameter values are unknown, as they would be in practice, and that we wish to make statements about the mean farm size on Rousay. We collect a single random sample of size $n = 30$ and calculate the statistics listed as sample 1 in table 7.4: $\bar{x}_1 = 298$, $S_1 = 626$. We would *not know* the value of σ to obtain $\sigma_{\bar{x}}$ from equation (7.32), however we can use the sample standard deviation, S, to estimate the standard error[8] as

$$\hat{\sigma}_{\bar{x}} = \frac{S}{\sqrt{n}} \cdot \sqrt{\frac{N-n}{N-1}}. \tag{7.33}$$

Therefore

$$\hat{\sigma}_{\bar{x}_1} = \frac{626}{\sqrt{30}} \cdot \sqrt{\frac{44-30}{44-1}} = 65.$$

where the hat symbol $\wedge$ denotes an estimate for an unknown population value (see table 7.5). The central limit theorem tells us that $\bar{x}_1$ is an independent observation taken from a general normal distribution centred on the unknown μ (the population mean) with standard deviation of $\hat{\sigma}_{\bar{x}}$. It is perfectly logical to reverse this argument and think of the unknown value μ being normally distributed around $\bar{x}_1$ with a standard error of $\hat{\sigma}_{\bar{x}}$ (figure 7.7c). This reversal of the central limit theorem permits us to make statements about the probability of the unknown value of μ falling within intervals on the general normal sampling distribution around $\bar{x}_1$. For example, there is a $p = 0.96$ probability that μ lies within the range $[\bar{x}_1 - 2\hat{\sigma}_{\bar{x}_1}, \bar{x} + 2\hat{\sigma}_{\bar{x}_1}] = [168\ 428]$ (the shaded area in figure 7.7c). Similarly, there is a $p = 0.68$ chance that μ lies within the range $[\bar{x}_1 - \hat{\sigma}_{\bar{x}_1}, \bar{x} + \hat{\sigma}_{\bar{x}_1}] = [233\ 363]$. Notice that the range of error becomes smaller as the probability of the statement being true declines. These statements illustrate the meaning of the term statistical inference. Because sample means are normally distributed we are able to infer population characteristics on the evidence of sample statistics.

Substitution of any sample observation, x_i, in the z transformation (see formula 7.31, p230)

$$z = \frac{x_i - \bar{x}}{\hat{\sigma}_{\bar{x}}}, \qquad (7.34)$$

where, as usual, x is the sample mean and $\hat{\sigma}_{\bar{x}}$ is the estimated standard error of the sampling distribution of sample means, allows us to obtain z values which express the observation x_i as a standard normal variable where the sample mean has a z value of zero and the standard error has a z value of one (figure 7.7(c)). For the first (Rousay) farm size random sample, if we substitute the value $x = 428$ into formula (7.34), we obtain the result

$$z = \frac{428 - 298}{65} = 2.$$

This example transformation indicates that the value $x = 428$ lies 2 standard errors to the right of the sample mean; that is $428 = 298 + (2 \times 65)$. Notice, in figure 7.7(c), that the transformation $z = \pm 2$ encloses $p = 0.96$ of the area under the normal distribution for the estimated value of the population mean $\hat{\mu}$, around the sample mean, $\bar{x}_i$. Similarly, the transformation $z = \pm 1$ encloses $p = 0.68$ of the area under this normal distribution. Therefore, the z values obtained from formula (7.34) can be attached to normal probabilities which express the likelihood of the unknown population mean occurring within a defined range of error around the sample mean.

The z transformation described by formula (7.34) can be used to construct concise inferential statements about random sampling experiments. For any single random sample the estimation of the population mean is written symbolically as

$$\hat{\mu} = \bar{x} \pm z \cdot \hat{\sigma}_{\bar{x}}. \qquad (7.35)$$

For the Rousay farm size problem when $z = 2$, formula (7.35) gives the result

$\hat{\mu} = 298 \pm 2 \times 65$

$= 298 \pm 130, \qquad p = 0.96.$

This statement is read as, 'there is a $p = 0.96$ probability that the unknown population mean occurs in the range of values of x between 168 and 428 acres'. The z value in formula (7.35), which we defined in formula (7.34), is termed the *confidence coefficient* or *critical value* and its value is chosen by the experimenter. To be of any practical value, inferential statements must have a high probability of being correct, therefore it is usual for experimenters to select z values (confidence coefficients) which are attached to normal probabilities whose value is close to one (unity)[9]. In practice, the confidence coefficient is usually given a value between ±2 and ±3. The coefficient $z = \pm3$ is attached to a normal probability of $p = 0.99$ and substituting this value in formula (7.35) gives the result

$$\hat{\mu} = 298 \pm 3 \times 65$$

$$= 298 \pm 195.$$

This statement is interpreted as, 'there is a $p = 0.99$ probability the population mean falls within the range 103 to 493 acres.' Notice that the inferential statement is less accurate for $z = \pm3$ because the range of error is greater, although there is a higher probability of the statement being correct. This trade-off between accuracy and the probability of correctness is decided by the experimenter *before* the random sample is collected.

In most statistical textbooks the probability

$$\alpha = 1 - p, \tag{7.36}$$

where p is the normal probability attached to the preselected confidence coefficient, is termed the *significance level*, or *probability of type I errors*. Here α is the probability that an inferential statement is incorrect. For example, in the Rousay problem when $z = \pm2$, then $p = 0.96$ and therefore $\alpha = 0.04$, which indicates that there are four chances in a 100 of the unknown population mean lying *outside* the range 168 to 428 acres. Notice that once the confidence coefficient has been selected the values of p and α are automatically fixed.

The chi-squared distribution and test

Careful thought is required to understand the probability theory which underpins the problem of estimating the value of a population mean, but the form of the problem is quite simple. The inferential statements are concerned only with the likely accuracy of a particular sample mean. However, the subject of inferential statistics, which is founded on relative frequency measurement, contains many tests and techniques which are intrinsically of far greater practical value to the geographer. One of the most useful of these tests is the chi-squared (χ^2) test which is founded on the properties of the χ^2 probability distribution, and follows the same type of reasoning by which we estimated a population mean.

To illustrate the geographical application of the χ^2 test, we will return to the office relocation problem which we discussed in the section on contingency tables (pp217—219). Remember in that section we used the multiplication axiom for independent events to interpret the relationship between type of move and distance of the office relocation from central London. However, one factor we ignored in that discussion was that all the data listed in table 7.2(a) were collected by random sampling and therefore our entire analysis of the problem was subject to sampling error. The question the χ^2 test is designed to answer is whether the differences between the observed, absolute frequencies (table 7.5a) and the expected, independent, absolute

Table 7.5. Office Relocation from Central London: calculation of χ^2 statistic.

(a) Observed frequencies, O_i (see table 7.2(a), p216).

(b) Expected independent frequencies, E_i. ($p_{ij} \times 519$, see table 7.2(c), p216).

Type of move	Distance moved (miles)		
	<20	20–39	>40
Complete	217	64	57
Partial	88	44	49

197	72	69
108	36	37

(c) $O_i - E_i$

20	−8	−12
−20	8	12

(d) $(O_i - E_i)^2/E_i$

2·03	0·89	2·09
3·70	1·78	3·89

$$\text{calc } \chi^2_{(2 \text{ df})} = 2\cdot03 + 0\cdot89 + 2\cdot09 + 3\cdot70 + 1\cdot78 + 3\cdot89$$
$$= 14\cdot38.$$

frequencies (table 7.5b) are due either to random sampling error or are the result of a genuine relationship between the two nominal variables. Moreover, the application of the χ^2 test to contingency tables is used to exemplify the reasoning that is common to the geographical application of most statistical techniques which are based on the probability theory of relative frequency measurement.

In statistical practice, when we examine the relationship between the two nominal variables in a contingency table, we begin by establishing what is termed the *research hypothesis*. This hypothesis is a symbolic, statistical statement which is designed to encapsulate the idea we are interested in testing. For the contingency table problem where the nominal variables are type of move and distance, the research hypothesis is written as

$$H_1 : \text{freq}(O_i) \neq \text{freq}(E_i). \tag{7.37}$$

This statement is read as, 'the observed frequencies, freq (O_i), are significantly different from ($\neq$) the expected, independent frequencies, freq (E_i) (see table 7.5a and b). By *significantly different from* ($\neq$) we mean that the differences between the individual cells of the two contingency tables are unlikely to be a result of errors due to random sampling. Therefore, the research hypothesis establishes the idea that we think the two nominal variables should be related to (dependent upon) each other in some unspecified way.

The second stage of statistical testing is to establish the *null hypothesis, H_0*, which is the antithesis (opposite) of the research hypothesis. Again, the null hypothesis is a statistical statement which is written symbolically as

$$H_0 : \text{freq}(O_i) = \text{freq}(E_i), \qquad \alpha = 0\cdot05 \tag{7.38}$$

This statement is read as, 'there is *no* significant difference between the observed and expected frequencies'. Similarly, *by no significant difference* (=) we mean that any differences between the observed and expected frequencies are caused solely by random sampling errors. The experimenter also has to qualify the null hypothesis with a probability, α. Remember (from p000) that we defined α as the significance level which represented the probability that a statistical statement was incorrect. The value of α is selected by experimenter before the application of the test, and for the office relocation example we have decided to set $\alpha = 0.05$[10]. Because the null hypothesis is designed to be the antithesis of the research hypothesis, at the conclusion of the test the experimenter is hoping to *reject* the null hypothesis, H_0, and *accept* the research hypothesis, H_1. But the rules of statistical reasoning are designed so that the experimenter can only reject H_0 if he is sure there is only a probability α of his being *incorrect*. Therefore, the value selected for α must always be close to zero so that the experimenter only rejects H_0 when he is virtually certain that the differences between the observed and expected frequencies (table 7.5c) are genuine and are not the result of random sampling errors. The probability theory of statistical testing makes it mathematically impossible to set α equal to zero and therefore, in any test, there must always be some probability that the rejection of H_0 is incorrect. Thus statistical tests are never capable of either proving or disproving a statistical hypothesis with certainty, they merely allow the experimenter to make statements, at the conclusion of the test, which have a high probability of being correct.

The third stage of statistical testing is the collection of data by random sampling. The observed frequency data (table 7.5a) are a random sample of size $n = 519$ drawn from the population of $N = 790$ offices that were relocated from central London during the period 1963 to 1969. The experimenter must obey two rules when collecting sample frequency data for contingency tables which are to be analysed by the χ^2 test. Firstly, the sample size, n, must be at least 50 and, secondly, when the expected frequencies are calculated each individual cell should possess an expected value of at least five. In problems where this second rule is not fulfilled the class boundaries of either one or both of the nominal variables must be redefined (in practice, classes are usually grouped together) until all the expected frequencies are greater than five.

The fourth stage of the analysis is the calculation of the χ^2 statistic for the problem in hand. The calculated value of χ^2 is obtained from the equation

$$\text{calc } \chi^2_{[(g-1).(m-1)\text{df}]} = \sum_i^k \frac{(O_i - E_i)^2}{E_i}, \tag{7.39}$$

where O_i = the value of the ith observed frequency, E_i = the value of the ith expected frequency, g = number of rows in the contingency table (2), m = number of columns in the contingency table (3), $k = g.m$ = number of cells in the contingency table (6),

$$\text{df} = (g-1)(m-1) = \text{degrees of freedom} = (2-1)(3-1) = 2. \tag{7.40}$$

For the office relocation data the calculation of formula (7.39) to obtain the result of calc $\chi^2_{(2\text{df})} = 14.38$ is demonstrated in table 7.5. The calculation of the χ^2 statistic is made with respect to a defined number (formula 7.40) of *degrees of freedom*. These are related to the number of cells in the contingency table[11]. Degrees of freedom appear in the calculation of most statistical tests because the interpretation of the calculated value of any test statistic, in this case χ^2, varies according to the amount of data that have been collected.

Inspection of formula (7.39) will reveal that calc χ^2 can theoretically take on any value between zero and infinity. When calc $\chi^2 = 0$, the observed frequencies, O_i, will be equal to the expected frequencies, E_i. Similarly, as the differences between O_i and E_i increase so will calc χ^2. Because of its numerical properties, calc χ^2 is termed a *goodness-of-fit statistic*. This term arises because, for contingency table problems, the value of calc χ^2 will tend to zero when there is no relationship between the two nominal variables; that is, the expected, independent frequencies are equal to the observed absolute frequencies, $O_i = E_i$. Alternatively, when a genuine relationship (dependency) between the variables exists, calc χ^2 will tend to some large value depending on the strength of the relationship.

The final, and most important, stage of statistical testing is the experimenter's decision as to whether he should reject or accept the null hypothesis. Because differences between O_i and E_i can occur purely as a result of the random sampling errors inherent in the values O_i it is possible for quite large values of calc χ^2 to be observed in a single experiment when, in reality, there is no relationship between the variables. In such experiments, calc χ^2 is merely reflecting a large amount of random sampling error in the experiment. Thus the statistical decision made by the experimenter centres on whether a single observed value of calc χ^2 is sufficiently large for him to be able to reject the null hypothesis with only the small probability (the significance level) of his being incorrect. The experimenter is always able to make this decision because it can be proved mathematically that in an infinite number of repetitions of the same contingency table experiment the variations in χ^2 which are due solely to random sampling error obey the continuous χ^2 probability distribution[1,2].

The previous statement makes it possible to interpret the results of the χ^2 test for our office relocation problem. In an infinite number of repetitions of the office relocation problem, where the contingency table has two degrees of freedom, it is known that the frequency distribution of χ^2 values which are due solely to random sampling error is defined by the curve of the χ^2 probability density function with two degrees of freedom. This curve, illustrated in figure 7.8, is termed the sampling distribution of χ^2 with two degrees of freedom. Also illustrated in figure 7.8 is the fact that $p = 0.95$ of

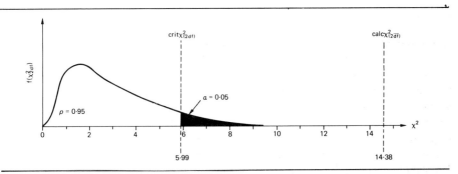

Figure 7.8. The sampling distribution of χ^2 with two degrees of freedom.

Table 7.6. Significance test for the office relocation problem.

H_1: freq(O_i) $\neq$ freq(E_i)
H_0: freq(O_i) = freq(E_i), α = 0·05, crit $\chi^2_{(2\ df)}$ = 5·99
calc $\chi^2_{(2\ df)}$ = 14·38
Decision: Reject H_0, accept H_1.

the area under the χ^2 curve lies to the left of the value χ^2 = 5·99, while α = 0·05 of the area under the curve lies to the right of this value. The probability α = 0·05 is the significance level we selected for testing the null hypothesis, and the value χ^2 = 5·99 which defines the significance level) is termed the *critical value of χ^2*. Crit $\chi^2_{(2df)}$ = 5·99 is interpreted as the value of χ^2 that would be exceeded in a proportion of 0·05 of an infinite number of repetitions of the office relocation experiment if there was no genuine relationship between distance and type of move. In fact, crit $\chi^2_{(2df)}$ = 5·99 is the value of χ^2 that would be exceeded, solely as a result of random sampling error, in all contingency table experiments with two degrees of freedom. In practical applications of the χ^2 test, critical values of the statistics are read from statistical tables which list critical values for both different numbers of degrees of freedom and significance levels (see Appendix D).

The significance test for the office relocation problem is written formally in table 7.6 and depicted graphically in figure 7.8. Because calc χ^2 = 14·38 is greater than crit χ^2 = 5·99, we *reject* the null hypothesis and accept our research hypothesis that there is a genuine relationship between distance and type of move[13]. The decision to reject the null hypothesis is made because calc χ^2 falls outside the range of values of χ^2 between 0 and 5·99 where we are uncertain theoretically whether calculated values are indicative of either random sampling errors or a genuine relationship. However, it must be remembered that in accepting our research hypothesis we do so with a probability α = 0·05 that in our decision is incorrect. This qualification arises because we know theoretically that in 0·05 of the repetitions of our experiment the critical value of χ^2 = 5·99 could be exceeded solely as a result of random sampling errors. In other applications of the χ^2 test where the critical value was found to exceed the calculated value, the experimenter would be forced to accept the null hypothesis and reject his research hypothesis.

General comments on statistical tests

In the subsequent chapters on geographical developments in probability theory some of the statistical tests more frequently used by geographers, such as analysis of variance and correlation coefficients will be discussed either in detail or in passing as a footnote. Although the arithmetical procedures for calculating these tests are different, the tests themselves all follow the relative frequency pattern of statistical reasoning. This reasoning process has been exemplified by our discussion of the χ^2 test: firstly, state the research hypothesis for the statistic in question; secondly, state the null hypothesis and desired significance level; thirdly, collect data by random sampling in accordance with the form of the calculated statistic; fourthly, identify the critical value of the statistic by referring to the sampling distributions describing variations in the statistic that are due solely to random sampling error; and finally, compare the critical and calculated values of the statistic in order to make decisions about the null and research hypotheses.

Notes

1   Formally, mathematicians say $p(E)$ tends to ($\rightarrow$) zero when $p(E)$ is impossible, and $p(E)$ tends to one (unity) when $p(E)$ is certain.

2   In a recent text on probability theory (Whittle 1976) the theory is deduced from the axiom of the *average*, and not the rules of addition and multiplication of probabilities expounded in this text.

3   Mutually exclusive means two events which, by nature, cannot occur together in a single trial on an experiment.

4   Factorial numbers like $r!$ are defined on p166. It should be remembered that both 0! and 1! are given defined values of 1.

5   Notice that the variance of the Poisson, rp, differs from the variance of the binomial, rpq. This occurs because to derive the Poisson we assumed $n \rightarrow \infty$, and from this assumption it follows that $p \rightarrow 0$ and $q = (1 - p) \rightarrow 1$. Therefore q is omitted from the Poisson's variance because it does not influence its value.

6   The probability density function for the general normal distribution is given by

$$f(x) = \frac{1}{\sigma\sqrt{2\pi}} \cdot e^{-[1/2(x-\mu/\sigma)^2]} \qquad -\infty \leqslant x \leqslant +\infty$$

where μ and σ are the mean and standard deviation of x, resp ectively. When $\mu = 0$ and $\sigma = 1$ this formula reduces to equation (7.27) for the standard normal probability density.

7   The central limit theorem is formally stated in terms of standard normal density as, 'let $\bar{x}_n$ denote the mean of the first n of a sequence of independent observations from a population with mean μ and variance σ^2, then the random variable

$$z = \frac{\bar{x}_n - \mu}{\sigma/\sqrt{n}}$$

has a distribution that approaches the standard normal distribution as n becomes infinite.'

8   Where N is infinite the standard error or the sample mean is estimated as

$$\hat{\sigma}_{\bar{x}} = S/\sqrt{n}.$$

9   It is impossible for the experimenter to select a z value (confidence coefficient) which is attached to a normal probability of $p = 1$ (unity), because the potential range of any normal variable is $\pm\infty$.

10   In geography it is common practice to set $\alpha = 0.05$, although values of 0.01 and 0.10 are quite common, depending on the degree of precision required for a particular experiment.

11   *Degrees of freedom in contingency tables.* The exact meaning of the term degrees of freedom is fairly easy to illustrate using the office relocation problem. In essence, degrees of freedom are the number of expected, independent frequencies that need to be calculated before the observed column and row frequencies (a) can be used for the calculation of the remaining expected

frequencies. In this example, once the two expected frequencies illustrated in (*b*) have been calculated by the procedures indicated in table 7.5 (*b*), the values of the remaining cells are automatically fixed by the values of the row and column frequencies. For example, the value of the starred cell is obtained from the information in (*b*) as 305 – 197 = 108. The remaining expected frequencies can be calculated in a similar manner. Therefore, in this example, once two expected frequencies have been calculated the remaining values are automatically fixed by the column and row frequencies, and so we say the values of the expected frequencies have 2 degrees of freedom to vary.

(*a*) Observed frequencies

| 217 | 64 | 57 | 338 |
| 88 | 44 | 49 | 181 |

305   108   106

(*b*) Expected frequencies

| 197 | 73 | | 338 |
| * | | | 181 |

305   108   106

12   *The chi-squared probability distribution.* The probability density function for this distribution, $f(x^2)$, controls the behaviour of the random variable x^2 between zero and plus infinity, and is given by

$$f(x^2) = \frac{1}{2^{t/2}\Gamma(t/2)} \cdot e^{-(x^2/2)} \cdot (x^2)^{(t-2)/2}, \qquad 0 \leqslant x^2 \leqslant +\infty, \tag{i}$$

where t = number of degrees of freedom, e = 2·71828, and the numbers of the form $\Gamma(t)$ are defined as

$$\Gamma(t) = (t - 1)!$$

For example, $\Gamma(3) = 2! = 1 \times 2 = 2$.

Although the x^2 probability density function is quite complex, the shape of the graph of formula (*i*) varies solely with the function's single parameter t, the number of degrees of freedom. Graphs of $f(x^2)$ for $t = 5$ and $t = 6$ are illustrated below.

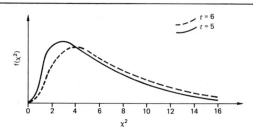

13   This test confirms the interpretation we gave to the relationship between distance and type of move on p218.

CHAPTER 8

Spatial Probability Models

In Part I of this book we discussed a number of models which predicted the behaviour of various types of spatial system. To construct these models we assumed that the spatial system was controlled by a known law, and thus the model predictions were simply the mathematical realization of the law. For example, the model of heat flow in a soil profile was a realization of Fourier's law of heat conduction (Chapter 4), while the model of shopping expenditure flows in a city was based upon the gravity propositions (Chapter 5). But for many spatial systems the processes controlling the occurrence of events are either not fully understood, or are too complex to be represented in a law-like manner. For instance, the exact location of a new factory in a city or the spread of a disease through a human population owe more to chance than any exact collection of laws. Nevertheless, a good deal of information about the past behaviour and spatial distribution of such systems is often available, and in this chapter we will explore the ways in which probability models can be constructed to help identify any trends, patterns or relationships that may be present in these data.

For the most part, this chapter will concentrate on the analysis of spatial variables, that is, the observed values, x_i, some phenomenon takes on over a set of $i = 1, 2, \ldots,$ n regions. The simplest spatial variables are obtained from the study of dot maps or point patterns. Most geographical objects, whether they be cities, factories, landforms, or a species of plant, may be depicted as a dot map which can be converted into a spatial variable simply by laying a grid of n equal-sized squares (quadrats) over the study region and then counting the number of points, x_i, found in each square. This chapter begins by describing how the methods of quadrat analysis are used to unravel the processes which generate the patterns of points we observe in reality. Quite often, sources of geographical information such as national censuses provide data describing the values of two or more variables distributed over the same set of regions, such as the census tracts in a city. Sometimes there will be good geographical reasons for suspecting that two variables will be related to one another in some way, and in the second section we describe how the techniques of correlation analysis are used to measure the strength and direction of such relationships. In this section we shall pay particular attention to the ways in which the size and shape of the regional units can prejudice our interpretation of geographical relationships. Finally, we will discuss the spatial autocorrelation problem which is concerned with understanding how a single variable is related to itself in space.

QUADRAT ANALYSIS

Patterns and processes

The term quadrat analysis embraces a variety of probability models which are used to understand the mathematical properties of observed point patterns. Here we will consider only the simpler methods and problems that arise in quadrat analysis. For more advanced treatments of the subject the reader is referred to the textbook by Greig-Smith (1964) and the monographs by Rogers (1974) and Thomas (1977).

Most probability models for spatial point patterns take their assumptions from some qualitative ideas about the geographical processes which control the evolution of a pattern both through time and space. The six patterns shown in figure 8.1 illustrate the continuum of possible point patterns which ranges from the totally clustered pattern (figure 8.1*a*), to the totally uniform pattern. Clustered patterns are usually controlled by *contagious* processes whereby the location of a point in space and time attracts subsequent point locations to its immediate vicinity (figure 8.1*b*). For example, if we were to study point patterns, representing the people infected with smallpox in an epidemic in a large city it is likely that we would observe clustered patterns

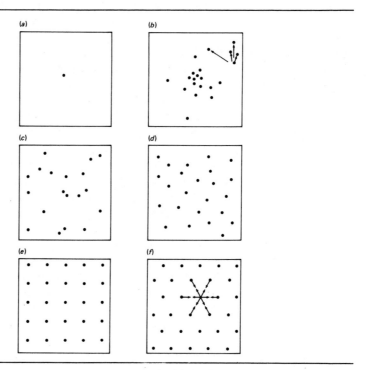

Figure 8.1. The continuum of spatial pattern. (*a*) Totally clustered. (*b*) More clustered than random. (*c*) Random. (*d*) More uniform than random. (*e*) Uniform (grid-iron). (*f*) Totally uniform (triangular). Arrows indicate contagion in (*b*) and competition in (*f*).

of infected people. Our expectation of clustering arises because the process by which smallpox spreads through time and space is person-to-person contact; therefore, patterns of infected people are likely to cluster around the original person who contracted the disease. In fact, many seemingly dissimilar geographical entities are found to exhibit clustering when mapped as point patterns. The distribution of karst depressions in limestone regions show a tendency for depressions to cluster in zones of structural weakness (McConnell and Horn 1972), while a similar clustered pattern is displayed by the distribution of people who voted for the winning candidate in Indianapolis Mayorial Elections (Reynolds 1974). The extreme case of a clustered pattern is where all the points are attracted to the same location. This theoretical limiting case of a clustered point pattern is assumed to occur when the contagious process controlling the clustering changes from a probabilistic to a deterministic nature (figure 8.1*a*).

Random patterns are the result of an independent probabilistic process where each point is located in the plane independently of existing, or subsequent, point locations. The operation of an independent, spatial process usually results in the haphazard distribution of points illustrated in figure 8.1(*c*) where, visually, there is no underlying structure to the pattern. Observed random point patterns are of no geographical interest because independent process are chance-like and therefore are not explainable in a geographical sense. However, theoretical random process are simple to represent mathematically and for this reason simple random models such as the binomial and Poisson distributions are used to classify observed patterns into one of the three pattern types: clustered, random or uniform.

Uniform patterns are created by competitive processes where, during the evolution of the pattern, the points compete for space in the plane (figure 8.1*d–f*). The distribution of settlements in predominantly agricultural regions often tend to uniformity (Dacey 1964) because the settlements compete with one another for market areas (figure 8.1*f*). Therefore, in a competitive probabilistic process, the location of one point will tend to repel the location of other points in its immediate vicinity. The upper limit of uniform patterns is where the points are located on the vertices of an imaginary grid of equilateral triangles (figure 8.1*f*). This limiting uniform pattern is created by a deterministic competitive process because, as soon as one point has been located and the distance between neighbouring points established, the location of all other points is immediately predetermined. For the same reason the grid-iron pattern where the points are located at the vertices of a square lattice is also caused by a deterministic competitive process. But the triangular pattern is considered the more uniform because, when the distance between neighbouring points is identical, more points can be packed into the same area with an equilateral grid than with a square grid.

Measurement and data collection

Quadrat analysis uses probability theory to model the *frequency* distribution of an observed point pattern. By frequency we mean the way in which the density of points varies over the study area. The simplest approach in quadrat analysis is to compare the observed frequency distribution of points with the theoretical random frequency distribution predicted by either the binomial or Poisson distributions (see Chapter 7, pp221–227. The second stage of the analysis is to decide whether there is a close enough correspondence between the observed and random frequency distributions to infer that the observed pattern is controlled by an independent spatial process. If the degree of correspondence is poor, we can then use theoretical properties of the Poisson distribution to decide whether the observed frequency distribution is clustered and therefore controlled by a contagious spatial process, or alternatively, uniform and controlled by a competitive spatial process. The essence of this method is that by analysing the frequency distribution of a point pattern we can infer the type of probabilistic process that controls the evolution of the pattern.

The design of any quadrat experiment begins with defining the boundaries of the study region. Clearly, the definition chosen depends on the particular nature of the objects that form the points, although the following guidelines provide a general background to this problem. Firstly, every part of the study area should be a possible location for a point. Thus, if we are studying the frequency distribution of karst depressions, the study area should be exclusively a limestone region. The inclusion of other rock types in the study area would create bias in any subsequent data collection technique designed to estimate the observed frequency distribution. To illustrate the quadrat method we have chosen the distribution of vehicle factories in Merseyside in 1966. It was decided to delimit the study area boundary as the nearest straight line equivalent to the boundary of the continuous built-up urban area of Merseyside in 1966 (figure 8.2*a*) because the potential locations for such factories are normally vacant spaces *within* the urban area. Notice that the delimitation of the study area boundary automatically fixes the *scale of resolution* of the problem, so

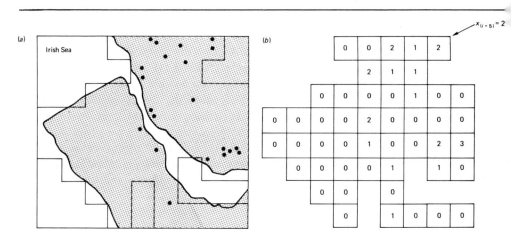

- Factory locations
- Nearest straight line approximation to the boundary of the Merseyside built-up urban area, 1966

Figure 8.2. The distribution of vehicle factories in Merseyside, 1966. (Adapted from R W Thomas 1977, p4.)

that any subsequent analysis can only make statements about vehicle factory location at the intra-urban scale.

The frequency distribution for the observed point pattern is obtained by either *quadrat sampling* or *quadrat censusing*. Quadrat sampling involves randomly placing a quadrat of predetermined size n times over the study area, each time counting the number of points, x_i, that fall within the quadrat. Quadrat sampling gives a random sample estimate of the observed frequency distribution and the technique is used either when r, the total number of points in the pattern, is extremely large, or when the points are difficult to map in the field, for example, the patterns of plant types that occur in biogeography. Quadrat censusing is the more usual technique in geographical applications and involves placing a continuous square grid containing n equal-sized cells over a map of the study area and counting the number of points falling within each cell, x_i. Thus censusing enables us to derive the frequency distribution for the whole population of points and the census of the vehicle factory pattern is shown in figure 8.2(b).

The information listed in table 8.1 demonstrates that once the census has been collected it is a straightforward procedure to convert the observed numbers of points per cell, x_i, into the observed frequency distribution which represents the number of cells containing exactly x points, n_x. In the factory problem there are 34 cells containing no points, eight cells containing one point, and so on, and the reader may wish to check these values in table 8.1 from the census in figure 8.2(b). Two useful summary statistics for the observed census are its mean, $\bar{x}$, and variance, S^2, defined, as usual, by

$$\bar{x} = \sum_i^n x_i/n = r/n, \tag{8.1}$$

Table 8.1. Definition of terms for the observed frequency distribution (numerical examples taken from figure 8.2).

r = total number of points = 21
n = total number of cells = 48
x_i = number of points in the ith cell
$x_{(i=5)} = 2$ (see figure 8.2b)
n_x = the number of cells containing x points
$n_{(x=0)} = 34$
k = the largest observed value of $x_i = 3$

The observed frequency distribution of vehicle factories in Merseyside in 1966.

x	n_x
0	34
1	8
2	5
3	1

which, for the factory problem gives the result

$\bar{x} = 21/48 = 0{\cdot}44$ points per cell, and

$$S_x^2 = \sum_i^n (x_i - \bar{x})^2 /n. \tag{8.2}$$

The variance formula may be written in a simpler form for making calculations direct from the frequency distribution as

$$S_x^2 = \sum_{x=0}^{k} n_x(x - \bar{x})^2 /n \tag{8.3}$$

and this formula gives the result for the variance of the factory census on

$$S_x^2 = [34(0 - 0{\cdot}44)^2 + 8(1 - 0{\cdot}44)^2 + 5(2 - 0{\cdot}44)^2 + 1(3 - 0{\cdot}44)^2]/48$$
$$= 27{\cdot}81/48$$
$$= 0{\cdot}58.$$

The random frequency distribution

To obtain the theoretical random frequency distribution which will be compared with the observed frequency distribution, we use the properties of the Poisson distribution[1] described in Chapter 7, pp225–227. It may be recalled that the Poisson distribution predicts the probability, P_x, that a single specified cell contains exactly x points at the termination of an experiment where r points are placed independently, and one at a time, into a grid of n equal-sized cells and where each placement has a $p = 1/n$ probability of landing in the specified cell. The formula is given by

$$P_x = \frac{e^{-\lambda} \cdot \lambda^x}{x!} \qquad x = 0, 1, \ldots \tag{8.4}$$

The calculation of the Poisson probabilities requires an estimate of λ, which is the mean of the Poisson's random variables x. In practice, λ is estimated by the value of $\bar{x}$, the observed mean number of points per cell in the census. For the factory problem this estimation procedure gives the result

$$\lambda = \bar{x} \quad = 0{\cdot}48. \tag{8.5}$$

Therefore, the probability that any specified cell in the factory census contains $x = 0$ factories at the termination of the theoretical process of 48 factories being placed independently into the census, is obtained from formula (8.4) as

$$P_{x=0} = \frac{e^{-0{\cdot}44}0{\cdot}44^0}{0!} = 2{\cdot}71828^{-0{\cdot}44} = 0{\cdot}644.$$

The remaining Poisson probabilities for $x = 1, 2$, and 3 are given in table 8.2. Because the Poisson model assumes that all cells are equally likely as a location for a single random placement, we can convert the values P_x into frequencies by multiplying these

Table 8.2. Vehicle factory frequency distributions.

x	Poisson P_x†	$(P_x)_n$	Observed n_x
0	0·644	30·9	34
1	0·283	13·6	8
2	0·062	3·0	5
3	0·009	0·4	1
$\geqslant 4$		0·1‡	

†Usually the Poisson probabilities (P_x) become extremely small as x increases in value and in practice need not be calculated for values of x where $(P_x)\, n < 0·5$.
‡This value may be obtained quickly using the approximation:

$$n - \sum_{x=0}^{k} (P_x)\, n = 48 - 47·9 = 0·1.$$

probabilities by n (the total number of cells). Therefore, the values $(P_x)n$ give the expected number of cells containing x points in a random pattern. For example, the result (see table 8.2)

$$(P_{x=0})n = 0·644 \times 48 = 30·9$$

indicates that we would expect random factory frequency distribution to contain 30·9 empty cells.

Goodness-of-fit

Recall that goodness-of-fit tests are statistical procedures for deciding whether a model prediction gives a close enough representation of the observed data for the assumptions of the model to be accepted as an adequate explanation for the processes controlling the system. Remember that, in Chapter 7, we used the χ^2 test to establish the goodness-of-fit between observed and expected contingency tables. To decide whether the Poisson prediction gives a close enought fit to the observed frequency distribution (table 8.2) for us to infer an independent spatial process the most appropriate goodness-of-fit statistic[2] is based on the ratio between the variance and mean of the Poisson's random variable x.

For the observed frequency distribution the variance/mean ratio is defined by

$$S_x^2/\bar{x} = \text{formula (8.3)/formula (8.1)} \tag{8.6}$$

which, for the vehicle distribution, gives the result

$$0·58/0·44 = 1·32.$$

The ratio calculated from formula (8.6) is used to construct a goodness-of-fit test for assessing whether the observed frequency distribution was generated by an independent spatial process. The test is derived from the theoretical property of the Poisson distribution that the mean, $\mu(x)$ of its random variable is always equal to the variance, σ_x^2 of the random variable. Therefore, if a perfect correspondence existed between the observed frequences, n_x, and the predicted Poisson frequencies, $(P_x)n$, the value of the observed variance/mean ratio (from formula 8.6) would be equal to one and we should infer the operation of an independent spatial process. Alternatively, for problems where the fit between the two frequency distributions was poor, the value of the observed variance/mean ratio would tend to zero as observed patterns tend to the triangular uniform distribution (figure 8.1*f*), and would tend to infinity as observed patterns tend to total clustering (figure 8.1*a*). These three statements are written symbolically as

$S^2/\bar{x} = 1{\cdot}0$: pattern random

$S^2/\bar{x} \rightarrow 0{\cdot}0$: pattern uniform

$S^2/\bar{x} \rightarrow \infty$: pattern clustered.

Clearly, the statistical problem here is to decide whether the observed variance/mean ratio is significantly different from one for the experimenter to conclude that the differences are due to either a competitive or contagious spatial process. Therefore, our research hypothesis is written as

$$H_1 : S_x^2/\bar{x} \neq 1{\cdot}0 \qquad\qquad (8.7)$$

and states that we are hoping to infer that the pattern is controlled by either a competitive or contagious spatial process. The research hypothesis predetermines the null hypothesis as

$$H_0 : S_x^2/\bar{x} = 1{\cdot}0, \qquad \alpha = 0{\cdot}05 \qquad\qquad (8.8)$$

which reads, 'there is no significant difference (=) between the observed and predicted random frequency distributions other than that which could be due to random sampling error[3] at the $\alpha = 0{\cdot}05$ significance level.' It can be proved mathematically that the sampling distribution of $S_x^2/\bar{x}$ around the Poisson prediction of one is approximated by a continuous probability density function known as the Student's t distribution[4]. The difference between the observed variance/mean ratio and the Poisson prediction of one is converted into a calculated value of t using the transformation

$$\text{calc } t_{(n-1,\,\text{df})} = \frac{(S_x^2/\bar{x}) - 1{\cdot}0}{\sqrt{2/(n-1)}}, \qquad\qquad (8.9)$$

which, for the vehicle factory problem, gives the result

$$\text{calc } t_{(48-1,\,\text{df})} = \frac{1{\cdot}32 - 1{\cdot}0}{\sqrt{2/(48-1)}}$$

$$\text{calc } t_{(47,\,\text{df})} = 1{\cdot}55.$$

For the vehicle factory problem, the critical value of t with 47 degrees of freedom which encloses 0·95 of the area under the sampling distribution of the variance/mean ratio around the Poisson prediction of one, is obtained from two-tailed t-tables (see

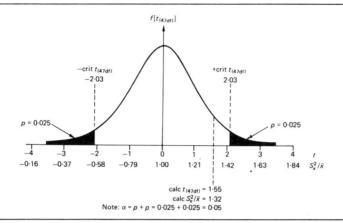

Figure 8.3. The *t*-test for vehicle factory variance/mean ratio.

Appendix E, p320) as, crit $t_{(47 \, df)}$ = ±2·03. The term *two-tailed* refers to the fact that in the research hypothesis (formula 8.7) we stated that we were interested in calculated values of the variance/mean ratio that were *either* significantly greater than one, *or* significantly less than one. Therefore, because the *t*-distribution is symmetrical around the value $t = 0$ $(S_x^2/\bar{x} = 1\cdot0)$, we have to select a critical value of t that leaves $p = 0\cdot025$ of the sampling distribution exposed beyond its negative value (−2·03) and $p = 0\cdot025$ exposed beyond its positive value (+2·03) in accordance with our significance level of $\alpha = 0\cdot05$ (figure 8.3). Thus two-tailed means that both tails of the sampling distribution are involved in the test decision, unlike the χ^2 test (p239) where only one tail of the sampling distribution was involved in the test decision.

Table 8.3 summarizes the result of the *t*-test on the vehicle factory variance/mean ratio. Because the calculated value of t is less than the critical value we accept the null hypothesis and conclude that at the $\alpha = 0\cdot05$ significance level it is likely that the difference between the observed variance/mean ratio of 1·32 and the Poisson prediction of one is due to random sampling error. Since there is no significant difference between the observed frequency distribution and the Poisson frequencies, we infer that the factory point pattern is the result of an independent spatial process. This interpretation makes good geographical sense because vehicle factories belong to large independent organizations which are unlikely to exhibit strong locational preferences for factory sites at the intra-urban scale. The vehicle factory location

Table 8.3. *t*-test for vehicle factory variance/mean ratio.

$H_1: S_x^2/\bar{x} \neq 1\cdot0$
$H_0: S_x^2/\bar{x} = 1\cdot0, \quad \alpha = 0\cdot05$
crit $t_{(47 \, df)}$ = ±2·03
calc $t_{(47 \, df)}$ = +1·55
Decision: Accept H_0, reject H_1.

decision is predominantly of the form 'in which town do we locate new plant', not 'in which part of a particular town.'

In other problems, where the calculated value of *t* exceeded the critical value of *t* in either a positive or negative direction, we should reject the null hypothesis and accept the research hypothesis. In such cases we should infer either a clustered or uniform pattern and then examine geographically the case for either a contagious or competitive process as a control upon the point locations.

The scale problem: Grieg-Smith's hierarchial analysis of variance

One of the major technical problems affecting the interpretation of quadrat experiments is the influence of cell size on the analysis of data. The selection of a cell size for either quadrat sampling or quadrat censusing is a subjective decision made by the experimenter, and it is often true that the *scale* of analysis, which is determined by the cell size, may influence the subsequent interpretation of results. For obvious reasons the influence of cell size upon the analysis is known as the *scale problem*.

The point pattern illustrated in figure 8.4 is a hypothetical distribution of a plant species in a square field, and is designed to illustrate the scale problem. If this pattern is subjected to a 4 x 4 quadrat census (figure 8.4a) and a Poisson frequency analysis, we come to the conclusion that the variance/mean ratio of 3 is indicative of clustering. We

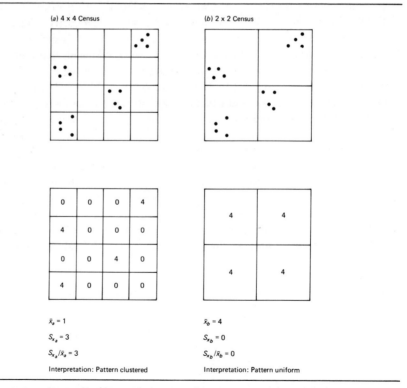

Figure 8.4.   Hypothetical censuses of a plant community in a square field.

might then suggest that the observed clustering was the result of contagion due to one plant propagating the three other plants in each of the four-point clusters. But frequency analysis of the *same* pattern by a 2 x 2 census leads to a variance/mean ratio of 0 and the inference of a uniform pattern controlled by a competitive process. At this larger scale our interpretation would probably be that each cell can only support four plants, and, once this limit is exceeded, subsequent plants die out. Therefore, at the 2 x 2 level we find spatial competition between the clusters of plants. Neither of these interpretations is incorrect; they differ simply because different spatial processes are dominant at the two scales of analysis.

The lesson to be learned from our hypothetical plant species is that, in any quadrat experiment the pattern must be analysed at a variety of different scales before any process interpretation is even attempted. A most ingenious treatment of scale effects is the hierarchical analysis of variance for scale effects devised by Greig-Smith (1952). The term *analysis of variance* refers to a whole range of statistical techniques which are founded on a body of statistical probability theory developed by R A Fisher. The Greig-Smith treatment of scale effects in quadrat experiments is a useful illustration of one of the simpler analyses of variance designs.

Analyses of variance techniques are methods by which the total variation in a set of data is separated into components associated with possible sources of variability whose relative importance we wish to assess. The Greig-Smith scheme involves assigning the variation among the observations, x_i, in a square quadrat census to a variety of scales (sources of variability) and then testing the variation due to each scale (cell size) for randomness. The whole scheme is based on the property of the Poisson distribution that its mean equals its variance. Figure 8.5(b) presents a hypothetical quadrat census of a plant community which we shall use to illustrate the Greig-Smith scheme. The test is always based on a square census where the number of cells on each axis is some power of 2, and the example shown in figure 8.5(b) is a 4 x 4 census (2^2). The number of points located within each cell of the basic 4 x 4 census are denoted by the symbol $x_j(i)$. Moreover, by grouping pairs of adjacent cells into blocks (figure 8.5b) and summing the observations in the adjacent cells, we obtain a second census for a scale which contains eight blocks. The subscript j refers to the number of basic cells contained in a block and i identifies the ith block (figure 8.5a). For example, $x_{j=1}(i = 11)$ identifies the eleventh cell in the basic 4 x 4 census, where $j = 1$ indicates one basic cell per block. The value of $x_1(11) = 3$ represents the three points located within the eleventh cell. Similarly, $x_2(6) = 5$ refers to the five points located in the sixth block of the $j = 2$ census where two cells are aggregated to form each block. The arithmetic of the aggregation procedure can be defined by the following equation

$$x_j(i) = x_{j/2}(2i - 1) + x_{j/2}(2i) \tag{8.10}$$

which, for the sixth cell in the $j = 2$ census, gives

$$x_2(6) = x_{2/2}(12 - 1) + x_{2/2}(12)$$
$$= x_1(11) + x_1(12)$$
$$= 3 + 2 = 5$$

Obviously, formula (8.10) does not apply to the cells of the $j = 1$ census. In a 4 x 4

(a) Subscripts ij **(b) Observations** $x_j(i)$ **(c) Cell means (scale j)** $x_j(i)/j$ **(d) Cell means (scale $2j$)** $x_{2j}(i)/2j$ **(e) Squared Deviations** **(f) Sums of squares** SS_j

$j = 1$

(a) Subscripts ij

1	3	5	7
2	4	6	8
9	11	13	15
10	12	14	16

(b) Observations $x_1(i)$

0	0	2	0
0	1	0	1
0	3	2	0
2	2	1	2

(c) $x_1(i)/1$

0·00	0·00	2·00	0·00
0·00	1·00	0·00	1·00
0·00	3·00	2·00	0·00
2·00	2·00	1·00	2·00

(d) $x_2(i)/2$

0·00	0·50	1·00	0·50
0·00	0·50	1·00	0·50
1·00	2·50	1·50	1·00
1·00	2·50	1·50	1·00

(e) $[x_1(i)/1 - x_2(i)/2]^2$

0·00	0·25	1·00	0·25
0·00	0·25	1·00	0·25
1·00	0·25	0·25	1·00
1·00	0·25	0·25	1·00

(f) $SS_1 = 8$

$j = 2$

(a) Subscripts ij

1	2	3	4
5	6	7	8

(b) Observations $x_2(i)$

0	1	2	1
2	5	3	2

(c) $x_2(i)/2$

0·00	0·50	1·00	0·50
0·00	0·50	1·00	0·50
1·00	2·50	1·50	1·00
1·00	2·50	1·50	1·00

(d) $x_4(i)/4$

0·25	0·25	0·75	0·75
0·25	0·25	0·75	0·75
1·75	1·75	1·25	1·25
1·75	1·75	1·25	1·25

(e) $[x_2(i)/2 - x_4(i)/4]^2$

0·0625	0·0625	0·0625	0·0625
0·0625	0·0625	0·0625	0·0625
0·5625	0·5625	0·0625	0·0625
0·5625	0·5625	0·0625	0·0625

(f) $SS_2 = 3$

$j = 4$

(a) Subscripts ij

1	2
3	4

(b) Observations $x_4(i)$

1	3
7	5

(c) $x_4(i)/4$

0·25	0·25	0·75	0·75
0·25	0·25	0·75	0·75
1·75	1·75	1·25	1·25
1·75	1·75	1·25	1·25

(d) $x_8(i)/8$

0·50	0·50	0·50	0·50
0·50	0·50	0·50	0·50
1·50	1·50	1·50	1·50
1·50	1·50	1·50	1·50

(e) $[x_4(i)/4 - x_8(i)/8]^2$

0·0625	0·0625	0·0625	0·0625
0·0625	0·0625	0·0625	0·0625
0·0625	0·0625	0·0625	0·0625
0·0625	0·0625	0·0625	0·0625

(f) $SS_4 = 1$

$j = 8$

(a) Subscripts ij

1
2

(b) Observations $x_8(i)$

4
12

(c) $x_8(i)/8$

0·50	0·50	0·50	0·50
0·50	0·50	0·50	0·50
1·50	1·50	1·50	1·50
1·50	1·50	1·50	1·50

(d) $x_{16}(i)/16$

1·00	1·00	1·00	1·00
1·00	1·00	1·00	1·00
1·00	1·00	1·00	1·00
1·00	1·00	1·00	1·00

(e) $[x_8(i)/8 - x_{16}(i)/16]^2$

0·25	0·25	0·25	0·25
0·25	0·25	0·25	0·25
0·25	0·25	0·25	0·25
0·25	0·25	0·25	0·25

(f) $SS_8 = 4$

basic census this method of aggregation can be continued to produce a census containing four blocks ($j = 4$ cells per block), a two-block census ($j = 8$ cells per block) and finally, a single-block census which in this example encloses all 16 points ($j = 16$ cells per block), $x_{16}(1) = 16$, which is not illustrated in figure 8.5(b).

Let n equal the number of cells in the basic census (16); then the *total* variation (or sums of squares) in the basic census is obtained from

$$TSS = \sum_{i=1}^{n} (x_1(i) - \bar{x}_1)^2 \qquad (8.11)$$

where

$$\bar{x}_1 = \sum_{i=1}^{n} x_1(i)/n. \qquad (8.12)$$

The total sums of squares is simply the variance of the basic census (formula 3.2) multiplied by n, and is obtained from the sum of the squared differences (deviations) between the observed number of points per cell, $x_1(i)$, and the mean number of points per cell for the whole census, $\bar{x}_1$. For the example census, where $\bar{x}_1 = 16/16 = 1$, formula (8.11) gives the total sums of squares as

$$TSS = (x_1(1) - \bar{x}_1)^2 + (x_1(2) - \bar{x}_1)^2 + \ldots, + (x_1(16) - \bar{x}_1)^2$$
$$= (0 - 1)^2 + (0 - 1)^2 + \ldots, + (2 - 1)^2$$
$$= 16.$$

The variance of the basic census is $16/16 = 1$. Therefore, the variance/mean ratio for the basic census is $S_x^2/\bar{x}_1 = 1/1 = 1$, indicating that the frequency distribution of the pattern is random at this scale.

The idea of the Greig-Smith scheme is that we split the total sums of squares into components which measure the proportion of the variation that is caused by each scale in the census. For any scale j, we calculate the proportion of the total sums of squares, SS_j, that is accounted for by differences between the observations $x_j(i)$ from the expression

$$SS_j = \sum_{i=1}^{n/2j} 2j[x_j(2i - 1)/j - x_{2j}(i)/2j]^2. \qquad (8.13)$$

For the scale $j = 1$ (the cells of the basic 4 x 4 census) in the example illustrated in figure 8.5, formula (8.13) gives the result

$$SS_1 = \sum_{i=1}^{8} 2[x_1(2_{i-1})/1 - x_2(i)/2]^2$$

$$= 2[x_1(1)/1 - x_2(1)/2]^2 + 2[x_1(3)/1 - x_2(2)/2]^2$$
$$+ 2[x_1(5)/1 - x_2(3)/2]^2 + 2[x_1(7)/1 - x_2(4)/2]^2$$
$$+ 2[x_1(9)/1 - x_2(5)/2]^2 + 2[x_1(11)/1 - x_2(6)/2]^2$$
$$+ 2[x_1(13)/1 - x_2(7)/2]^2 + 2[x_1(15)/1 - x_2(8)/2]^2$$
$$= 2[0\cdot00 - 0\cdot00]^2 + 2[0\cdot00 - 0\cdot50]^2 + 2[2\cdot00 - 1\cdot00]^2$$
$$+ 2[0\cdot00 - 0\cdot50]^2 + 2[0\cdot00 - 1\cdot00]^2 + 2[3\cdot00 - 2\cdot50]^2$$
$$+ 2[2\cdot00 - 1\cdot50]^2 + 2[0\cdot00 - 1\cdot00]^2$$
$$= 8.$$

Careful inspection of this arithmetic will show that each of the eight individual calculations in the summation $\Sigma_{i=1}^{8}$ represent the sum of the two squared differences between a pair of cell observations at the $j = 1$ scale and the mean number of points per cell, $x_1(i)/2$, obtained from the block the pair of cells form at the $j = 2$ scale. The summation of these eight terms gives the sums of squares (or variation) between cells (sixteenths) within pairs of cells (eighths). The procedure for calculating the variation *between* observations at the jth scale within observations at $(j + 1)$th scale will be fully understood if the reader works through the full set of calculations listed in figure 8.5 (c–f). In a 4 x 4 census we assign the total sums of squares to four sources (scales): the variation between sixteenths within eighths, the variations between eighths within quarters, the variation between eighths within quarters, the variation between quarters within halves, and the variation between halves. The values of these sums of squares for the example data are listed in figure 8.5(f) and table 8.4.

The calculation of the terms SS_j are made with respect to a given number of degrees of freedom. In analyses of variance experiments the degrees of freedom are related to the number of observations used in the calculation of a particular sum of squares and therefore, as usual in statistical tests, degrees of freedom vary with the size of the experiment. For the sums of squares associated with any scale j the number of degrees of freedom are given by

$$\text{degrees of freedom at scale } j = n/2j \qquad (8.14)$$

and these values for the example data are listed in table 8.4. The sums of squares is converted into a variance estimate, ms_j, of the amount of variance associated with a particular scale by dividing its value by the number of degrees of freedom:

$$ms_j = 2j\,SS_j/n. \qquad (8.15)$$

Again, the variance estimates for the example problem are given in table 8.4.

Table 8.4. Variance table for census given in figure 8.5(b).

Source of variation, j	Sums of squares (SS_j)	Degrees of freedom ($n/2j$)	Variance estimate (ms_j)	Variance ratios (F_j)
Between halves ($j = 8$)	4	1	4·00	4·00
Between quarters within halves ($j = 4$)	1	2	0·50	0·50
Between eighths within quarters ($j = 2$)	3	4	0·75	0·75
Between cells within eighths ($j = 1$)	8	8	1·00	1·00
Total	16	–	–	–

The simplest way of interpreting the Greig-Smith analysis is to calculate variance ratios, F, for each scale from the formula

$$F_j = ms_j/ms_1. \tag{8.16}$$

The four F ratios from the example census are listed in the final column of table 8.4. Because we know from the variance/mean ratio that the frequency distribution of the 16 cells in the $j = 1$ census is random, the variance estimate $ms_1 = 1$ must also be random variance associated with the cells of the $j = 1$ census varying within the blocks of the $j = 2$ census. When ms_1 is known to be random variance, the ratios F_j may be interpreted in the same way as variance/mean ratios: when F_j tends to one, this indicates that the observations at the jth scale are varying randomly within the blocks of the $(j + 1)$th scale; when F_j tends to infinity the observations at the jth scale are clustering within the blocks of the $(j + 1)$th scale; and when F_j tends to zero, the observations at the jth scale are uniformly distributed within the blocks of the $(j + 1)$th scale. Figure 8.6 shows that for the example data the major clustering takes place between halves ($j = 8$), although at the $j = 2$ and $j = 4$ scales the observations are tending to be evenly distributed within the blocks of the $j = 4$ and $j = 8$ scales, respectively.

Two major problems are associated with the Greig-Smith scheme. Firstly, the values $x_j(i)$ will vary according to the direction in which cells are aggregated into blocks. This aggregation effect will not be particularly serious for scales containing a large number of blocks, but the effect can be quite marked for the higher scales which contain only two or four blocks[5]. In practice, it is usual to apply the Greig-Smith scheme to basic censuses containing at least 256 (16^2) cells and, therefore, aggregation effects are only likely to influence higher scales. However, to avoid erroneous interpretation, it is sensible in practice to conduct a variety of analyses of the same basic census using

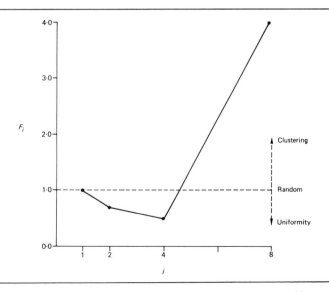

Figure 8.6.   Graph of scale effects for the hypothetical census (figure 8.5*b*).

alternative aggregation schemes. Secondly, although statistical significance tests for the F_j ratios produced by the Greig-Smith scheme do exist, they are beset by numerous technical problems which restrict their general applicability. The interested reader is referred to papers by Mead (1974) and Zahl (1974) for discussions of these tests.

Alternative approaches

So far we have only considered methods of determining observed departures from randomness. However, probability models can be derived from assumptions about non-random point processes. One of the most widely applied models is the negative binomial distribution[6] which is based on the assumption that the probability a point is located in the specified cell increases linearly with the number of points previously placed in the cell. This assumption leads to the prediction of a clustered frequency array and the negative binomial has been fitted to a number of point patterns including the distributions of farmers adopting agricultural innovations (Harvey 1966) and the distribution of shops in urban areas (Rogers 1965).

Another feature of quadrat analysis is that clustered frequency distributions are much more prevalent that either random or uniform distributions. A possible explanation for this occurrence has been put forward by Thomas and Reeve (1976). If the principle of entropy-maximization (Chapter 5, p155) is used to define the most likely frequency array, the prediction is a clustered array. Indeed, quite different interpretations are given to point patterns if we adopt entropy-maximizing procedures as our interpretive yardstick in place spatial independence. Further developments on this topic are reported in papers by Webber (1976), and Liebetrau and Karr (1977).

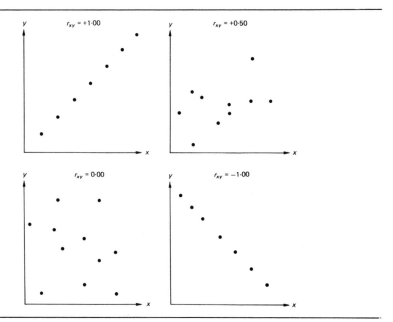

Figure 8.7.   Scatter diagrams illustrating some example correlations, r_{xy}.

SPATIAL CORRELATION ANALYSIS

Correlation coefficients are a deceptively simple methods of statistical analysis which measure the degree of association between two variables. Geographical journals abound with applications of correlation analysis which, recently, have ranged from the study of relationships between soil characteristics in Zambia (King 1975) to the analysis of social relationships between house type and tenure in Exeter (Morgan 1976). Correlation coefficients measure the extent to which two variables, x and y, tend towards a *linear* relationship. Correlation coefficients, r_{xy}, have the property that their values range from +1·00 when there is a perfect positive correlation between the two variables (one variable increasing in value with the value of the second variable) to −1·00 when there is a perfect negative correlation (one variable increasing in value in exact linear proportions with decreases in the value of the second variable). Between these limits lie the imperfect relationships of correlation analysis which will tend to a value of $r_{xy} = 0·0$ when no relationship exists between the variables. Illustrative examples of all these relationships are shown in figure 8.7.

Calculations for Pearson's product—moment coefficient

The formula for Pearson's correlation coefficient is the most general expression for the degree of association between two variables. The data used in correlation analysis are always a set of *paired* observations which consist of either, a random sample, n, or a population (N) of objects, where observations on two variables, x and y, have been taken on each object. For example, the data listed in table 8.5 are a random sample of $n = 4$ engineering companies in Manchester where for each company we have collected x — the number of employees, and y — the value of the company output in millions of £ per annum.

Table 8.5. Calculation of the covariance for the companies problem.

$\bar{x}$ = mean employment = 5·000					
$\bar{y}$ = mean output = 6·000					
$S_x = 2·236$					
$S_y = 2·916$					
$S_{xy} = 26/4 = 6·5$					

Companies i	Employment (000s) x	Output y	$x_i - \bar{x}$	$y_i - \bar{y}$	$(x_i - \bar{x}) \cdot (y_i - \bar{y})$
1	2	2	−3	−4	12
2	4	5	−1	−1	1
3	6	7	1	1	1
4	8	10	3	4	12
Totals, $n = 4$					26

Pearson's formula[7] is the ratio of the covariance of the two variables and the total amount of variance in both variables, and is given by

$$r_{xy} = \frac{S_{xy}}{S_x S_y},$$ (8.17)

where S_{xy} is the covariance of x and y defined by

$$S_{xy} = \frac{1}{n} \sum_{i=1}^{n} (x_i - x)(y_i - y),$$ (8.18)

where n = number of pairs of observations and, as usual, S_x and S_y are the standard deviations of x and y.

$$S_x = \sqrt{\sum_{i=1}^{n} (x_i - \bar{x})^2 / n}, \qquad S_y = \sqrt{\sum_{i=1}^{n} (y_i - \bar{y})^2 / n}.$$ (8.19)

The calculation of the covariance for the companies problem is demonstrated in table 8.5. In theory the value of the covariance can vary between $\pm\infty$; in practice, its actual value will depend on two factors. Firstly, the strength of the relationship (or tendency towards linearity) between the variables, which causes the covariance to be large and positive if the points within quadrants II and III of figure 8.8 are tending to linearity, and to be large and negative if the points tend to negative linearity in quadrants I and IV. Secondly, the value of the covariance is influenced by the range of value of both the x and y observations – the larger the range the greater the value of the covariance. For this second reason the covariance is converted into a standard comparable measure by dividing its value by the product of the two standard deviations $(S_x.S_y)$ which measures the total variance present in both variables. Therefore, for the companies problem, the correlation coefficient is calculated from formula (8.17) as

$$r_{xy} = \frac{6 \cdot 500}{2 \cdot 236 \times 2 \cdot 916} = +0 \cdot 997,$$

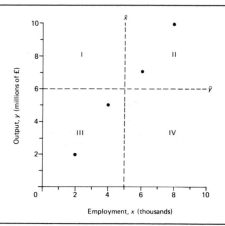

Figure 8.8.   Scatter diagram for the companies problem.

which is indicative of a strong posivitive relationship between numbers employed and the value of a company's output.

Significance tests for correlation

The simplest research hypothesis in correlation analysis is that we expect either a significant positive or negative correlation between the variables[9], which is written symbolically as

$$H_1 : r_{xy} \neq 0.0. \tag{8.20}$$

The null hypothesis states that there is no significant difference between the observed correlation and zero, and is written as

$$H_0 ; r_{xy} = 0.0, \qquad \alpha = 0.05.$$

If a number of data requirements are met, and the observations belong to a random sample where n is reasonably large, then it can be proved that the sampling distribution of the correlation coefficient around the expected random value of zero is approximated by the Student's t distribution. The observed correlation coefficient is converted into a calculated t value using the transformation

$$\text{calc } t_{(n-2, \text{df})} = \frac{r_{xy}\sqrt{n-2}}{\sqrt{1 - r_{xy}^2}}. \tag{8.21}$$

Although the observed correlation of $r_{xy} = +0.997$ in the companies problem does not conform to the assumptions of the test, we will use its value in the test to illustrate the statistical reasoning. Substituting in formula (8.21) gives

$$\text{calc } t_{(2\text{df})} = \frac{+0.997\sqrt{2}}{\sqrt{1 - 0.997^2}}$$

$$= +18.216.$$

The critical value of t with 2 degrees of freedom at the $\alpha = 0.05$ significance level is obtained from the two-tailed t-table (Appendix E, p320) as ± 4.303. For this problem, because the calculated value of t exceeds the critical value, we reject the null hypothesis and accept the research hypothesis that there is a statistically significant relationship between employment and output.

Unfortunately, the data restrictions on this significance test are so severe that the test is rarely appropriate for geographical correlation experiments. To be reasonably large, the random sample size must be at least $n \leqslant 30$, and more restrictive is the requirement that the histograms of the frequency distributions of both the x and y variables must be approximated by the normal curve.

The coefficient of determination

Because the t-test for correlation coefficient is usually inappropriate in geography, the geographer is often forced to take a qualitative approach to the interpretation of correlation which makes use of an index known as the coefficient of determination, R^2, which is defined by the formula

$$R^2 = 100r_{xy}^2. \tag{8.22}$$

For the companies problem, the value of the coefficient of determination is obtained as

$$R^2 = 100(0.997^2) = 99.40,$$

which is interpreted as 99·40 per cent of the variation in the values of x (employment) that is explained (or paralleled) by the variation in y (output) or vice versa. The 'vice versa' indicates that in correlation analysis we do not assume that one variable *causes* variation in the second variable, we merely analyse the strength of the association between the two.

A word of caution is necessary about the interpretation of coefficients of determination. Research workers who apply simple correlation techniques are, at the conclusion of the analysis, apt to declare gleefully that a correlation coefficient of say +0·8 has given a 64 per cent explanation of their problem. However, the coefficient of determination is a susceptible statistic which often gives a grossly misleading picture of a relationship. Some of the more obvious spurious relationships are illustrated in figure 8.9. In the example shown in figure 8.9(a) there is no correlation between the main group of points, and the linearity is due solely to the position of the rogue point with exceptionally large values of x and y. Similarly, the value of $R^2 = 80$ per cent in figure

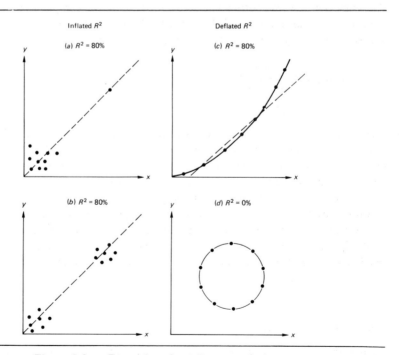

Figure 8.9.   Examples of spurious correlations.

8.9(*b*) also gives an exaggerated impression of the actual relationship. The points belonging to each of the two groups are uncorrelated within the group and therefore in figure 8.9(*b*) we are looking at two different statistical populations. Figures 8.9(*c–d*) illustrate deterministic relationships where there are a perfect nonlinear relationships between the points but, because correlation analysis only tests for linearity, the value of R^2 under-estimates the true 100 per cent nonlinear correlation. The more obvious cases of spurious correlation can easily be detected by careful examination of the scatter diagram.

Scale effects and ecological correlations

One of the most illuminating published debates about the interpretation of correlation coefficients is the analysis of the relationship between race and illiteracy in the USA in 1930. The original research on this subject appears in a remarkable paper by Robinson (1950) and this work has more recently been extended and clarified by Alker (1969). The relationship has obvious political importance because, if a positive correlation between blacks and illiteracy is established statistically, it provides ammunition for segregationalist political policies. Historically, it is also important to remember that the analysis is based on 1930 US Census data. The geographical and statistical arguments surrounding this problem will become clearer as we unravel the relationship.

The first relationship Robinson analysed was the individual correlation between race ($x = 0$ if white, $x = 1$ if black) and illiteracy ($y = 0$ if literate, $y = 1$ if illiterate) for the $N = 97\ 272\ 000$ persons in the USA, 1930 over 10 years old (table 8.6). We define the variables x_{ij} and y_{ij} as the paired observations on the ith individual person who is resident in the jth US Census division (figure 8.10). For example $x_{6\ 702\ 001,2} = 1$, $y_{6\ 702\ 002,2} = 1$ (table 8.6) is the 6 702 002nd individual in the Census list who is a black illiterate living in the Middle Atlantic Census division. The individual correlation is calculated, irrespective of Census division, j, from the standard correlation formula (8.17) which in the present notation is

$$r_{xy} = \frac{\sigma_{xy}}{\sigma_x \sigma_y}, \tag{8.23}$$

where the covariance is defined as

$$\sigma_{xy} = \frac{1}{N} \sum_{i=1}^{N} [x_{ij} - \bar{x}] [y_{ij} - \bar{y}]. \tag{8.24}$$

Entering the whole list of $N = 97\ 272\ 000$ paired observations of zeros and ones into formulae (8.23 and 8.24), Robinson obtained the individual correlation[10] between race and illiteracy as +0·203. The corresponding value of $R^2 = 4$ per cent indicates that only 4 per cent of the variation in race is explained by variation in illiteracy. Such a low level of explanation does not provide clear evidence for segregationalist arguments.

The predominant aim in Robinson's paper was to illustrate the fallacious arguments that can be developed from correlations which are based on regions as the object of analysis rather than individuals. When regions are the objects of correlation analysis the resulting coefficients are termed *ecological* or *spatial correlations*. The second correlation calculated by Robinson was the ecological correlation between race and

264 *Probability Models*

Table 8.6. A summary of Robinson's (1950) data. Source: W S Robinson (1950), tables 1 and 2, p353.

Regional subscripts	Individual subscripts	Race Black, $x = 1$ White, $x = 0$	Illiteracy Illiterate, $y = 1$ Literate, $y = 0$
$j = 1, 2, \ldots, J$	$i = 1, 2, \ldots, n_1, \ldots,$ $n_2, \ldots, N$	x_{ij}	y_{ij}
New England $(j = 1)$	1 $\quad(i \in j)$ 2 $\vdots$ 6 702 000 $(n_{j=1})$	0 0 $\vdots$ 1	1 0 $\vdots$ 0
Regional totals, $x_{(j=1)}, y_{(j=1)}$		76 000	244 000
Middle Atlantic $(j = 2)$	6 702 001 $(i \in 2)$ 6 702 002 $\vdots$ 28 247 000 (n_2)	0 1 $\vdots$ 0	0 1 $\vdots$ 0
Regional totals, x_2, y_2		868 000	751 000
$\ldots j = 3, 4, \ldots, 8$	$\vdots$	$\vdots$	$\vdots$
Pacific $(J = 9)$	90 792 000 $(i \in 9)$ 90 792 001 $\vdots$ 97 272 000 (N)	0 0 $\vdots$ 0	0 0 $\vdots$ 0
Regional totals x_9, y_9		77 000	73 000
Grand totals $\sum_{i=1}^{N} x_{ij}, \sum_{i=1}^{N} y_{ij}$		9 292 000	3 918 000

Figure 8.10.   Major census divisions of the USA.

illiteracy based on the nine major Census divisions in the USA (figure 8.10). To obtain the nine paired observations for the ecological correlations we simply aggregate the individual observation x_{ij} over each region j. The total number of blacks in the jth region, x_j, and the total number of illiterates in the jth region, y_j, are obtained from the following aggregation formulae

$$x_j = \sum_{i \in j}^{n_j} x_{ij}, \qquad y_j = \sum_{i \in j}^{n_j} y_{ij}, \qquad (8.25)$$

where $i \in j$ defines the value of the subscript i that identifies the first individual in the set of individuals living in the jth region, and n_j defines the subscript value i for the last individual in the set belonging to the jth region. For example, the summations given in formula (8.25) take the following values for the Middle Atlantic region ($j = 2$, table 8.6):

$$x_2 = \sum_{6\,702\,001}^{28\,247\,000} x_{i,2}$$

$$= x_{6\,702\,001,2} + x_{6\,702\,002,2} +, \dots, + x_{28\,247\,000,2}$$

$$= \quad 0 \quad + \quad 1 \quad +, \dots, + \quad 0$$

$$= 868\,000 \text{ blacks in Middle Atlantic; and,}$$

$$y_2 = y_{6\,702\,001,2} + y_{6\,702\,002,2} +, \dots, + y_{28\,247\,000,2}$$

$$= \quad 0 \quad + \quad 1 \quad +, \dots, + \quad 0$$

$$= 751\,000 \text{ illiterates in Middle Atlantic.}$$

Obviously, we have not listed all the $x_{i,2}$ observations for reasons of space.

If J denotes the total number of regions in the aggregation (in this problem $J = 9$), then the ecological correlation coefficient is defined by

$$r_{x_J y_J} = \frac{\sigma_{xy_J}}{\sigma_{x_J}\sigma_{y_J}}, \tag{8.26}$$

where the ecological covariance is

$$\sigma_{x_J y_J} = \frac{1}{J} \sum_{j=i}^{J} (x_j - \bar{x}_J)(y_j - \bar{y}_J), \tag{8.27}$$

the ecological mean is

$$\bar{x}_J = \sum_{j=i}^{J} x_j / J, \tag{8.28}$$

and the ecological standard deviation is

$$\sigma_{x_J} = \sqrt{\sum_{j=1}^{J} (x_j - \bar{x}_J)^2 / J}. \tag{8.29}$$

For the relationship between race and illiteracy the ecological correlation obtained from formulae (8.26–8.29) is +0·946, which gives a corresponding R^2 value of 89 per cent. For the segregationalist this is an impressive result. Indeed, Robinson shows that, when the ecological unit is changed from the Census division to the more detailed spatial level of the State, the relationship persists with an ecological correlation of +0·773.

To understand how the individual and ecological correlations give such vastly different results we must consider the *within-region correlation* which provides the final link in the interpretation of aggregation effects on correlation coefficients. The within-region correlation coefficient is the individual correlation between race and illiteracy for the people residing in the jth division. Thus the present problem produces nine within-region coefficients, one for each division. Using the notation we have established for individual and ecological correlations, the within-region correlation coefficient for the jth region is obtained from

$$r_{x_j y_j} = \frac{\sigma_{x_j y_j}}{\sigma_{x_j}\sigma_{y_j}}. \tag{8.30}$$

If $m_j = n_j - n_{(j-1)}$, then the m_j is the number of people residing in the jth region. For example, from table 8.6 the number of people living in Middle Atlantic ($j = 2$) are $m_2 = n_2 - n_1$, $m_2 = 28\ 247\ 000 - 6\ 702\ 000 = 21\ 545\ 000$. We can now calculate the within-region mean from

$$\bar{x}_j = \sum_{i \in j}^{n_j} x_{ij} / m_j \tag{8.31}$$

the within-region standard deviation from

$$\sigma_{x_j} = \sqrt{\sum_{i \in j}^{n_j} [x_{ij} - \bar{x}_j]^2 / m_j}, \tag{8.32}$$

and the within region covariance for the jth region from

$$\sigma_{x_j y_j} = \frac{1}{m_j} \sum_{i \in j}^{n_j} [x_{ij} - \bar{x}_j] \cdot [y_{ij} - \bar{y}_j]. \tag{8.33}$$

The within-region correlation coefficients[11] together with the coefficients of determination for the three regions listed in table 8.6 are given in table 8.7. All the coefficients of determination, R^2, listed in table 8.7 indicate the complete absence of a relationship between race and illiteracy within each of the three regions. Fortunately, the conflicting interpretation obtained from the individual, ecological and within-region correlations can be reconciled using theoretical arguments. A result known as the *covariance theorem* states that, 'the individual covariance is the sum of the ecological covariance and the sum of the J within-region covariances.' Symbolically, we write this theorem as

$$\sigma_{xy} = \sigma_{x_J y_J} + \sum_{j=1}^{J} \sigma_{x_j y_j}. \tag{8.34}$$

This theorem asserts that the individual correlation is derived from two distinct sources of covariation. Firstly, the ecological covariance measures the importance of the relationship *between* the J regions, and secondly, the sum of the within-region covariances measures the strength of the relationship *within* the regions. Clearly, for the relationship between race and illiteracy there is a strong between-region component and a negligible within-region component.

How do we interpret these results? It is important to stress that we cannot begin to interpret an ecological correlation without knowing the values of all the covariances in the covariance theorem. Ecological data are easily available to geographers in the form of census variables collected over census regions, however the corresponding individual and within-region data are rarely available. Therefore, from the geographer's point of view, the covariance theorem takes the form

$$\sigma_{x_J y_J} = \sigma_{xy} - \sum_{j=1}^{J} \sigma_{x_j y_j}, \tag{8.35}$$

with the individual and within-region covariances being unknown quantities. Given this situation, it is *impossible* to infer that a strong correlation exists at the individual level on the basis of a strong ecological correlation. Similarly, ecological and within-region correlations cannot be inferred from a strong individual correlation.

Alker's (1969) interpretation of the race and illiteracy relationship uses all three terms in the covariance theorem. He begins by arguing that the system of census division illustrated in figure 8.10 acts as a third variable in the analysis. Broadly, the

Table 8.7. Within-region correlation statistics for selected divisions.

Regions	New England	Middle Atlantic	Pacific
j	1	2	9
$r_{x_j y_j}$	+0·009	+0·002	+0·015
R^2 (per cent)	0·0	0·0	0·0

regional system is a nominal variable which is a surrogate measure of the distribution of urbanization/industrialization in the USA. The Northern and Western Divisions are more industrialized and urbanized than the Southern and Central Divisions. Further, he suggests that the regional system is negatively correlated with both race and illiteracy. Thus increasing urbanization/industrialization is associated with both decreasing illiteracy rates and decreasing numbers of blacks, and similarly, low levels of urbanization/industrialization are related to large black populations and poor educational standards. It follows that the large ecological correlation for race and illiteracy is spurious and the strong correlation arises because both these variables are correlated with 'regional urbanization levels'. For the race/illiteracy problem, we know from our analysis that the individual covariance (σ_{xy} in formula 8.35) is composed almost entirely of ecological covariance because the within-region covariances are all close to zero in value. Therefore, the individual correlation is also unreliable because the individual covariance in formula (8.35) is composed almost entirely of the spurious ecological covariance. Thus the only reliable information we possess are the within-region correlations which indicate a complete absence of a relationship between race and illiteracy.

Alker's interpretation of Robinson's problem is subjective because he uses information not contained in the aggregation data to refute the relationship between race and illiteracy. The geographical properties of the US Census divisions are not represented quantitatively in the problem. At first sight, the individual relationship between race and illiteracy would not appear to contain a geographical component, and it is only when the within-region correlations (these are geographical entities) are considered that a complete statistical interpretation of an important social relationship can be attempted.

Spatial correlation analysis has been one of the mainstream traditions during the recent evolution of quantitative and theoretical geography. Although geographers have often indicated an awareness of the interpretive problems posed by ecological correlations, there are numerous published examples of ecological correlations where no reference is made to either individual or within-region data. Perhaps the most notorious example of such research is the tradition of factorial ecology in urban social geography which has been reveiwed in a social geographical context by Jones and Eyles (1977), and in a statistical geographical context by Taylor (1977). Factorial ecology involves intercorrelating large numbers of social variables based on the census divisions of a city. The purpose of the analysis is to identify the major groups of intercorrelated variables and, in so doing, isolate the major 'socio-spatial dimensions' of the city. More often than not these dimensions have names like 'social class', 'housing', and 'stage in the life cycle'. However, such analyses can be extremely misleading because, without individual and within-region information, they are totally dependent on the artibtrary system of census regions which acts as an unspecified, rogue variable on the analysis and interpretation of a complex set of social relationships. The reader who wishes to pursue the topic of multiple correlation analysis is referred to the excellent textbook by Mather (1976)

SPATIAL AUTOCORRELATION ANALYSIS

Definition

So far, our discussion of correlation analysis has centred on the analysis and interpretation of relationships between a pair of geographical variables. Recently, however, geographers and statisticians have begun to tackle a more fundamental idea known as the problem of *spatial autocorrelation*. This idea involves analysing the way in which a single variable, measured over a set of regions, is correlated with *itself* in space. Essentially, spatial autocorrelation is concerned with establishing whether the presence of a variable in one region in a regional system makes the presence of that variable in neighbouring regions *more*, or *less*, likely. If the variable is more likely to occur in neighbouring regions we say that it is *positively* spatially autocorrelated and, alternatively, if the variable is less likely to occur in neighbouring regions we say it is *negatively* spatially autocorrelated. For example, a casual glance at the arrangement of values in census B (figure 8.11) will suggest the presence of positive spatial autocorrelation because all the large values of x are grouped around the value $x = 3$ in the top left-hand corner of the lattice. The two types of spatial autocorrelation are synonymous with the continuum of point patterns in quadrat analysis (figure 8.1). Positive spatial autocorrelation is akin to clustering, while negative spatial autocorrelation is similar to uniformity. Also, the strategy of spatial autocorrelation analysis is similar in structure to the quadrat method. The original papers on the subject of spatial autocorrelation by Moran (1948, 1950) were concerned with constructing significance tests which summarized the behaviour of random patterns. Random spatial autocorrelation occurs when the value of the variable in any one region does not influence the variable's value in any other region. The null hypothesis

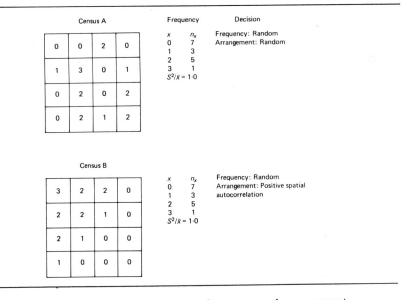

Figure 8.11.   The distinction between frequency and arrangement.

for applications of these tests states that there is no significant difference between the observed autocorrelation and the expected autocorrelation in a random pattern. Thus observed cases of positive or negative spatial autocorrelation are measured against a predefined random spatial autocorrelation probability model. This is the same idea as the quadrat method, where observed uniformity and clustering are measured against the random Poisson frequency model.

The fundamental distinction between the quadrat method and spatial autocorrelation analysis is that the former measures the frequency distribution of a pattern whereas the latter measures the arrangement of a pattern. The distinction between frequency and arrangement is made in the two quadrat censuses illustrated in figure 8.11. Analysis of both census A and B by quadrat methods would lead us to the conclusion that both were controlled by an independent spatial process because in both cases $S^2/\bar{x} = 1\cdot0$. However, inspection of the arrangement of the censuses gives the impression that the values in A are unrelated to one another (random spatial autocorrelation), whereas B is a highly clustered arrangement indicative of positive spatial autocorrelation. This ambiguous interpretation arises because, at present, there is no spatial statistical test which is capable of measuring all the geometrical properties of a geographical variable[12].

The specification of simple autocorrelation processes

Spatial autocorrelation in a geographical variable can arise in many different ways and, therefore, if we analyse the arrangement of a geographical variable, we must first specify the type of spatial autocorrelation process that is appropriate for the variable. This specification is made from our existing knowledge of the spatial behaviour of the variable. For the simpler autocorrelation tests the specification is represented by a contiguity matrix whose elements are denoted by δ_{ij}. To illustrate the simpler types of autocorrelation process we will use the analysis of a 4×4 chessboard presented in Cliff and Ord (1973).

Figure 8.12 illustrates the distribution of a nominal variable over the $n = 16$ cells of the 4×4 grid. When x_i takes on the value 0, the cell has been coloured white, and when $x_i = 1$ the cell is coloured black. Therefore, from figure 8.12 it can be seen that $x_1 = 0$ and $x_2 = 1$, etc. Because spatial autocorrelation analysis measures the arrangement of a pattern an important feature of the analysis is the contiguity matrix whose elements δ_{ij} specify the type of autocorrelation process that is appropriate for the variable. In the first instance we will define contiguity between cells as edge-to-edge contact (the Rook's case). These contacts are recorded in an $n \times n$ contiguity matrix. If cell i is contiguous with cell j, then $\delta_{ij} = 1$ and if cell i is non-contiguous with cell j then $\delta_{ij} = 0$. For example, cell 1 is contiguous only with cells 2 and 5, therefore, $\delta_{1,2} = 1$, and $\delta_{1,5} = 1$ while the remaining values in row 1 of the matrix are all set equal to 0 (figure 8.12a). Notice also that the contact of a cell with

---

Figure 8.12. Spatial autocorrelation on a chessboard. (Adapted from A D Cliff and J K Ord 1973, p16.) (*a*) The Rook's case (edge-to-edge contact). (*b*) The Bishop's case (vertex-to-vertex contact). (*c*) The Queen's case (edge-to-edge and vertex-to-vertex contact).

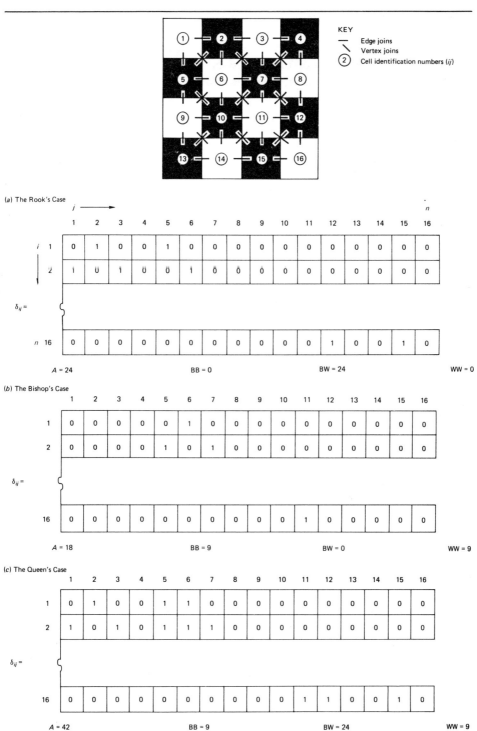

itself, $\delta_{i=j}$ always has a defined value of zero. Therefore, in the part of the contiguity matrix illustrated in figure 8.12(a) the elements $\delta_{1,1}$, $\delta_{2,2}$, and $\delta_{16,16}$ are all given the value 0. By making various summations over the contiguity matrix we can identify the main numerical characteristics of the system of contacts. The total number of contacts, or joins between cells, A, is obtained from

$$A = \tfrac{1}{2} \sum_i^n \sum_j^n \delta_{ij}, \tag{8.36}$$

which is an instruction to add all values of the elements (δ_{ij}) together and divide the result by two. For the Rook's case on the 4 x 4 chessboard formula (8.36) gives

$$A = \tfrac{1}{2} \times 48 = 24.$$

The summation of the elements δ_{ij} is divided by two because each join is recorded twice in the contiguity matrix. For example, in figure 8.12(a) the join between cell 1 and cell 2 is recorded both as $\delta_{12} = 1$ and $\delta_{21} = 1$.

To discover whether the chessboard is either positively, negatively or randomly spatially autocorrelated, we simply classify the total number of joins into three groups. These three types are a black cell being in contact with a black cell (BB), a white cell being in contact with a white cell (WW), and a black cell being in contact with a white cell (BW). The sum of the total number of 36 joins belonging to each of these three groups is A, the total number of contacts between cells[13]. If the reader classifies and counts the edge-to-edge contacts on the chessboard (figure 8.12) he will find the pattern is composed entirely of 24 BW joins. There are no BB or WW edge to edge joins in the pattern.

The frequency distribution of join types indicates the form of the autocorrelation process. For the Rook's case, the values of the join counts are BB = 0, BW = 24 and WW = 0, which is indicative of perfect negative spatial autocorrelation. The Rook's case on the chessboard is the deterministic limiting case of negative autocorrelation processes because every cell containing the maximum value of x (black, $x = 1$) is a neighbour on its edges to cells containing the minimum value of x (white, $x = 0$) and vice versa. Therefore, if a cell is coloured black, there is a probability of unity that its edge neighbours are white. Thus the values of the three join counts are diagnostic statistics for the autocorrelation process. For observed patterns, where the arrangement of colours is unlikely to follow the perfect chessboard pattern, negative spatial auto-correlation will be indicated by more BW and less BB and WW joins than one would expect in a random arrangement of colours[14].

If our definition of a join between cells is changed from edge-to-edge contact (the Rook's case) to vertex-to-vertex contact (the Bishop's case), then the pattern on the chessboard looks totally different. A portion of the vertex-to-vertex contiguity matrix is shown in figure 8.12(b) and, with this new definition of a join, the total number of cell contacts, A, is 18. The total number of BW joins is counted as 0. The number of BB and WW joins may be counted by eye from the chessboard as BB = 9 and WW = 9. This join count frequency distribution of BB = 9, BW = 0 and WW = 9 is indicative of positive spatial autocorrelation. Every black cell is a vertex neighbour with another black cell and therefore, if a cell is black there is a probability of one that is vertex neighbours are also black. The same is true for white squares. Thus the Bishop's case on the chessboard is the opposite extreme to the Rook's case.

Finally, we analyse the contiguity matrix which defines joins between cells as both edge-to-edge and vertex-to-vertex (the Queen's case, figure 8.12c). The autocorrelation process specified as the Queen's case gives the join counts BB = 9, BW = 24 and WW = 9, which is indicative of random spatial autocorrelation. The random spatial autocorrelation inferred from the Queen's case arises because the definition of contiguity in the Queen's case is the sum of the contiguity in the Rook's case and the contiguity in the Bishop's case. Therefore the negative autocorrelation in the Rook's case cancels out the positive autocorrelation in the Bishop's case to produce a random arrangement of colours over the Queen's joins.

The chessboard pattern of the nominal variable x is unlikely to be observed in reality, but the purpose of this discussion has been to illustrate how the same pattern can be given totally different interpretations depending on the definition of contact in the contiguity matrix. When we define a join we specify the spatial process which controls the arrangement of the variable. The join counts will indicate whether the specified process is positively, negatively or randomly autocorrelated, but clearly a single set of join counts will not be capable of indentifying the type of autocorrelation associated with alternative definitions of contiguity. Therefore, in practical applications of spatial autocorrelation analysis, it is extremely important for the experimenter to be able to specify an appropriate definition of contiguity from his knowledge of the spatial processes that influence the arrangement of the variable over the system of regions.

Moran's autocorrelation coefficient

The most commonly used statistical test for indentifying spatial autocorrelation was derived by Moran (1950)[15]. His test involves calculating a spatial autocorrelation coefficient, r_{xx}, which is synonymous with Pearson's coefficient, r_{xy}, for testing relationships between two variables. The test is appropriate for regular (e.g. a square grid), and irregular (for example, the map in figure 8.13) regional systems. Moreover, although the test was designed for an interval or ratio geographical variable, nominal and ordinal (ranked) variables may also be tested for spatial autocorrelation by applying the Moran technique.

Moran's spatial autocorrelation coefficient is calculated from a variable x which is composed of the observations x_i for each of n contiguous regions. The coefficient is defined by the ratio

$$r_{xx} = \frac{S_{xx}}{S_x^2},\qquad(8.37)$$

where

S_{xx} = the spatial autocovariance of x

S_x^2 = the variance of x

$$= \sum_i^n (x_i - \bar{x})^2/n.$$

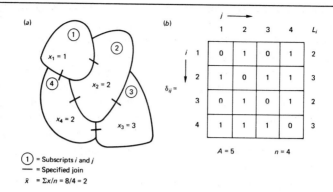

(a) = Subscripts *i* and *j*
— = Specified join
$\bar{x} = \Sigma x/n = 8/4 = 2$

(c)

i	*j*	x_i	x_j	δ_{ij}	$(x_i - \bar{x})(x_j - \bar{x})\delta_{ij}$ = Total	
1	1	1	1	0	(1 − 2) (1 − 2) 0	0
1	2	1	2	1	(1 − 2) (2 − 2) 1	0
1	3	1	3	0	(1 − 2) (3 − 2) 0	0
1	4	1	2	1	(1 − 2) (2 − 2) 1	0
2	1	2	1	1	(2 − 2) (1 − 2) 1	0
2	2	2	2	0	(2 − 2) (2 − 2) 0	0
2	3	2	3	1	(2 − 2) (3 − 2) 1	0
2	4	2	2	1	(2 − 2) (2 − 2) 1	0
3	1	3	1	0	(3 − 2) (1 − 2) 0	0
3	2	3	2	1	(3 − 2) (2 − 2) 1	0
3	3	3	3	0	(3 − 2) (3 − 2) 0	0
3	4	3	2	1	(3 − 2) (2 − 2) 1	0
4	1	2	1	1	(2 − 2) (1 − 2) 1	0
4	2	2	2	1	(2 − 2) (2 − 2) 1	0
4	3	2	3	1	(2 − 2) (3 − 2) 1	0
4	4	2	2	0	(2 − 2) (2 − 2) 0	0

Auto *SS* = 0

(d) Autocovariance, S_{xx} = Auto *SS*/2A = 0/10 = 0

Figure 8.13. Calculating the spatial autocovariance, S_{xx} (formula 8.42). (*a*) Example map pattern. (*b*) Contiguity matrix. (*c*) Work table for the auto-sums-of-squares. (*d*) Autocovariance.

The important quantity in formula (8.37) is the spatial autocovariance, S_{xx}, which is defined by

$$S_{xx} = \frac{1}{2A} \sum_i^n \sum_j^n (x_i - \bar{x})(x_j - \bar{x})\delta_{ij}. \tag{8.38}$$

As usual, δ_{ij} refers to the elements of the contiguity matrix which are defined as zero when region i is not connected to region j, and 1 when the regions i and j are connected. Therefore, the individual terms $(x_i - \bar{x})(x_j - \bar{x})\delta_{ij}$ will be zero for all unconnected regions, because in such cases $\delta_{ij} = 0$. For pairs of contiguous regions (where $\delta_{ij} = 1$) the value of $(x_i - \bar{x})(x_j - \bar{x})\delta_{ij}$ will be relatively large and positive if the values of x in both region i and j are large positive deviations from variable's mean, $\bar{x}$. Similarly, the same terms will be large and negative if the value of x in region i is a large positive deviation from the mean and the value in region j is a large negative deviation from the mean, or vice versa. Thus the terms in the autocovariance measure the extent to which x is positively or negatively associated with itself across the joins in the regional system. The sum of the terms $(x_i - \bar{x})(x_j - \bar{x})\delta_{ij}$ is divided by twice the number of joins, $2A$[16], to give the autocovariance. The autocovariance will be large and positive if the variable is positively spatially autocorrelated, and large and negative when the variable is negatively spatially autocorrelated.

The calculation of the autocovariance for an example variable distributed over $n = 4$ regions is shown in figure 8.13. The autocovariance is converted into a standard spatial autocorrelation coefficient by dividing its value by the variance of x (see formula 8.37). The spatial autocorrelation coefficient, r_{xx}, will tend to a value of plus one when the variable is perfectly positively spatially autocorrelated, and minus one when the variable is perfectly negatively spatially autocorrelated. These limiting cases of the Moran coefficient are illustrated in figure 8.14 where the graphs show the plots of values x_i against x_j for neighbouring regions where, by definition, $\delta_{ij} = 1$.

The value of the variance for the example map pattern (figure 8.13a) is calculated, as usual, as

$$S_x = [(1-2)^2 + (2-2)^2 + (3-2)^2 + (2-2)^2]/4$$
$$= 0{\cdot}5.$$

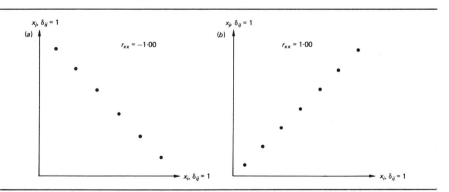

Figure 8.14. The limits of Moran's spatial autocorrelation coefficient. (a) Perfect negative spatial autocorrelation. (b) Perfect positive spatial autocorrelation.

Therefore, the value of the example spatial autocorrelation coefficient is

$$r_{xx} = \frac{S_{xx}}{S_x} = \frac{0 \cdot 0}{0 \cdot 5} = 0 \cdot 0.$$

Significance tests for spatial autocorrelation

Before we can construct a significance test for the Moran coefficient, it is necessary to define exactly what we mean by a random pattern which is characterized by the absence of spatial autocorrelation. The simplest approach[17] to this problem is taken by Moran (1950). He defines a random map pattern for a contiguous group of n regions as 'a set of n independent values of the standard normal random variable x' (see Chapter 7, p227). He then proves that the mean value of the spatial autocorrelation coefficient, $\mu(r_{xx})$, for an infinite number of sets of standard normal values each containing n observed values is given by

$$\mu(r_{xx}) = -1/(n-1) \tag{8.39}$$

Therefore, the expected mean value of r_{xx} for the example map pattern, where $n = 4$, is

$$\mu(r_{xx}) = -1/(4-1)$$
$$= -0 \cdot 333.$$

Unlike the significance text for Pearson's coefficient, r_{xy}, where the expected value of a random correlation is always zero, the random value of the spatial autocorrelation will always be negative and its exact value depends on n, the number of regions. By inspecting formula (8.39) it should be clear that as n tends to infinity ($n \to \infty$) so the random value of r_{xx} tends to zero ($\mu(r_{xx}) \to 0 \cdot 0$).

From our definition of random spatial autocorrelation it should be clear that $\mu(r_{xx})$ is the mean value of the sampling distribution of r_{xx}. The shape of the graph of this sampling distribution depends on n. Moran proves that, so long as n is reasonably large ($n \geqslant 30$ is the usual definition of reasonably large), the sampling distribution of r_{xx} follows a general normal distribution with a mean of $\mu(r_{xx})$ and standard deviation (or standard error) given by

$$\sigma(r_{xx}) = \sqrt{\left[\frac{4An^2 - 8n(A+D) + 12A^2}{4A^2(n^2-1)} \right] - \mu^2(r_{xx})}, \tag{8.40}$$

where A = total number of joins (formula 8.36), n = number of regions, and

$$D = \tfrac{1}{2} \sum_{i}^{n} L_i(L_i - 1), \tag{8.41}$$

where L_i is the number of regions joined to the ith region (figure 8.13b). D is a value which occurs in the mathematical derivation of formula (8.40) and for the example pattern (figure 8.13) its value is

$$D = [(L_1(L_1 - 1)) \times (L_2(L_2 - 1)) \times (L_3(L_3 - 1)) \times (L_4(L_4 - 1))]/2$$
$$= [(2 \times 1) \times (3 \times 2) \times (2 \times 1) \times (3 \times 2)]/2$$
$$= 16/2 = 8.$$

The standard deviation of the sampling distribution for the example pattern where $n = 4$ and $A = 5$ is calculated[18] as

$$\sigma(r_{xx}) = \sqrt{\left[\frac{(4 \times 5 \times 4^2) - [8 \times 4(5 + 8)] + (12 \times 5^2)}{4 \times 5^2(4^2 - 1)}\right] - (-0.333)^2}$$

$$= \sqrt{[204/1500] - 0.111} = \sqrt{0.136 - 0.111} = \sqrt{0.025} = 0.158.$$

Because the sampling distribution of r_{xx} is generally normally distributed, we can test any observed value of r_{xx} for significance by calculating the z transformation

$$\text{calc } z = \frac{r_{xx} - \mu(r_{xx})}{\sigma(r_{xx})} \tag{8.42}$$

which, substituting the values in the example map pattern, gives

$$\text{calc } z = \frac{0.000 - (-0.333)}{0.158}$$

$$= \frac{0.333}{0.158} = +2.108.$$

The significance test for a Moran coefficient is based on the sampling theory presented above. The research hypothesis for the example problem is that we expect either positive or negative spatial autocorrelation. We write this symbolically as

$$H_1 : r_{xx} \neq \mu(r_{xx}). \tag{8.43}$$

The null hypothesis states that there is no significant difference between the observed value of the coefficient and the expected value in a random pattern, and this statement is written as

$$H_0 : r_{xx} = \mu(r_{xx}) \qquad \alpha = 0.05. \tag{8.44}$$

The significance level of $\alpha = 0.05$ defines a *critical* value of $z = \pm 1.96$, which encloses $p = 0.95$ of the area under the normal curve. For the example problem, the *calculated* value of $z = +2.108$ falls outside the range of the critical value and so, in this case, we reject the null hypothesis and accept the research hypothesis of positive (calc z is positive) spatial autocorrelation (or clustering) among the arrangement of the example variable x.

The distribution of Labour voters in Wales

Here we present a spatial autocorrelation analysis of Labour voting in the Welsh constituencies (figure 8.15). The analysis is designed to illustrate the practical application of the Moran technique to a geographical variable.

There is abundant evidence in the literature of electoral geography (for example, Cox 1969), to suggest that the voting behaviour of an elector is partly influenced by the political attitudes of his friends and neighbours. Many electors tend to swim with the tide and vote for the dominant political party in a constituency. Thus when we analyse maps of the voting patterns for a political party at the constituency level we find a wide variation in the level of support for the party. A quick glance at figure 8.15

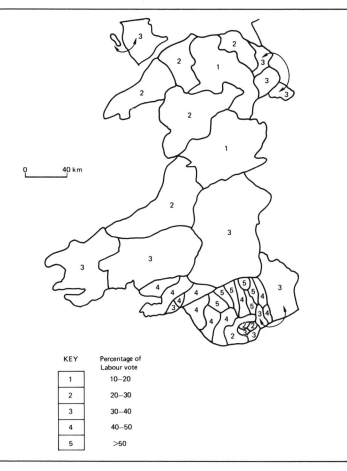

Figure 8.15.   Welsh constituencies, 1974, UK general election: percentage of constituency electorate voting Labour.

which shows the percentage of the constituency electorate voting Labour in the 1974 UK General Election, suggests the Labour Party enjoys above-average support in South Wales and below average support in North Wales. It is probable that political contacts between electors occur over quite short distances, and therefore we might expect neighbouring constituencies to support the same party. These ideas would lead us to propose a research hypothesis of positive spatial autocorrelation for the pattern of Labour voting in Wales. We write the research hypothesis, which specifies only positive[19] spatial autocorrelation, as

$$H_1: r_{xx} > \mu(r_{xx}) \tag{8.45}$$

and we shall test the null hypothesis of no spatial autocorrelation (formula 8.44) at the $\alpha = 0.05$ significance level.

Defining contiguity as $\delta_{ij} = 1$ when the ith constituency shares a common boundary with the ith constituency and $\delta_{ij} = 0$ otherwise, the application of the Moran test to

Table 8.8. Moran test for the Labour voting problem.

$A = 76$ $\qquad$ $D = 321$ $\qquad$ $n = 36$

$r_{xx} = +0 \cdot 540$ $\qquad$ $\mu(r_{xx}) = -0 \cdot 029$ $\qquad$ $\sigma_{r_{xx}} = 0 \cdot 104$

$$\text{calc } z = \frac{0 \cdot 540 - (-0 \cdot 029)}{0 \cdot 104} = +5 \cdot 47$$

crit $z = +1 \cdot 64$ $\qquad$ $\alpha = 0 \cdot 05$

Decision: reject H_0, accept H_1

the percentage of electors voting Labour in each of the $n = 36$ Welsh constituencies gives the results listed in table 8.8.

These results are depicted graphically in figure 8.16. Notice that, because the research hypothesis specifies that we are only interested in the case where the observed spatial autocorrelation coefficient is positive, the test is one-tailed. Therefore, the critical z value $+1 \cdot 64$ leaves $\alpha = 0 \cdot 05$ of the area under the normal sampling distribution exposed to the right of its position. Because calc z exceeds crit z, we reject the null hypothesis and accept our research hypothesis of positive spatial autocorrelation. Thus it would seem that neighbouring Welsh constituencies have similar Labour voting tendencies.

The use of non-contiguous weights

Our previous analysis of Labour voting is restricted by the definition of contiguity. If two constituencies were contiguous, δ_{ij} was set equal to one irrespective of the size of the constituency or the length of the common boundary. This situation will apply to any autocorrelation analysis of a variable distributed over irregularly shaped regions. Since the values of δ_{ij} represent our numerical specification of the spatial process controlling the arrangement of the pattern, the zero—one representation of contiguity restricts the amount of detail contained in the specification. Indeed, even with regular regions the same values of δ_{ij} may be obtained from different regional systems (figure 8.17).

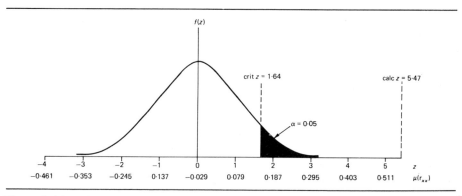

Figure 8.16. One-tailed z-test for the Labour voting spatial autocorrelation coefficient.

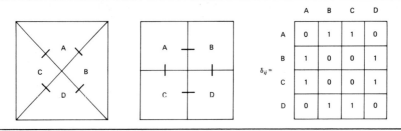

Figure 8.17.   Different regional systems with the same join structure.

To circumvent these problems, Cliff and Ord (1973) derived a version of the Moran coefficient where the zero—one elements, δ_{ij} were replaced by generalized weights, w_{ij}. These generalized weights represent the hypothesized influence of zone i on zone j and can take on any value so long as all the weights for a single problem are comparable. Generalized weights allow us to test ideas about spatial processes that operate between regions that are not necessarily neighbours. For example, when we tested the idea that the pattern of Labour voting was influenced by the amount of social contact between electors we only considered neighbouring constituencies. Moreover, we ignored both the areal extent of the constituency and the number of electors in each constituency. Clearly, using generalized weights we can given a more detailed numerical representation of these ideas. In the absence of observational information on the amount of social contact between electors living in Welsh constituencies we could quantify the amount of contact theoretically using a simple gravity model. If E_i is the size of the electorate in the ith constituency, E_j the size of the electorate in the jth constituency, and d_{ij} the distance between the ith constituency and the jth constituency, then the generalized weight for the influence of any constituency i on any constituency j is defined from the simple gravity model (Chapter 5, p133 as

$$w_{ij} = E_i E_j d_{ij}^{-1}. \tag{8.46}$$

The values $w_{i=j}$, which measure the region's influence upon itself, are always set equal to zero in the generalized weights matrix. Using these gravity model weights in an autocorrelation test would enable us to represent the ideas, which led to our research hypothesis about Labour voting patterns, in greater numerical detail than was possible with the elements δ_{ij}.

It is often the case in autocorrelation analysis that insufficient information is known about the spatial processes controlling the arrangement of the variable for us to specify a set of generalized weights that are particular to that variable. For example, Geary's (1954) data for the number of milch cows per 1000 acres over the Counties of Eire has been subjected to autocorrelation analysis by Cliff and Ord (1973). Intuitively, no one set of weights seems particularly appropriate as an explanation for the arrangement of milch cow densities over the Irish counties. There are many geographical variables of this nature, and in such cases Cliff and Ord suggest a set of standard weights that may be used to obtain an approximation of the unknown auto-correlation process. Standard weights are defined by

$$w_{ij} = \beta_{i(j)}/d_{ij}, \tag{8.47}$$

where $\beta_{i(j)}$ is the proportion of the boundary of region i (excluding the portion of i's perimeter that is the study area boundary and therefore non-contiguous with any other region) that is shared with region j, and d_{ij} is the distance between the centroid of region i and the centroid of j. The calculation of standard weights for a hypothetical three-region system is illustrated in figure 8.18. For example, the weight w_{12} for the map shown in figure 8.18(a) is calculated in the following way. The value of $\beta_{i(j)}$ is

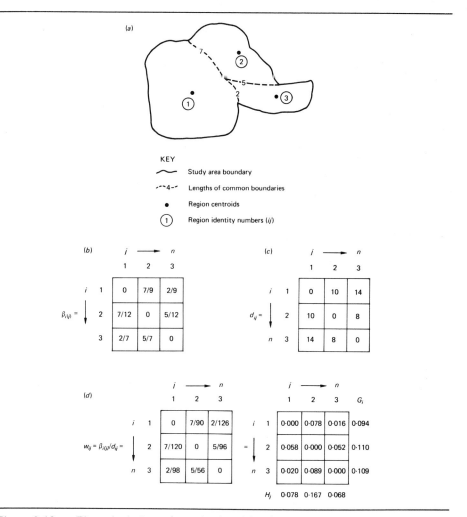

Figure 8.18. The calculation of standard weights for a three-region system. (a) Map. (b) The matrix of $\beta_{i(j)}$. (c) The matrix of d_{ij}. (d) The matrix of standard weights, w_{ij}.

defined as the ratio

$$\beta_{i(j)} = \frac{\text{length of boundary of region } i \text{ common with } j}{\text{length of common boundaries of region } i},$$

which for

$$\beta_{1(1)} = \frac{7}{7+2} = 7/9.$$

The distance between the centroids of regions 1 and 2 is measured as $d_{12} = 10$, and therefore, $w_{12} = 7/9 \times 1/10 = 7/90$. The idea incorporated in the definition of standard weights is that the influence of region i on region j will be large when j is a close neighbour to i and shares a large proportion of i's common boundary, and the influence will be small in the reverse situation. Notice that, unlike binary weights, δ_{ij}, $w_{ij} \neq w_{ji}$ because the lengths of the common boundaries of region i will normally be different from those of j. Also standard weights are calculated only for neighbouring regions and so preclude the specification of autocorrelation processes which involve non-contiguous regions.

Haggett (1976) has shown how weights can be used to construct spatial autocorrelation tests which take into account other geographical attributes of the counties which might influence the distribution of the spatial variable. For example, figure 8.19(a) records a weekly infection rate during a measles epidemic in the $n = 5$

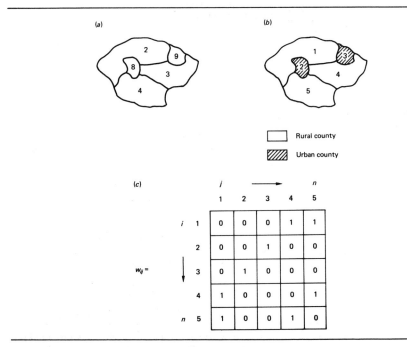

Figure 8.19.   Weighting attributes of regions. (a) Incidence of measles, x_i (cases per thousand per week). (b) Region numbers (ij). (c) Weighting matrix.

counties. Notice that counties 2 and 3 have been classified as urban and the remainder as rural. To test the idea that the incidence of measles is high in urban counties and low in rural counties, weights are given values of one when both i and j are either rural or urban, and zero when i is rural and j urban, or vice versa (figure 8.19c). The incidence of measles can then be tested for positive spatial autocorrelation between urban and rural counties using the Cliff and Ord coefficient which is described in the following section.

Cliff and Ord's autocorrelation coefficient

Once the appropriate weights have been specified for the test variable the calculation and significance tests for the Cliff and Ord coefficient are virtually identical to the Moran coefficient. For this reason, we have not included a worked example. The weighted autocorrelation coefficient, wr_{xx}, is defined by Cliff and Ord as

$$wr_{xx} = \frac{wS_{xx}}{S_x^2} \qquad (8.48)$$

where, as usual, S_x^2 is the variance of the geographical variable x distributed over the n regions, and wS_{xx} is the weighted spatial autocovariance given by

$$wS_{xx} = \frac{1}{W} \sum_i^n \sum_j^n (x_i - \bar{x})(x_j - \bar{x})w_{ij}. \qquad (8.49)$$

The divisor W is simply the sum of all the elements w_{ij} which form the specified weighting matrix; therefore

$$W = \sum_i^n \sum_j^n w_{ij}. \qquad (8.50)$$

The weighted autocovariance is simply an extension of Moran's autocovariance given by formula (8.38). Whereas in Moran's autocovariance for zero—one join definitions the divisor is $2A$ (the sum of the elements δ_{ij}), so the divisor in the weighted autocovariance is W, the sum of the elements w_{ij}. Unlike Moran's autocorrelation coefficient, r_{xx} (formula 8.37), which takes on values only between minus one and plus one, the weighted autocorrelation coefficient can take on any value between plus and minus infinity. The extended range of values of the weighted coefficient arises because the weights w_{ij} can be given any value depending on the nature of the variable x. Therefore, the value of wS_{xx} will often exceed that of S_x^2. Despite this difference, the significance test for the weighted coefficient is virtually identical to that for the Moran coefficient.

The expected mean value of the weighted coefficient in a random arrangement (see p276 for the precise definition) of the values of x over the n counties is

$$\mu(wr_{xx}) = -1/(n-1), \qquad (8.51)$$

which is identical to the random expectation of Moran's coefficient, formula (8.39). The standard deviation of the normally distributed sampling distribution of wr_{xx}

around $\mu(wr_{xx})$ is given by

$$\sigma_{(wr_{xx})} = \sqrt{\left[\frac{n^2 h - ng + 3W^2}{W^2(n^2 - 1)}\right] - \mu^2(wr_{xx})}. \tag{8.52}$$

In this formula the previously undefined terms h and g are obtained from the weights w_{ij} as

$$h = \tfrac{1}{2} \sum_i^n \sum_j^n (w_{ij} + w_{ji})^2 \tag{8.53}$$

and

$$g = \sum_i^n (G_i - H_i)^2, \tag{8.54}$$

where

$$G_i = \sum_j^n w_{ij}, \quad \text{and} \quad H_j = \sum_i^n w_{ij}. \tag{8.55}$$

To illustrate the calculation of formulae (8.51) and (8.52) we will use the weighting matrix given in figure 8.18(d). For this example, n, the number of regions is three, therefore

$$\mu(wr_{xx}) = -1/(3 - 1) = -0.5000$$

is the expected random value of the weighted coefficient. From formula (8.53) and figure 8.19(d) we obtain the values

$$h = [(w_{11} - w_{11})^2 + (w_{12} - w_{21})^2 + (w_{13} + w_{31})^2$$
$$+ (w_{21} - w_{12})^2 + (w_{12} - w_{22})^2 + (w_{23} - w_{32})^2$$
$$+ (w_{31} - w_{13})^2 + (w_{32} - w_{23})^2 + (w_{33} - w_{33})^2]/2$$
$$= [(0.000 - 0.000)^2 + (0.078 - 0.058)^2 + (0.016 - 0.020)^2$$
$$+ (0.058 - 0.078)^2 + (0.000 - 0.000)^2 + (0.052 - 0.089)^2$$
$$+ (0.020 - 0.016)^2 + (0.089 - 0.052)^2 + (0.000 - 0.000)^2]/2$$
$$= 0.004/2 = 0.002.$$

From formula (8.55) and figure 8.18(d) we see that G_i is simply the sum of the weights in row i of the weighting matrix, and H_j is the sum of the weights in column j. Therefore, from formula (8.54) we obtain

$$g = (G_1 - H_1)^2 + (G_2 - H_2)^2 + (G_3 - H_3)^2$$
$$= (0.094 - 0.078)^2 + (0.110 - 0.167)^2 + (0.109 - 0.068)^2$$
$$= 0.005.$$

From formula (8.50) we obtain W, the sum of all the weights, as

$$W = w_{11} + w_{12} + w_{13} + w_{21} + w_{22} + w_{23} + w_{31} + w_{32} + w_{33}$$
$$= 0.000 + 0.078 + 0.016 + 0.058 + 0.000 + 0.052 + 0.020 + 0.089 + 0.000$$
$$= 0.313.$$

Finally, we are now in a position to calculate $\sigma(wr_{xx})$, the standard error of the sampling distribution of wr_{xx}, from formula (8.52) as

$$\sigma_{(wr_{xx})} = \sqrt{\left[\frac{(3^2 \times 0\cdot002) - (3 \times 0\cdot005) + (3 \times 0\cdot313^2)}{0\cdot313^2(3^2 - 1)} - (-0\cdot500)^2\right]}$$

$$= \sqrt{[0\cdot297/0\cdot784] - 0\cdot25} = \sqrt{0\cdot379 - 0\cdot25} = \sqrt{0\cdot129} = 0\cdot359.$$

The null hypothesis of no spatial autocorrelation for the weighted spatial autocorrelation coefficient is stated in standard form as

$$H_0: wr_{xx} = \mu(wr_{xx}) \qquad \alpha = 0\cdot05. \tag{8.56}$$

The calculated value of z for any observed value of wr_{xx} is obtained from the standard normal transformation as

$$\text{calc } z = \frac{wr_{xx} - \mu(wr_{xx})}{\sigma(wr_{xx})}. \tag{8.57}$$

The decision whether to accept or reject the null hypothesis will be obtained by comparing the calculated value of z with the critical value of z defined by the research hypothesis and significance level for the problem in hand.

CONCLUDING REMARKS

The discussion in this chapter has been limited to processes and relationships which occur in space. However, more recent research in this field has extended the autocorrelation framework to the space—time domain. This style takes into account not only the spatial dependencies of a variable, but also its behaviour in the past. Cliff (1977) has proposed the following general model to describe the space—time scheme,

$$x_{i,t} = f(x_{i,t-k}, x_{j,t-k}) \qquad k = 1, 2, \ldots \tag{8.58}$$

Here, the subscript t denotes time and k is some pre-defined time interval. This model asserts that a variable's present value at someplace, $x_{i,t}$, is some function of its past value at that place, $x_{i,t-k}$, and also some function of its past value at some *other* place, $x_{j,t-k}$. For example, we might expect present unemployment levels in Manchester to be related not only to past unemployment levels in Manchester, but also past unemployment in a neighbouring cities such as Liverpool. The general structure of the space—time scheme is illustrated in figure 8.20. Implicit in this diagram is the idea that future values of the variable, $x_{i,t+k}$, may be forecast if the past space—time dependencies have been unravelled by the model. The reader is referred to the papers by Bennett (1975) on the population distribution in north-west England, and Martin and Oeppen (1975) on, among other things, the spread of fowl pest in East Anglia 1970—71, for operational examples of space—time probability models.

Talk of forecasting the geographical future must be tempered by some words of caution. To operationalize space—time probability models, many assumptions have to be made about our data which often cannot be justified. For this reason a number of

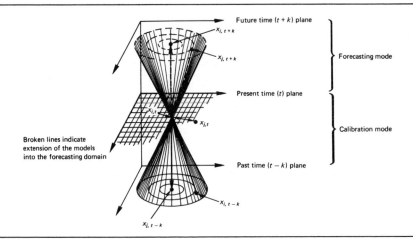

Figure 8.20.   The pattern of dependencies between regions in time and space. Reprinted with permission from P Haggett et al. 1977, *Locational Analysis in Human Geography* (2nd edition), Vol. 2, *Locational Methods*, Edward Arnold, fig. 11.1.

statisticians, amongst them Besag and McNeil (1976), have suggested that geographers would be well advised to spend more time understanding spatial data using simple exploratory techniques, and less time subjecting poorly understood data to complex forms of autocorrelation analysis. Indeed, in a marvellously inventive text, Tukey (1977) has devised numerous exploratory techniques designed to make data speak for themselves.

Notes

1   In practice, the binomial distribution should be chosen as the random model for data collected by quadrat censusing, while the Poisson is appropriate for quadrat sample data. However, in most geographical applications of quadrat analysis the binomial probability, $p = 1/n$ is sufficiently small for Poisson's approximation to the binomial to be close enough for the Poisson to be used as the random model for both data types (see Chapter 7, p225).

2   A version of the χ^2 test and a test known as the Kolmogorov–Smirnov D statistic, are often used as goodness-of-fit statistics in quadrat analysis (see Thomas 1977). However, for the style of quadrat analysis described in the text the variance/mean ratio test is preferred because it is derived directly from theoretical properties of the Poisson distribution.

3   Strictly speaking, the variance/mean ratio test is only appropriate when the observed frequency distribution has been collected by quadrat sampling because obviously, there is no sampling error inherent in the technique of quadrat censusing. However, in practice, the variance/mean ratio test is often applied to census data simply as a guide to interpretation.

4   The probability density function for the random variable t, which takes on values between $\pm\infty$, is defined by

$$f(t) = \frac{1}{\sqrt{\pi n}} \cdot \frac{\Gamma[(n+1)/2]}{\Gamma(n/2)} \cdot \left(1 + \frac{t^2}{2}\right)^{(-n+1)/2} \qquad -\infty \leqslant t \leqslant +\infty$$

where n = number of degrees of freedom and, as usual, numbers of the form $\Gamma(n)$ are defined as

$$\Gamma(n) = (n-1)!$$

5   The accompanying figure gives an alternative aggregation scheme for censusing the hypothetical plant community. For an illustration of how aggregation can alter the interpretation of scale effects, the reader is invited to apply hierarchical analysis of variance to this scheme and attempt an interpretation of the scale effects.

Subscripts (ij)   Observations ($x_j(i)$)

$j = 1$ $x_1(i)$

1	2	9	10
3	4	11	12
5	6	13	14
7	8	15	16

0	0	2	0
0	1	0	1
0	3	2	0
2	2	1	2

$j = 2$ $x_2(i)$

1	5
2	6
3	7
4	8

0	2
1	1
3	2
4	3

$j = 4$ $x_4(i)$

1	3
2	4

1	3
7	5

$j = 8$ $x_8(i)$

1	2

8	8

An alternative aggregation scheme.

6   The density function from the negative binomial is given by

$$P_x = \binom{k+x-1}{x}\left(\frac{p}{1+p}\right)^x \left(\frac{1}{1+p}\right)^k \qquad x = 0, 1, \ldots ,$$

where P_x is the probability that the specified cell contains x points, and the parameters p and k are

288   *Probability Models*

estimated as

$$\hat{k} = \bar{x}^2/(s_x^2 - \bar{x}),$$
$$\hat{p} = \bar{x}/\hat{k}.$$

As usual, $\bar{x}$ and s_x^2 are the observed mean and variance, respectively.

7   When the variables in correlation analysis are expressed on either ordinal or nominal scales Pearson's formula (8.17) can be simplified considerably to speed up calculations. The interested reader is referred to Siegel (1956, p203–5) for a discussion of these procedures for ordinal data, and note 11 for nominal data.

8   The symbols S_x, S_y, and S_{xy} denote that the calculation of the correlation coefficient is based on the random sample statistics for the standard deviation of x, the standard deviation of y, and the covariance of x and y. When the calculation is for a population, these symbols are σ_x, σ_y and σ_{xy}.

9   Often we possess evidence which enables us to specify the sign of the correlation coefficient in the research hypothesis as either,

$$H_1 : r_{xy} \geqslant 0.0 \text{ (positive correlation)}$$

or

$$H_1 : r_{xy} \leqslant 0.0 \text{ (negative correlation)}.$$

In such cases we test for significance using a one-tailed t-test. See p239 for an illustration of a one-tail test.

10   This calculation can be greatly simplified using the methods discussed in note 11.

11   The individual values of x and y for both race and illiteracy are measured on a nominal scale (0 or 1). When both variables are nominal the lengthy calculations implied by formulae (8.30–8.33) for the within-region correlation coefficient can be simplified by arranging the data in a 2 × 2 contingency table. For the New England division ($j = 1$) this contingency table takes the values

	Negro ($x = 1$)	White ($x = 0$)	Totals
Illiterate ($y = 1$)	4000	240 000	244 000
Literate ($y = 0$)	72 000	6 386 000	6 458 000
Totals	76 000	6 626 000	6 702 000

These data are frequencies obtained from the list of m_1 = 6 702 000 paired observations for each individual living in New England. The row and column frequencies can be obtained by inspecting the information listed in table 8.6; however, for obvious reasons of space sufficient information is presented in table 8.6 to obtain the individual cell frequencies. If we represent the frequencies by the following notation

	Negro	White	Totals
Illiterate	a	b	$(a + b)$
Literate	c	d	$(c + d)$
Totals	$(a + c)$	$(b + d)$	

then the following formula gives the value of the within-region correlation coefficient, r_{xjyj}, as

$$r_{x_jy_j} = \frac{a.d - b.c}{\sqrt{(a+c)(b+d)(c+d)(a+b)}}.$$

Rounding the frequencies for the New England division ($j = i$) up to the nearest thousand, the within-region coefficient is calculated quickly as

$$r_{x_jy_j} = \frac{(4 \times 6386) - (240 \times 72)}{\sqrt{(76)(6626)(6458)(244)}}$$

$$= \frac{8264}{890\,793} = +0 \cdot 009.$$

The same procedure may be used to calculate the individual correlation, r_{xy}.

12   An alternative method of pattern analysis to quadrat and spatial autocorrelation techniques is nearest-neighbour analysis. This technique is founded on the sampling theory of the frequency distribution of distances measured between each point in the observed pattern and its nearest neighbouring point. However, too many theoretical and practical problems bedevil the interpretation of nearest-neighbour analysis to make it a reliable technique at the present time. The interested reader is referred to King (1969) and papers in *Area* 8 pp161–71, (1976) for a discussion of the method and its problems.

13   This statement is true because, although WB joins are recorded in the contiguity matrix as $\delta_{ji} = 1$. They are simply double recordings of the equivalent BW joins $\delta_{ij} = 1$. For example, in figure 8.12(a) the BW join between cell 2 and cell 1 is recorded as $\delta_{21} = 1$, but the same join is recorded as WB between cell 1 and cell 2 as the element $\delta_{12} = 1$.

14   The observed pattern of a nominal variable taking on only values $x = 0$ (absent in a region) and 1 (present in a region) can be tested for spatial autocorrelation using Moran's coefficient (formula 8.37). The values x_i and δ_{ij} are substituted in the formula and the value of r_{xx} is tested for significance against the null hypothesis of random autocorrelation in the usual way.

15   A number of autocorrelation tests have been proposed, and the interested reader is referred to Cliff and Ord (1973) for a discussion of their relative merits.

16    $2A$ is the denominator in formula (8.38) because in the calculations for the autocovariance each join is counted twice, once for δ_{ij} and once for δ_{ji}.

17   A more complex definition is termed randomization. For this we calculate the mean value of the spatial autocorrelation coefficient for the $n!$ rearrangements of the observed values of x_i around the county system. The reader is referred to Cliff and Ord (1973) for a complete discussion of this approach.

18   Although n should be greater than 30 for the sampling distribution to be normal, we are using this example where $n = 4$ for the purpose of numerical illustration only.

19   The research hypothesis for negative spatial autocorrelation is

$$H_1 : r_{xx} < \mu(r_{xx})$$

and is tested using a negative, one-tail, critical z value.

CHAPTER 9

Geographical Decision Models

MAKING GEOGRAPHICAL DECISIONS

In his introductory textbook on the mathematics of decision-making, Lindley (1971) states that decision theory is about the logical processes that are used in order to arrive at a rational decision. Conversely, he argues that decision theory is *not* about the ways people actually take decisions. At first sight, this argument might appear a rather lame excuse; instead of becoming personally involved in the human conflicts that precede even the most trivial actions, the mathematician prefers to construct an abstract theory which tells the unsuspecting individual what actions he ought to be taking if only he were perfectly rational. On reflection, however, the reader should find Lindley's argument quite comforting. It would be most unfortunate if mathematicians felt they could predict human behaviour and so eliminate the uncertainties and irrationalities that are such an integral part of man's progress.

How, then, do mathematicians define a rational decision-making procedure? Again, it is instructive to quote Lindley's (1971) succinct definition of the procedures a sensible decision-maker ought to follow:

> First, the uncertainties present in the situation must be quantified in terms of values called *probabilities* Secondly, the various *consequences* of the courses of *action* must be similarly described in terms of *utilities*. Thirdly, the decision must be taken which is expected — on the basis of the calculated probabilities — to give the greatest utility.

We shall set Lindley's terminology into a more formal mathematical framework later in this chapter. However, at this stage it is worth commenting on the notions of probability and utility as they occur in decision theory. Probability is used in the context of subjective probability which we described in Chapter 7, p210. For instance, we shall take the decision-maker to be a farmer who has to decide which crop combinations to plant in the face of uncertain weather conditions. The decision theorist will assume that this farmer is capable of assigning subjective, prior probabilities to the outcome of different weather types in a single growing season. Further, it will be assumed that the farmer can quantify a utility to measure the attractiveness for the consequence of planting each crop in association with each possible weather type. Many consequences (for example, winning an argument as opposed to watching your favourite television programme) are difficult to quantify as ultilities. However, decision theorists suggest that, because many other consequences involve monetary expenditure, a decision-maker can easily express his utilities as monetary value judgments each measuring the relative benefits of different consequences. Typically, our farmer would probably measure the utility of different crop combinations as some function of the potential revenue obtained from their ultimate sale. Thus to operationalize his rational decision-making framework the mathematician assumes that individuals solve problems by their ability to quantify their uncertainties into subjective judgments. In addition, to incorporate some simple aspects of human personality into the framework, decision theorists allow *different* individuals facing the *same* decision problem to quantify *different* numerical utilities for the consequences of their actions. For example, an extroverted farmer would probably assign higher numerical values to his utilities than would an introverted farmer.

To understand the present position of decision theory in geography it is necessary to sketch the subject's historical development. Since the publication of Bayes' theorem in the late eighteenth century, mathematicians and philosophers of science alike have puzzled over the paradoxes it posed (see Chapter 7, p219). However, it is since 1945 that the subject of decision theory has made substantial progress. Savage (1961) cites the mathematical statistician, Neyman, as the major source of inspiration. Neyman (1947) suggested that the behaviour of the statistician in relation to his data was a more fertile field of study than the inferential tests developed by statisticians who followed the relative frequency view of probability. In essence, Neyman was arguing for a mathematical framework which allowed individual statisticians to express their own personalities as statistical decision-makers. In constrast, the relative frequency theory attempted to eliminate personality effects by the development of unbiased and repeatable statistical tests. At this point it is worth mentioning a concomitant theory which developed in the field of mathematical economics. Von Neumann and Morgenstern (1944), proposed a theory of games which also assumed the notions of subjective probability and utility among is basic structures. However, the subject matter of games theory was two or more individuals competing with one another in a uniform environment. This uniform environment was defined by the rules of the game. As such, games theory has given some simple insights about the conflicts that arise between, for example, businessmen competing for the same market, nations competing for territory, or gamblers playing games of chance. Hence, games theory is a study of small *group* behaviour in a *uniform* environment, whereas decision theory is the study of individual behaviour in an uncertain environment.

One of the first geographers to appreciate the potential of gaming and decision frameworks was Gould (1963). He adapted games theory to identify optimum crop combinations for Ghanaian farmers facing uncertain weather conditions. Unfortunately Gould's analysis violated the basic tenets of games theory because he assumed the farmer played against the natural environment in an attempt to gain high returns and not, as is proper, playing against another farmer. This misunderstanding led to unnecessary ambiguities appearing in Gould's analysis and, although subsequent writers such as Lloyd and Dicken (1977) have pointed to the error, little attempt has been made to rectify the problem.

Curry (1966), however, has suggested quite correctly that it would be fruitful to study the crop combination problem in a purely decision theoretical framework. Accordingly, the major part of this chapter is devoted to illustrating the geographical potential of decision theory using the farmer and his crop combinations as the example. The farming problem used throughout this chapter has been adapted from a teaching problem described in an elementary textbook on decision theory by Chernoff and Moses (1959). Their problem was about a man who has to decide which clothes to wear in the face of uncertain daily weather conditions, and this problem has the same theoretical form as that of the farmer having to decide which crops to plant in the face of uncertain weather conditions during the growing season.

THE STRUCTURE OF A DECISION PROBLEM

We begin by describing the basic structures which mathematicians assume characterize most rational decision problems. The farming problem is used to demonstrate the meaning of the following elements of a decision model: states of nature, which refer to the uncertain environment; actions available to the decision-maker (the farmer); losses of utility which measure the consequences of the various actions; frequency information which is conditional upon the true state of nature; and the strategies, or the plans of action which are open to the decision-maker.

States of nature

The first stage in any decision problem is to categorize the uncertainties posed by the environment as various 'states of nature'. For our farmer this categorization requires him to define discrete, mutually exclusive events which describe all the different possible weather conditions that could occur during the growing season. For the sake of simplicity, we will assume that the climatic effects on crop returns can be adequately classified as either wet weather effects or dry weather effects. Symbolically, these two possible states of nature are denoted by

E_1: a wet season,

E_2: a dry season.

In order to distinguish between these two categories the farmer would use climatic information that is relevant to crop yields such as the mean seasonal rainfall or the mean number of daily sunshine hours. Generally, the precise number of states of nature chosen by the decision-maker will depend on his accumulated knowledge of the major environmental effects on his particular problem. As a rule, it is unwise to specify large numbers of states of nature each representing quite minor environmental effects, because the states of nature are intended to simplify environmental uncertainties and too many categories can make the subsequent analysis unncessarily complicated.

Notice that states of nature are expressions of environmental uncertainties and not environmental locations. An alternative decision problem might involve a business organization trying to decide between a number of competing, spatial investment programmes. In this problem, environmental uncertainties would be the economic uncertainties affecting the various programmes. The economic climate would be represented by states of nature describing the different, possible, national inflation rates (or deflation rates) likely to occur during the period of the investment programme. The spatial aspect of this problem is the ultimate location of the investment programme, and in the case of the farmer is the final decision to plant a particular crop.

Available actions

Any decision problem involves the decision-maker in taking action, and the set of discrete actions {a} simply specifies all the different actions the decision-maker must eventually choose between. We will allow our farmer three possible courses of action, defined as

a_1: plant rice,

a_2: plant barley,

a_3: plant wheat.

Again, we have assumed for simplicity that ultimately the farmer will plant all his land with one of the three listed crops. The problem could easily be made more realistic by adding additional actions; for example, we could define a_4 as plant 50 per cent of the land under rice and 50 per cent under wheat, etc.

Losses of utility

As we noted previously, it is assumed that the decision-maker is capable of assigning numerical utilities to express his preference for the consequence of each action occurring in conjunction with each state of nature. It was suggested that the farmer uses knowledge of the financial returns he is likely to receive from planting a specific crop in given climatic condition to construct a set of utilities such as those listed in table 9.1. It is a convention in decision theory that utilities are measured negatively as losses of utility. Accordingly, the most favourable combination of an action, a, occurring in conjunction with a state of nature, E, is assigned a loss of utility of zero. All other combinations (E, a) are then assigned a loss, $l(E, a)$, which can be compared with the best, zero combination. The bigger the loss of utility, the less favourable the comparison with the best combination. From table 9.1 it can be seen that the most favourable outcome is planting rice in a wet season, $l(E_1, a_1) = 0$, and the least favourable outcome is when rice is planted in a dry season, $l(E_2, a_1) = 5$. The use of

Table 9.1. The farmer's loss of utility, $l(E, a)$.

E \ a	a_1 (rice)	a_2 (barley)	a_3 (wheat)
E_1 (wet)	0	1	3
E_2 (dry)	5	3	2

losses instead of actual utilities is a mathematical convenience to avoid the problem of negative utilities. It is typical of the conservative traditions amongst statisticians that a pessimistic loss scale is preferred to an optimistic utility scale.

Information conditional upon the true state of nature

The final element of a decision problem is information about the occurrence of the true state of nature. For our farmer we will assume this information is contained in a long-term weather forecast made by the local Meteorological Office about the likely weather in the coming growing season. At the beginning of each season the Met. Office issues one of the following statements:

x_1: wet weather likely,

x_2: dubious,

x_3: dry weather indicated.

The category 'dubious' is issued when the climatic evidence contained in the Met. Office's data is not decisive enough to make a definite prediction.

The reader may have noticed that, if the weather forecast was always correct, the farmer would be able to minimize his loss of utility whenever either x_1 or x_3 was the forecast. For example, he would plant rice, a_1, whenever x_1 (wet) was forecast, and plant wheat, a_3, whenever x_3 (dry) was forecast (table 9.1). Unfortunately, such accuracy is never the case and we must regard the set of forecasts $\{x\}$ as a random variable which is conditional (dependent) upon the true state of nature. To illustrate this statement we must consider the Meteorological Office's past record of accuracy. Table 9.2 gives two conditional probability distributions, $p(x = x \mid E)$, which measure the previous success rates of the Meteorological Office's forecasts; firstly, when a wet season subsequently occurred, and secondly, when a dry season occurred. For example, the probability $p(X = x_1 \mid E_1) = 0.60$ indicates that for 60 per cent of the previous wet seasons the Meteorological Office actually forecast wet weather, x_1. However, in 25 per cent $[p(X = x_2 \mid E_1) = 0.25]$ of the previous wet seasons the office gave the forecast dubious, x_2, and in 15 per cent of the wet seasons an incorrect forecast of dry weather was given. The Office's forecasting record in dry season is poorer than wet seasons because in only a proportion of $p(X = x_3 \mid E_2) = 0.50$ of previous dry seasons was the correct forecast, x_3, given. Therefore, at the beginning of a growing season, the particular forecast X given to the farmer is a random observation which is dependent upon the ultimate occurrence of one or other of the states of nature.

Table 9.2. Probability of obtaining a weather forecast x when E is the true state of nature; $p(X = x \mid E)$.

	x_1	x_2	x_3
E_1	0·60	0·25	0·15
E_2	0·20	0·30	0·50

The reader may be wondering if it is sensible for our farmer to base his actions upon information which may eventually turn out to be inaccurate. However, inspection of the probabilities listed in table 9.2 will reveal that the forecast is more use than no information, because the correct forecast is always made more often than the incorrect forecast. Notice too that, so far, we have assumed that the decision-maker does *not* know the past frequency of occurrence of the difference states of nature.

Strategies

The farmer now has to decide how he will react to the information provided by the weather forecast. His plan of reaction is termed a *strategy*. A strategy is a list of actions which together form a response to each of the different forecasts, x. For example, one possible strategy is $s(a_1, a_2, a_2)$; this list denotes that the farmer will plant rice, a_1, when wet weather, x_1 is forecast, plant barley, a_2, when 'dubious', x_2, is forecast, and plant barley, a_1, again when dry weather, x_3, is the forecast. Table 9.3 lists all the strategies that are possible reactions to the three weather forecasts. For any decision problem the total number of possible strategies may be quickly calculated from the formula n^r, where n is the number of available actions and r is the number of observations forming the random variable x. Therefore, our farmer has to choose between $3^3 = 27$ possible plans of action.

Before taking our analysis further, it is worth evaluating some of the possible strategies on purely intuitive grounds. At first sight some of the strategies are quite ridiculous while others seem quite reasonable. For instance, the strategies $s_1(a_1, a_1, a_1)$, $s_{14}(a_2, a_2, a_2)$ would seem quite unreasonable because they effectively ignore the weather forecast. The farmer who employs one of these strategies is determined to plant a single crop irrespective of the information provided by the Met. Office. Conversely, the strategy $s_6(a_1, a_2, a_3)$ seems quite a reasonable plan of action because the farmer will be combining the information provided by the weather forecast with his losses of utility (table 9.1). He will plant rice when wet weather is forecast, barley when dubious is forecast and wheat when dry is forecast.

Table 9.3. List of all pure strategies involved in the farming problem.

Strategy, s / Possible observation, x	s_1	s_2	s_3	s_4	s_5	s_6	s_7	s_8	s_9	s_{10}	s_{11}	s_{12}	s_{13}	s_{14}
x_1	a_1	a_1	a_1	a_1	a_1	a_1	a_1	a_1	a_1	a_2	a_2	a_2	a_2	a_2
x_2	a_1	a_1	a_1	a_2	a_2	a_2	a_3	a_3	a_3	a_1	a_1	a_1	a_2	a_2
x_3	a_1	a_2	a_3	a_1	a_2	a_3	a_1	a_2	a_3	a_1	a_2	a_3	a_1	a_2

	s_{15}	s_{16}	s_{17}	s_{18}	s_{19}	s_{20}	s_{21}	s_{22}	s_{23}	s_{24}	s_{25}	s_{26}	s_{27}
x_1	a_2	a_2	a_2	a_2	a_3	a_3	a_3	a_3	a_3	a_3	a_3	a_3	a_3
x_2	a_2	a_3	a_3	a_3	a_1	a_1	a_1	a_2	a_2	a_2	a_3	a_3	a_3
x_3	a_3	a_1	a_2	a_3	a_1	a_2	a_3	a_1	a_2	a_3	a_1	a_2	a_3

OPTIMAL STRATEGIES

Expected Losses

The subject matter of decision theory is intended primarily to provide the decision-maker with rational statistical arguments which allow him to select one of the possible strategies as the 'best' plan of action. In other words, the subject is about finding optimal strategies for an individual decision-maker. To identify optimal strategies it is first necessary to define the *expected loss of utility* that each strategy is likely to generate.

The expected loss of utility accruing to a particular strategy is calculated using the notion of an *action probability*, and these probabilities are estimated for all actions on each possible strategy for each state of nature. An action probability is an estimate of *the proportion of the time a particular action is taken when a particular state of nature occurs*. For example, consider the action probabilities associated with the strategy $s_5(a_1, a_2, a_2)$ in wet seasons. Firstly, calculating the action probability of planting rice, a_1, it can be seen from table 9.2 that in 0·60 of all wet seasons, E_1, the farmer will receive the wet weather forecast, x_1, and plant rice. Therefore, the action probability for rice, p_1, on strategy s_5 in wet seasons is defined as

$$p_1(E_1, s_5) = p(x_1 \mid E_1)$$
$$= 0·60.$$

The corresponding action probability for barley, $p_2(E_1, s_5)$, is again calculated with reference to table 9.2. Notice that, by continual application of s_5, the farmer will receive the forecast 'dubious', x_2, in 0·25 of all wet seasons and plant barley. Also, in 0·15 of all wet seasons, he will have received the incorrect forecast 'dry', x_2, and again planted barley as a response. Therefore, the action probability for barley on s_5 in wet seasons is obtained as

$$p_1(E_1, s_5) = p(x_2 \mid E_1) + p(x_3 \mid E_1)$$
$$= 0·25 + 0·15 = 40.$$

Finally, s_5 never requires the farmer to plant wheat, a_3, and so this final action probability is zero, that is,

$$p_3(E_1, s_5) = 0·00.$$

By using the idea of an action probability we can now calculate the average loss of utility, L, to the farmer applying a particular strategy, s, when j is the state of nature, E_j. The general formula for calculating this average loss of utility, $L(E_j, s)$, is given by

$$L(E_j, s) = \sum_{i=1}^{n} p_i(E_j, s) \cdot l(E_j, a_i), \qquad (9.1)$$

where i refers to the ith action and n is the total number of available actions, Applying formula (9.1) to operate strategy s_5 in all wet seasons, $E_{j=1}$, we obtain the result

$$L(E_1, s_5) = [p_1(E_1, s_5) \times l(E_1, a_1)] + [p_2(E_1, s_5) \times l(E_1, a_2)]$$
$$+ [(p_3(E_1, s_5) \times l(E_1, a_3)]$$
$$= [0·60 \times 0] \times [0·40 \times 1] \times [0·00 \times 3]$$
$$= 0·40.$$

Inspection of these calculations will reveal that, on the average, the application of s_5 in all wet seasons, E_1, will entail the farmer planting rice, a_1, in $p_1(E_1, s_5) = 0.60$ of all wet seasons, which makes the average loss of utility on rice $0.60 \times 0 = 0.00$ (see table 9.1 for the losses $l(E, a)$). Similarly, in 0.40 of all wet seasons he will plant barley with $l(E_1, a_2) = 1$, and so incur the average loss on barley of $0.40 \times 1 = 0.40$. Finally, he never plants wheat, a_3, and so his average loss of utility on wheat is $0.00 \times 3 = 0$. Summing these average losses for each of the i actions gives a total average loss of utility to the farmer applying s_5 in wet seasons as 0.40 units of loss.

A second average loss of utility is also incurred by the farmer who applies s_5. This penalty is the average loss in all dry seasons, $E_{(j=2)}$. If the reader inspects the probabilities in table 9.2 he can check that the action probabilities for $s_5(a_1, a_2, a_3)$ in all dry seasons are

$$p_1(E_2, s_5) = p(x_1 \mid E_2) = 0.20$$
$$p_2(E_2, s_5) = p(x_2 \mid E_2) + p(x_3 \mid E_2) = 0.30 + 0.50 = 0.80.$$
$$p_3(E_2, s_5) = 0.00$$

Applying formula (9.1) to the individual losses, $l(E_2, a_i)$, listed in table 9.1 we obtain the average loss of utility for s_5 in dry seasons as

$$L(E_2, s_5) = [p_1(E_2, s_5) \times l(E_2, a_1)] + [p_2(E_2, s_5) \times l(E_2, a_2)]$$
$$+ [p_3(E_2, s_5) \times l(E_2, a_3)]$$
$$= [0.20 \times 5] + [0.80 \times 3] + [0.00 \times 2]$$
$$= 3.40.$$

The development of arguments which allow the decision-maker to select an optimal strategy require him to compare the two average losses, one for each state of nature, for all his possible plans of action. These average losses accruing to each of the 27 strategies in the farming problem are listed in table 9.4. The reader may wish to check

Table 9.4. Expected losses of utility, $L(E, s)$, for the farming problem.

State of nature, E	s_1	s_2	s_3	s_4	s_5	s_6	s_7	s_8	s_9
E_1	0·00	0·15	0·43	0·25	0·40	0·70	0·75	0·90	1·20
E_2	5·00	4·00	3·50	4·40	3·40	2·90	4·10	3·10	2·60

s	s_{10}	s_{11}	s_{12}	s_{13}	s_{14}	s_{15}	s_{16}	s_{17}	s_{18}
E_1	0·60	0·75	1·05	0·85	1·00	1·30	1·35	1·50	1·80
E_2	4·60	3·60	3·10	4·00	3·00	2·50	3·70	2·70	2·20

s	s_{19}	s_{20}	s_{21}	s_{22}	s_{23}	s_{24}	s_{25}	s_{26}	s_{27}
E_1	1·80	1·95	2·25	2·05	2·20	2·50	2·55	2·70	3·00
E_2	4·40	3·40	2·90	3·80	2·80	2·30	3·50	2·50	2·00

one of these pairs of average losses in order to familiarize himself with the notions of action probability and average loss.

Because the strategies in the farming problem involve two expected losses, one average loss corresponding to each state of nature, each strategy can be represented as a point on a graph where the axes define the expected loss in each state of nature. The values of the expected losses listed in table 9.4 are plotted on figure 9.1 for all the 27 possible strategies. We can use this graph to illustrate the problem of identifying optimal strategies. From inspecting the diagram, it should be apparent to the reader that the only strategies worth considering as possible plans of reaction to the weather forecast are those strategies which lie on the lower lines connecting strategy s_1 to s_{27}; that is, s_1, s_2, s_5, s_6, s_{15}, s_{18} and s_{27}. A fundamental theorem in decision theory asserts that only strategies positioned on these lower lines are *admissible* plans of reaction. Such strategies are said to be admissible because their position on the line denotes the fact that one or other of their expected loss values *must* be lower than all the remaining *inadmissible* strategies which are positioned *above* the lower lines. For example, a quick

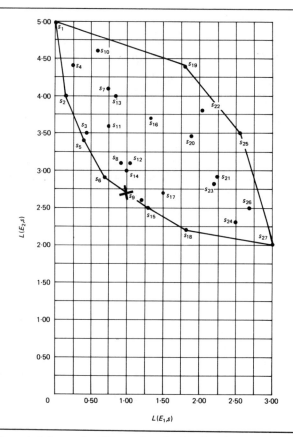

Figure 9.1. Expected loss of utility points in the farming problem. x Randomized strategy obtained by selecting s_6 with $p = 0.5$ and s_{15} with $p = 0.5$. (Adapted from H Chernoff and L E Moses 1959, p124.)

glance at the values in table 9.4 might lead the farmer to conclude that s_3, with expected losses $L(E_1, s_3) = 0.45$ and $L(E_2, s_3) = 3.5$, is quite a reasonable strategy because the expected losses on both states of nature are relatively low. However, inspection of figure 9.1 reveals immediately that s_3 is an inadmissible strategy, because strategy s_5 has the expected losses $L(E_1, s_5) = 0.4$ and $L(E_2, s_5) = 3.4$. Hence, no matter which state of nature ultimately occurs, s_5 will yield a smaller expected loss than s_3, and for this reason the admissible s_5 is said to *dominate* the inadmissible s_3.

So far our arguments have reduced the initial 27 strategies to the seven admissible strategies positioned on the lower limits. However, the problem is complicated by the fact that other strategies are open to the farmer in addition to the 27 already described. These additional strategies are termed *randomized* or mixed strategies. For example, suppose our farmer decides he likes both s_6 and s_{15} but cannot decide between them. In such circumstances he could operate both strategies over a number of growing seasons selecting either s_6 or s_{15} with probability 0.50 at the beginning of each season. To operationalize such a randomized strategy the farmer would simply toss an unbiased coin at the beginning of the season and select s_6 if heads showed, and s_{15} if tails showed. The expected losses for this randomized strategy are easily evaluated. In wet growing seasons, E_1, the farmer would select s_6 with probability 0.5 and incur the expected loss of utility $L(E_1, s_6) = 0.70$ (table 9.4). Similarly, with probability 0.5, he would select s_{15} and incur the expected loss, $L(E_1, s_{15}) = 1.30$. Therefore, the expected loss in wet seasons for this mixed strategy is given by $(0.5 \times 0.70) + (0.5 \times 1.30) = 1.0$. In the same way, the expected loss for this randomized strategy in dry seasons is obtained as $(0.5 \times 2.9) + (0.5 \times 2.5) = 2.7$. These two expected loss points have been plotted on figure 9.1 and it can be seen that this randomized strategy (marked 'x') lies half-way along the line segment connecting s_6 with s_{15}. This mid-point position reflects the fact that the two strategies have been randomized with probability 0.5.

Alternatively, the farmer can incur the same expected losses by operating the randomized strategy as a *mixed* strategy. For example, s_6 comprises the actions (a_1, a_2, a_3), while s_{16} comprises the actions (a_2, a_2, a_3). Notice that the only difference between the two strategies occurs when wet weather, x_1, is forecast; then s_6 requires the farmer to plant rice, a_1, while s_{15} requires the planting of barley a_2. If either dubious, x_2, or dry, s_3, are forecast, both strategies require the farmer to take identical actions. Hence, in order to mix s_6 and s_{15} with probability 0.5 in a *single* season, the farmer can take the following actions: firstly, if x_1 is forecast he plants 50 per cent of his land in rice and 50 per cent in barley; secondly, if x_2 is forecast he plants barley; and finally, if x_3 is forecast he plants wheat. This mixed plan of action gives the same expected losses (1.0 and 2.7) as the randomized plan involving selecting *either* s_6 *or* s_{15} with probability 0.5. Our interpretation of the mixed strategy has added a further spatial dimension to our decision problem because mixing allows the farmer to plant more than one crop in a single season. In geographical decision problems, mixing is preferable to randomization because mixing eliminates the additional, and artificial, experiment of tossing a coin which is needed to randomize two strategies with probability 0.5.

The mixed strategy we have just described is one of an infinitely large number of mixed strategies open to the decision-maker. For example, the farmer could use a strategy which selects s_6 with probability 0·67 and s_{15} with probability 0·33. Such a mixture would result in the expected losses (table 9.4) of

$$0·67(0·70) + 0·33(1·30) = 0·90$$

and

$$0·67(2·9) + 0·33(2·5) = 2·76,$$

which, if plotted on figure 9.1, locates a point a third of the distance from s_6 towards s_{15}. Notice that the location of the randomized strategy on the line between s_6 and s_{15} is *inversely* proportional to the probability of selecting each pure strategy. Therefore, because s_{15} is selected with $p = 0·33$ it makes a lower contribution to the expected losses on the mixed strategy than s_6. It is for this reason that the mixed strategy is located on the line segment two-thirds of the distance from s_{15} towards s_6.

Before taking our discussion further we need to define carefully the two types of strategy open to the decision-maker. The initial 27 strategies are termed *pure* strategies. Formally, a strategy is said to be pure if it assigns a single action to each of the observations, x. A *mixed* or randomized strategy is a choice between two or more pure strategies and will require the decision-maker to select among actions using an appropriate mixing or randomizing device. Notice that our definition of a mixed strategy allows the decision-maker to operate more than two pure strategies each with a defined probability of selection. This definition allows the decision-maker tremendous freedom to concoct mixed strategies, and such freedom makes it necessary to consider the range of average losses the decision-maker is likely to incur by operating different mixed strategies. The range of possible average losses is obtained by delimiting what is commonly termed the *feasible region*. For a decision problem involving just two states of nature the feasible region is identified by drawing lines between the points representing the outermost pure strategies on the average loss graph. The feasible region obtained from this procedure in the farming problem is shown on figure 9.1. The feasible region is so named because the decision-maker can incur any pair of losses within, or on the boundaries, of this region by operating either a pure or mixed strategy. Conversely, it is *not* possible for the farmer to devise a strategy which incurs a pair of average losses located outside the boundary of the feasible region.

The idea of a mixed strategy does not alter our previous discussion about admissible and inadmissible strategies. It is still true that the only strategies worth considering as possible plans of action are those whose loss points are located on the lower line segments (s_1 to s_{27} on figure 9.1) of the feasible region. The introduction of mixed strategies simply makes it possible for the decision-maker to incur the average losses defined by *any* point on the lower line segments by mixing two pure, admissible strategies with the appropriate probabilities. Obviously all the mixed strategies located above the lower line segments are inadmissible because they are each dominated by one or more of the admissible strategies.

Bayes' strategies

So far, we have not considered the possibility that the decision-maker possesses knowledge which would enable him to estimate the probability of occurrence for each state of nature. Suppose the farmer used past climatic records to obtain the previous frequency of occurrence of wet and dry seasons. He could then estimate the probability of occurrence for each state of nature as a relative frequency. These relative frequencies are termed prior probabilities because their value is known before the final decision is taken and we will refer to them by the symbols

$p(E_1)$: prior probability of a wet growing season,

$p(E_2)$: prior probability of a dry growing season.

Whenever the decision-maker can estimate the priors, he may use their values to identify an optimum strategy known as Bayes' strategy. Indeed, many statisticians would argue that, even in the absence of data to estimate the priors, the decision-maker is entitled to make an inspired guess at their value and use these subjective values to obtain the appropriate Bayes' strategy.

The optimal Bayes' strategy is found by using the priors to calculate a single value of total average loss for each admissible strategy. The formula for the total average loss $L(s)$ incurred by any strategy is defined by

$$L(s) = \sum_{j=1}^{m} p(E_j)L(E_j, s), \qquad (9.2)$$

where m is the total number of states of nature. For example, consider the admissible pure strategy s_6 with its two average losses of $L(E_1, s_6) = 0{\cdot}7$ in wet seasons and $L(E_2, s_6) = 2{\cdot}9$ in dry seasons. If the farmer estimates the prior probability of wet weather to be $p(E_1) = 0{\cdot}33$, and the prior probability of dry weather to be $p(E_2) = 0{\cdot}67$, he may use this information to calculate the total loss incurred by s_6 in any growing season. From formula (9.2) we obtain the result

$$L(s_6) = [p(E_1) \times L(E_1, s_6)] + [p(E_2) \times L(E_2, s_6)]$$
$$= [0{\cdot}33 \times 0{\cdot}7] + [0{\cdot}67 \times 2{\cdot}9]$$
$$= 0{\cdot}23 + 1{\cdot}94 = 2{\cdot}17.$$

From these calculations it can be seen that the total average loss is the sum of the average loss for each state of nature weighted by its prior probability of occurrence. Therefore, irrespective of the state of nature, in any single growing season the application of strategy s_6 is likely to result in a loss of $2{\cdot}17$ units of utility to the farmer.

One could imagine the farmer using the priors to calculate a total average loss for each admissible strategy and then choosing the admissible strategy with *minimum* total average loss to be his optimum plan of action. By selecting this minimum loss strategy, the farmer is said to be choosing the Bayes' strategy which corresponds to the particular priors $p(E_1) = 0{\cdot}33$ and $p(E_2) = 0{\cdot}67$. In general, for any set of priors, $\{p(E_j)\}$, corresponding to each of the m states of nature, the Bayes' strategy, s, is defined as that strategy which minimizes the value of the expression

$$\sum_{j}^{m} p(E_j)L(E_j, s) = L(s). \qquad (9.3)$$

If the farmer calculated the total average loss for all the admissible strategies, he would find that s_{18}, with an average total loss of $L(s_{18}) = 0{\cdot}33\ (1{\cdot}80) + 0{\cdot}67$

(2·20) = 2·07, was the Bayes' strategy which corresponded to his estimates of the priors. Clearly, calculating the total average expected loss for all admissible strategies is an extremely tedious method of finding the optimum Bayes' strategy. Fortunately, for decision problems involving just two states of nature, a simple graphical technique may be used to identify the minimum-loss Bayes' strategy. This technique uses the idea of a line of equal average total loss. Such a line may be drawn for any pair of prior probabilites, $p(E_1)$ and $p(E_2)$, and is a representation of the different combinations of average loss on each state of nature $L(E_1, s)$ and $L(E_2, s)$ which result in the *same* total average loss. The general formula for this equal average total loss line is

$$p(E_1) . L(E_1, s) + p(E_2) . L(E_2, s) = L, \tag{9.4}$$

where L is an arbitrary value given to the average total loss. For example, take the line for the priors $p(E_1) = 0·33$ and $p(E_2) = 0·67$ where the combinations of $L(E_1, s)$ and $L(E_2, s)$ each result in an average total loss of $L = 1$. This line is defined by the equation

$$0·33 \, L(E_1, s) + 0·67 \, L(E_2, s) = 1.$$

The simplest way to plot this equation is to find the value of $L(E_1, s)$ when $L(E_2, s)$ = 0, and then to find the value of $L(E_2, s)$ when $L(E_1, s) = 0$. Substituting these values in the equation we obtain the following coordinate values: when $L(E_2, s) = 0$, then

$$0·33 \, L(E_1, s) + 0·67 \times 0 = 1$$
$$0·33 \, L(E_1, s) = 1$$
$$\therefore \quad L(E_1, s) = 3;$$

and when $L(E_1, s) = 0$, then

$$0·33 \times 0 + 0·67 \, L(E_2, s) = 1$$
$$0·67 \, L(E_2, s) = 1$$
$$\therefore \quad L(E_2, s) = 1·5.$$

These pairs of coordinates have been used to plot the equation as line (iv) on figure 9.2(a). To convince himself of the equal average loss property of the line, the reader may wish to substitute the coordinate values, $L(E_1, s)$ and $L(E_2, s)$, of an additional point on the line into the example equation and calculate the value of L.

It is important to know how changes in the value of $p(E_1)$, $p(E_2) = 1 - p(E_1)$, and L affect the position of the line. Suppose we keep the priors constant at $p(E_1) = 0·33$ and $p(E_2) = 0·67$, but *increase* the arbitrary total average loss to $L = 1·5$. The equation for this new line is listed on figure 9.2 and plotted as line (ii). Notice that line (ii) is parallel to line (iv), but is located at a greater distance from the origin. Thus an increase in the average total loss causes an upward, parallel movement of the line. Indeed, figure 9.2(b) shows that the value L is proportional to the length of a line c which bisects the axes and connects the origin to the equal average loss line. Alternatively, when the values of the priors are changed but L is kept constant, a different picture emerges. Figure 9.2a, line (iii) illustrates the graph of the equation

$$0·5 \, L(E_1, s) + 0·5 \, L(E_2, s) = 1,$$

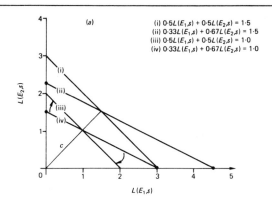

$$\text{(i)}\ 0{\cdot}5L(E_1,s) + 0{\cdot}5L(E_2,s) = 1{\cdot}5$$
$$\text{(ii)}\ 0{\cdot}33L(E_1,s) + 0{\cdot}67L(E_2,s) = 1{\cdot}5$$
$$\text{(iii)}\ 0{\cdot}5L(E_1,s) + 0{\cdot}5L(E_2,s) = 1{\cdot}0$$
$$\text{(iv)}\ 0{\cdot}33L(E_1,s) + 0{\cdot}67L(E_2,s) = 1{\cdot}0$$

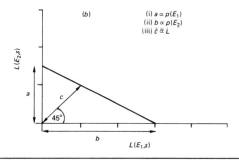

$$\text{(i)}\ a \propto p(E_1)$$
$$\text{(ii)}\ b \propto p(E_2)$$
$$\text{(iii)}\ \hat{c} \& L$$

Figure 9.2.   Lines of equal average total loss. (*a*) Some example lines. (*b*) Properties of the line of equal average total loss.

where the priors have been changed to $p(E_1) = 0{\cdot}5$ and $p(E_2) = 0{\cdot}5$. A comparison of equation (iv) with equation (iii) on figure 9.2(*a*) will demonstrate that the effect of increasing $p(E_1)$ from $0{\cdot}33$ to $0{\cdot}50$ is to *increase* the contribution of $L(E_1,s)$ to the arbitrary average total loss of $L = 1$ at the expense of the contribution of $L(E_2,s)$. Graphically, the consequence of increasing $p(E_1)$ is a *clockwise* rotation of the equal average loss line for the estimated priors is plotted for some arbitrarily low value of L. For example, equation (i) in figure 9.3 represents the equal average loss line for $p(E_1) =$ rotate *anti-clockwise* around the point where c intersects. Therefore, changing the values of the priors alters the points at which the equal average loss line intercepts the axes of the average loss graph. From figure 9.2(*b*) it can be seen that the intercept a is proportional to the value of $p(E_1)$ and the value of $p(E_2) = 1 - p(E_1)$ is proportional to the value of the intercept b^1.

Lines of equal average total loss are now used to locate graphically the Bayes' strategy which corresponds to a particular pair of prior probabilities. Firstly, an equal average loss line for the estimated priors is plotted for some arbitrarily low value of L. For example, equation (i) in figure 9.3 represents the equal average loss line for $p(E_1) = 0{\cdot}33$, $p(E_2) = 0{\cdot}67$ and $L = 1$. Notice that this line is positioned *below* the feasible region, indicating that no strategy exists which yields an average total loss of $L(s) = 1$.

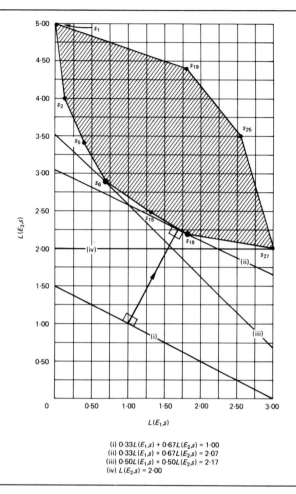

(i) $0.33L(E_1,s) + 0.67L(E_2,s) = 1.00$
(ii) $0.33L(E_1,s) + 0.67L(E_2,s) = 2.07$
(iii) $0.50L(E_1,s) + 0.50L(E_2,s) = 2.17$
(iv) $L(E_2,s) = 2.00$

Figure 9.3.   Locating the Bayes' strategy.

Secondly, we draw a line parallel to the initial line which is also *tangential* to the feasible region. This second parallel line is depicted as equation (ii) on figure 9.3. As we noted in the previous paragraph, moving an equal loss line upward and parallel has the effect of increasing the value of L. In this example (figure 9.3) the average total loss is increased from $L = 1$ to $L = 2.07$. The point at which this second line is tangential to the feasible region corresponds to an admissible strategy which the decision-maker can operate to incur the average total loss of $L(s) = 2.07$. The tangent is the Bayes' strategy for the estimated priors because its location corresponds to the admissible strategy where the average total loss is at a minimum. From figure 9.3 it can be seen that, for the priors $p(E_1) = 0.33$ and $p(E_2) = 0.67$, the Bayes' strategy is s_{18}. Graphically, the property of s_{18}, as the point of minimum average total loss is demonstrated by the fact that *all* other admissible strategies are positioned *above* the line of equal average loss and, therefore, *must* incur a greater total average loss to the farmer.

From the preceding discussion it should be apparent that the particular admissible strategy which is Bayes' optimum depends on the values estimated for the priors. For example, if the priors in the farming problem were estimated as $p(E_1) = 0 \cdot 5$ and $p(E_2) = 0 \cdot 5$, then the line of equal average loss would be rotated in a clockwise direction (see line (iii) on figure 9.3) and the tangent would locate the Bayes' strategy as s_6 with an average total loss of $L(s_6) = 2 \cdot 17$. Indeed, all the admissible strategies are potential Bayes' strategies for a particular pair of priors. Consider what happens to the Bayes' strategy in the limit, when one of the priors is zero and the other unity. Suppose wet seasons never occur, then $p(E_1) = 0, p(E_2) = 1$ and the line of equal average loss is defined by the equation

$$0[L(E_1, s)] + 1[L(E_2, s)] = L$$

$$L(E_2, s) = L.$$

This equation represents a set of horizontal lines each parallel to the $L(E_1, s)$ axis. The line $L(E_2, s) = 2 \cdot 00$ is tangential to the feasible region (figure 9.3(iv)) and identifies $s_{27}(a_3, a_3, a_3)$ as the Bayes' strategy. Previously, we discounted s_{27} as a possible plan of action because it entailed planting wheat irrespective of the information provided by the weather forecast. However, if wet seasons never occur, the weather forecast is rendered meaningless and the farmer can safely minimize his loss of utility by planting wheat each year (table 9.1). Statistically, this means that if the decision-maker knows E_1 is an impossible event he will adopt a strategy that minimizes $L(E_2, s)$ no matter how large $L(E_1, s)$. Similarly, if $p(E_1) = 1$, the feasible region will be tangential to a *vertical* line defined by the equation of the general form

$$L(E_1, s) = L.$$

For the farming problem the line $L(E_1, s) = 0$ identifies the Bayes' strategy as the end

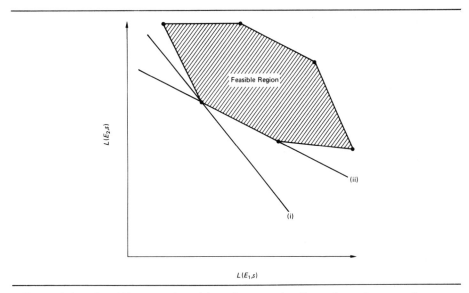

Figure 9.4. Possible general position of supporting lines of equal average loss.

point $s_1(a_1, a_1, a_1)$. Thus, knowing that dry seasons never occur, the farmer always plants rice and experiences zero loss of utility.

The fact that all admissible strategies are potentially Bayes' strategies leads to some further interesting results. Figure 9.4 illustrates that the line of equal average loss which supports the feasible region *must* either be tangential at a single point, case (i), or along a single line segment, case (ii). If the prior probabilities result in case (ii), then every point on the line segment is an optimum Bayes' strategy, including the two end points which are *pure* strategies. Therefore, at least one of the strategies which is Bayes' optimum will always be a pure strategy irrespective of the values estimated for the priors. Consequently the decision-maker who operates the Bayes' criterion can safely ignore randomized strategies which are infinite in number and cumbersome to operate.

Minimax strategies

The statistical criterion of minimum average loss is used to justify the selection of the Bayes' strategy as the optimal plan of action. However, the decision-maker may well feel that criteria other than the purely statistical must be taken into account before he decides upon a final plan of action. For instance, suppose our farmer tilled marginal land and simply could not afford to sustain high losses of utility without facing bankruptcy. Similarly, farmers working in a subsistence economy cannot sustain large losses of utility. The peasant farmer and his kin might well starve if the wrong crop is planted. Such farmers are likely to take a pessimistic approach to decision-making, and adopt strategies which avoid the possibility of large losses of utility. A pessimistic farmer inspecting the losses in table 9.1 would beware of any strategy which might result in his planting rice in a dry season, $l(E_2, a_1) = 5$.

An optimal solution to the pessimist's dilemma is to select a strategy defined by the *minimax* criterion. 'Minimax' is an abbreviation for the strategy which minimizes the maximum average loss. The minimax strategy is obtained from the following argument. The pessimist will assume that no matter what strategy he applies the state of nature, E_j, which occurs will be the one which has a maximum loss of utility $L(E_j, s)$. For example, if the farmer selects $s_1(a_1, a_1, a_1)$ with average losses $L(E_1, s_1) = 0$ and $L(E_2, s_1) = 5·0$, it will no doubt be a dry season and the *maximum* average loss of 5 units will be experienced. Similarly, if $s_{18}(a_2, a_3, a_3)$ is selected with average losses $L(E_1, s_{18}) = 1·80$ and $L(E_2, s_{18})$ 2·20, he fears it will again be a dry season and the larger of the two average losses will be incurred. However, if the pessimistic farmer was asked to choose between s_1 and s_{18} he would choose s_{18} because the maximum average loss with s_{18} (2·20) is *smaller* than the maximum average loss on s_1 (5·00). The minimax criterion extends this reasoning to suggest that the pessimist should select that admissible strategy where *maximum* average loss takes on a *minimum* value. Hence, if the worst state of nature *does* ultimately occur, the pessimist can be satisfied that he has selected a strategy which has kept his average loss to a minimum. Notice that, because the pessimist is only interested in the worst state of nature, the prior probabilities $p(E_1)$ and $p(E_2)$ are ignored in the minimax criterion. Thus the minimax strategy is selected by comparing the average losses $L(E_1, s)$ and $L(E_2, s)$ for each admissible strategy, unlike the Bayes' strategy, which is selected using *total* average losses, $L(s)$.

Again, for decision problems with only two states of nature, the minimax strategy can be found using a simple graphical technique. The minimax strategy is the point on the lower boundary of the feasible region where the average loss $L(E_1, s)$ is *equal* to the average loss $L(E_2, s)$. This property of the minimax strategy is illustrated in figure 9.5. The point marked s denotes the location of the minimax strategy where $L(E_1, s) = L(E_2, s)$. To understand the definition of minimax we will compare the properties of s with all other admissible strategies on the lower boundary of the feasible region. Firstly, all admissible strategies located to the right of s (direction 'a' on figure 9.5) possess larger values of $L(E_1, s)$ than point s. Therefore, the maximum loss on all these strategies is always larger than the maximum loss at s. Similarly, all the admissible strategies located to the left of s (direction 'b' on figure 9.5) possess larger values of $L(E_2, s)$ than s. Therefore, s is the minimax strategy because all strategies to the right and left of s have one or other of their average losses greater than those at s.

The graphical solution for the minimax strategy is quite straightforward. A line is constructed along which the coordinate values are equal, that is, $L(E_1, s) = L(E_2, s)$. This is the line which bisects the axes at an angle of 45 degrees. The point where this line intersects the lower boundary of the feasible region is the location of the minimax strategy[2] (see figure 9.5). The location of the minimax strategy for the farming problem is shown in figure 9.6. It can be seen that the minimax point is a *randomized*

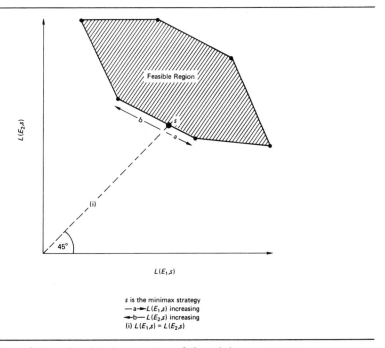

Figure 9.5. Graphical properties of the minimax strategy.

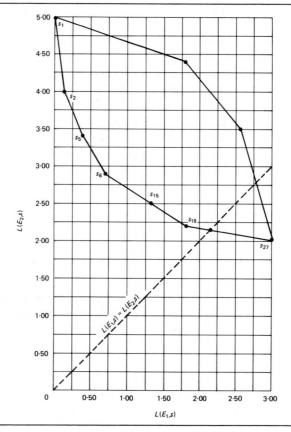

Figure 9.6.   Locating the minimax strategy in the farming problem.

strategy where the farmer selects s_{18} with probability $0·71$, and s_{27} with probability $p = 0·29$, yielding an average loss of $2·14$ on each state of nature. Notice that, if the pessimist adopts the minimax strategy with its mixture of s_{18} (a_2, a_3, a_3) and s_{27} (a_3, a_3, a_3), he will always avoid planting rice with its maximum loss of five units in a dry season.

Regrets and risks

Suppose the farmer planted rice and it turned out to be a dry growing season, his loss of utility would be $l(E_2, a_1) = 5$. However, he could rationalize this outcome in another way. When a dry growing season occurs the best action is to have planted wheat with a minimum loss of $l(E_2, a_3) = 2$. Because in his ignorance the farmer planted rice, his *regret* due to not having guessed the state of nature correctly can be measured as $L(E_2, a_1) - L(E_2, a_3) = 5 - 2 = 3$. Similarly, if barley had been planted his regret would be $L(E_2, a_2) - L(E_2, a_3) = 3 - 2 = 1$. More formally, the regret, $r(E_j, a_i)$, at not having guessed the jth state of nature correctly is the *difference* between the loss on the action that was taken and the action with the minimum loss,

Table 9.5. Losses and regrets for the farming problem.

		Losses $= l(E_j, a_i)$				Regrets $= r(E_j, a_i)$		
		$i \longrightarrow n$				$i \longrightarrow n$		
		a_1	a_2	a_3	min l_j		a_1	a_2	a_3
j	E_1	0	1	3	0	E_1	0	1	3
$\downarrow$									
m	E_2	5	3	2	2	E_2	3	1	0

min l_j, on the jth state of nature. Symbolically we write this definition as

$$r(E_j, a_i) = l(E_j, a_i) - \min l_j. \tag{9.5}$$

The full table of regrets for the farming problem is listed in table 9.5. Notice that the regrets at not having guessed a wet season correctly are the same as the losses, because min $l_1 = 0$. In what follows, we will examine the effects of using regrets instead of losses on the location of the Bayes' and minimax strategies.

Proceeding in the usual way, the decision-maker calculates the average regret, R, for the jth state of nature incurred by each pure strategy. Average regret is more commonly termed *risk* and is defined by the formula

$$R(E_j, s) = \sum_{i=1}^{n} p_i(E_j, s) \cdot r(E_j, a_i), \tag{9.6}$$

where, as usual, $\{p_i(E_j, s)\}$ is the set of action probabilities for the jth state of nature. The only difference between risk (formula 9.6) and average loss (see formula (9.1), p296) is that the regrets, $\{r(E_j, a_i)\}$ have replaced the losses, $\{l(E_j, a_i)\}$. Substituting the definition of regret (formula 9.5) in formula (9.6) we obtain

$$R(E_j, s) = \left[\sum_{i=1}^{n} p_i(E_j, s) \cdot l(E_j, a_i) \right] - \min l_j.$$

Since we have already calculated all the terms inside the square brackets as average losses, $L(E_j, s)$ (see formula 9.1), we may now write formula (9.6) more simply as

$$R(E_j, s) = L(E_j, s) - \min l_j. \tag{9.7}$$

Therefore, for any pure strategy or strategies, the risk for each state of nature is calculated by subtracting the minimum loss, min l_j, from the respective average loss (table 9.4). For example, the risks on s_1 are obtained from formula (9.7) as

$$R(E_1, s_1) = L(E_1, s_1) - \min l_1$$
$$= 0 \cdot 00 - 0 = 0$$
$$R(E_2, s_1) = L(E_2, s_1) - \min l_2$$
$$= 5 - 2 = 3.$$

The pairs of risks for each pure strategy are plotted on a risk graph in the same way as pairs of average losses. The feasible region obtained from this procedure for the farming problem is shown in figure 9.7. A comparison of the feasible risk region with the

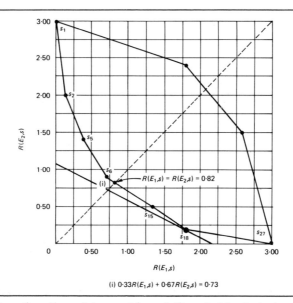

Figure 9.7.   Locating the minimax risk and Bayes' risk strategies for the farming problem. (Adapted from H Chernoff and L E Moses 1959, p152.)

average loss region will reveal that the regions are identical except that each risk point is located two units below its corresponding loss point on the dry season, E_2, axis. This result occurs because the only numerical difference between the problems is that each regret $r(E_2, a_i)$ is two units (min $l_2 = 2$) less than the corresponding loss $l(E_2, a_i)$.

The location of the Bayes' and minimax risk strategies are found using the graphical procedures described for the loss problem. For instance, the Bayes' risk strategy corresponding to any pair of priors $\{p(E_j)\}$ is located by finding the equal total risk, R, line, defined by the equation

$$p(E_1)R(E_1, s) + p(E_2)R(E_2, s) = R,$$

which is tangential to the lower boundary of the feasible risk region. Recalling that the farmer estimated the priors to be $p(E_1) = 0.33$ and $p(E_2) = 0.67$, then the Bayes risk strategy turns out to be s_{18}, which is tangential to the line $0.33\,R(E_1, s) + 0.67\,R(E_2, s) = 0.73$ (figure 9.7). The solution to minimizing total average loss with these priors was also s_{18}. Therefore, the location of Bayes' strategy is unaltered by replacing losses with regrets. However, this is not the case for the minimax criterion. Figure 9.7 illustrates that the line, with the property $R(E_1, s) = R(E_2, s)$, intersects the lower boundary between s_6 and s_{15}. The minimax risk strategy is a randomized strategy which mixes s_6 and s_{15} with probabilities 0.8 and 0.2, respectively. The risks for this strategy are $R(E_1, s) = 0.82$, $R(E_2, s) = 0.82$, while the expected losses are $L(E_1, s) = 0.82$, $L(E_2, s) = 2.82$. The minimax risk strategy is composed of the actions $s_6(a_1, a_2, a_3)$ and $s_{15}(a_2, a_2, a_3)$ and, therefore, if a wet growing season, x_1, is forecast the farmer will plant 80 per cent of his land in rice and 20 per cent in barley. Remember that the minimax average loss strategy only admitted the actions a_2 and a_3 and so avoided the catastrophic state a_1. The farmer's regret at planting wheat in a wet season,

$r(E_1, a_3) = 3$, is identical to that of planting rice in a dry season, $r(E_2, a_1) = 3$. For this reason, the minimax risk strategy often requires the farmer to plant the 'safe' crop barley as a hedge against uncertainty.

A comparison of the optimum strategies

Our discussion of optimal strategies has provided our farmer with a number of plans of action depending on his particular circumstances. Yet we have only considered the best known criterion of optimality. Decision theory abounds with different statistical and psychological criteria for identifying optimum plans of action; however, there is little agreement about the way a decision-maker should choose between these optimal strategies. To conclude this section we shall consider the general preferences of statisticians for the solution to the farming problem listed in table 9.6.

All statisticians are agreed that optimum strategies must be selected from the class of admissible strategies. Furthermore, it is generally agreed that, if the decision-maker is confident about his estimate of the priors, then the best plan of action is to select the corresponding Bayes' strategy and minimize total average loss. Statisticians are more sceptical about the minimax criterion because it is determined by psychological reasoning. The pessimistic decision-maker selects minimax to avoid the action which incurs a catastrophic loss. However, the penalty for this course of action is that many opportunities to make small losses are missed. Therefore, because the pessimist ignores the information contained in the priors, in the long term his selection of the minimax strategy will result in a greater total loss than the corresponding Bayes' strategy. Statisticians also agree that an analysis based on regrets is more rational than an analysis of losses. Accordingly, the minimax risk criterion, which takes into account the catastrophic action on each state of nature, is preferred to the minimax loss strategy.

To extend the argument we will consider the optimal strategy for the decision-maker who has no information with which to estimate the priors. It was suggested previously that, when the priors are unknown, the decision-maker is entitled to guess the priors and then select the appropriate Bayes' strategy. The problem can be solved

Table 9.6. Optimum strategies in the farming problem.

(a) *Prior probability of states of nature are known*

Bayes' optimum, $\{p(E_1) = 0.33, p(E_2) = 0.67\}$: $s_{18}(a_2, a_3, a_3)$

(b) **Priors ignored**

Minimax average loss: randomization of $s_{18}(a_2, a_3, a_3)$
with $p = 0.71$ and $s_{27}(a_3, a_3, a_3)$
with $p = 0.29$.

Minimax risk: randomization of $s_6(a_1, a_2, a_3)$
with $p = 0.80$ and $s_{15}(a_2, a_2, a_3)$
with $p = 0.20$.

(c) *Priors unknown*

Bayes' optimum, $\{p(E_1) = 0.5, p(E_2) = 0.5\}$: $s_6(a_1, a_2, a_3)$.

---

using the entropy-maximizing criterion expounded in Chapter 5, pp153–156. Recall that, in the absence of information about constraints, the most likely state of a system is the one where all events are equally likely to occur. Therefore, we assume the most likely probability distribution is the assignment of *equal* probabilities. According to the maximum entropy criterion the ignorant farmer will guess the priors to be $p(E_1) = p(E_2) = 0.5$. In such circumstances the appropriate Bayes' strategy is s_6.

Inspection of the list of optimum strategies in table 9.6 will show that they entail quite different plans of action. However, this variety of action is mitigated by the fact that the optimum strategies are all located within a relatively small proportion of the lower boundary of the feasible region. Indeed, s_6 and s_{18} each occur as the optimum strategy for two different criteria. This tendency for the different optimum strategies to incur similar risks is common to many decision problems and, fortunately, diminishes the importance of the final choice amongst them. More generally, decision theorists use this tendency to assert that their subject is more concerned with isolating a small finite range of sensible strategies than giving definitive solutions to decision problems.

CONCLUDING REMARKS

In this chapter we have taken a *normative* view of decision-making. That is, we have been concerned with the way spatial decision-makers ought to behave, rather than how they actually behave. It has been assumed decision-makers are capable of structuring problems in terms of probabilities and then acting rationally in accordance with their assignments. It is perhaps unfortunate that geographers have not realized the potential of this normative framework and, for this reason, geographical applications have been slow to develop. Typically, geographical studies of decision-making have simply described the unfolding history of sequences of actual decisions. For instance, the location decisions of business organizations is an enormous field of geographical research (see Lloyd and Dicken 1977). However, this non-mathematical approach to decision-making is outside the scope of this book.

Nevertheless, there are encouraging signs that things are changing for the better. Recently, a number of Dutch regional economists have made significant progress by applying the methods of 'multi-criteria decision analysis' to the evaluation of alternative regional planning policies. In their monograph, Delft and Nijkamp (1977), applied this framework to evaluating a set of proposed industrial development plans for the reclaimed Maasvalkte peninsula in the Rhine delta region. The five plans they evaluate are listed in table 9.7. For example, Plan 1 involves building a tank storage plant for chemicals, a container terminal, and a ship repair yard with a tanker cleaning yard. Application of a variety of decision-optimizing criteria showed that Plan 1 was preferred if utilities were expressed in terms of possible environmental damage, while Plan 4 was preferred if utilities were expressed in terms of employment and economic growth. Again, we see that decision methods do not provide unique solutions; rather, they focus attention on a limited range of preferable alternatives. Finally, the more mathematical readers are referred to the text by White (1976) for an impression of the present scope of decision models and their application.

Table 9.7. Planned activities for the Maasvalkte peninsula:
based upon Delft and Nijkamp (1977) p86.

Activity	Plan Numbers				
	1	2	3	4	5
a. Integrated steel works				*	
b. Steel works			*		*
c. Tank storage	*	*	*	*	*
d. Trans-shipment works		*		*	*
e. Container terminal	*	*	*		*
f. Ship and tanker cleaning	*	*	*	*	*

*Activity included in proposed plan.

Notes

1   The equation $p(E_1) . L(E_1, s) + p(E_2) . L(E_2, s) = L$ is an example of a derivation of the equation for a straight line which is written as

$$ax + by = c.$$

In this equation c is some arbitrary constant and x and y are the coordinate values of any point located on the line. The coefficients a and b are proportional to the respective contributions of any pair values of x and y to the arbitrary constant c. So long as a and b are positive, the line will possess a negative gradient.

2   For some decision problems involving two states of nature the feasible region will not be intersected by the line where $L(E_1, s) = L(E_2, s)$. For these eccentric feasible regions the minimax strategy will either be the Bayes' strategy with $p(E_1) = 1$ or the Bayes' strategy with $p(E_2) = 1$. These examples are illustrated in the figure.

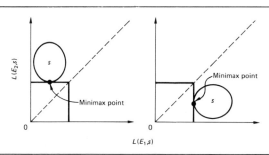

Locating the minimax strategy for eccentric feasible region, S.

APPENDIX A

Exponential Functions

x	e^x	e^{-x}	x	e^x	e^{-x}	x	e^x	e^{-x}	x	e^x	e^{-x}
0·02	1·0202	0·9802	0·30	1·3499	0·7408	1·0	2·7183	0·3679	3·5	33·115	0·0302
0·04	1·0408	0·9608	0·31	1·3634	0·7335	1·1	3·0042	0·3329	3·6	36·598	0·0273
0·06	1·0618	0·9418	0·32	1·3771	0·7261	1·2	3·3201	0·3012	3·7	40·447	0·0247
0·08	1·0833	0·9231	0·33	1·3910	0·7189	1·3	3·6693	0·2725	3·8	44·701	0·0224
			0·34	1·4050	0·7118	1·4	4·0552	0·2466	3·9	49·402	0·0202
0·10	1·1052	0·9048	0·35	1·4191	0·7047	1·5	4·4817	0·2231	4·0	54·598	0·0183
0·11	1·1163	0·8958	0·36	1·4333	0·6977	1·6	4·9530	0·2019	4·1	60·340	0·0166
0·12	1·1275	0·8869	0·37	1·4477	0·6907	1·7	5·4739	0·1827	4·2	66·686	0·0150
0·13	1·1388	0·8781	0·38	1·4623	0·6839	1·8	6·0497	0·1653	4·3	73·700	0·0136
0·14	1·1503	0·8694	0·39	1·4770	0·6771	1·9	6·6859	0·1496	4·4	81·451	0·0123
0·15	1·1618	0·8607	0·40	1·4918	0·6703	2·0	7·3891	0·1353	4·5	90·017	0·0111
0·16	1·1735	0·8521	0·41	1·5068	0·6636	2·1	8·1662	0·1225	4·6	99·484	0·0100
0·17	1·1853	0·8437	0·42	1·5220	0·6570	2·2	9·0250	0·1108	4·7	109·95	0·00910
0·18	1·1972	0·8353	0·43	1·5373	0·6505	2·3	9·9742	0·1003	4·8	121·51	0·00823
0·19	1·2092	0·8270	0·44	1·5527	0·6440	2·4	11·023	0·0907	4·9	134·29	0·00745
0·20	1·2214	0·8187	0·45	1·5683	0·6376	2·5	12·182	0·0821	5·0	148·41	0·00674
0·21	1·2337	0·8106	0·46	1·5841	0·6313	2·6	13·464	0·0743	5·1	164·02	0·00610
0·22	1·2461	0·8025	0·47	1·6000	0·6250	2·7	14·880	0·0672	5·2	181·27	0·00552
0·23	1·2586	0·7945	0·48	1·6161	0·6188	2·8	16·445	0·0608	5·3	200·34	0·00499
0·24	1·2712	0·7866	0·49	1·6323	0·6126	2·9	18·174	0·0550	5·4	211·41	0·00452
0·25	1·2840	0·7788	0·50	1·6487	0·6065	3·0	20·085	0·0498	5·5	244·69	0·00409
0·26	1·2969	0·7711	0·6	1·8221	0·5488	3·1	22·198	0·0450	5·6	270·43	0·00370
0·27	1·3100	0·7634	0·7	2·0138	0·4966	3·2	24·532	0·0408	5·7	298·87	0·00335
0·28	1·3231	0·7558	0·8	2·2255	0·4493	3·3	27·113	0·0369	5·8	330·30	0·00303
0·29	1·3364	0·7483	0·9	2·4596	0·4066	3·4	29·964	0·0334	5·9	365·04	0·00274
									6·0	403·43	0·00248

APPENDIX B

Random Numbers

57780	97609	52482	12783	88768	12323	64967	22970	11204	37576
68327	00067	17487	49149	25894	23639	86557	04139	10756	76285
55888	82253	67464	91628	88764	43598	45481	00331	15900	97699
84910	44827	31173	44247	56573	91759	79931	26644	27048	53704
35654	53638	00563	57230	07395	10813	99194	81592	96834	21374
46381	60071	20835	43110	31842	02855	73446	24456	24268	85291
11212	06034	77313	66896	47902	63483	09924	83635	30013	61791
49703	07226	73337	49223	73312	09534	64005	79267	76590	26066
05482	30340	24606	99042	16536	14267	84084	16198	94852	44305
92947	65090	47455	90675	89921	13036	92867	04786	76776	18675
51806	61445	46719	60281	03644	70024	07629	55805	44482	40004
16383	30577	74687	71227	72423	81307	75192	80443	22901	74351
30893	85406	42731	50249	74479	68273	78133	34506	61539	25717
59790	11682	71740	29429	99033	76460	36814	36917	05886	11205
06271	74980	96746	05938	43525	16516	26393	89082	96725	67903
93325	61834	27564	81744	17507	90432	50973	35591	72644	74441
46690	08927	21895	29683	83156	58597	88267	32479	22051	31743
82041	88942	01492	40778	43812	58483	43779	42718	40202	15824
14306	04003	55846	19271	62700	99408	72236	52722	25245	81239
63471	77583	14615	75196	37031	05819	90836	19530	26785	66830
68467	17634	77848	15755	92996	75644	82043	84157	97406	67988
94308	57895	87167	03106	65080	51928	74237	00449	53782	32412
52218	32502	73018	56511	79935	34620	37386	00243	92150	14737
46586	08309	29247	67792	06670	18796	74713	81632	21182	10765
07869	80471	17412	09161	33989	44250	79597	15182	43536	39705

APPENDIX C

Areas under the Normal Curve

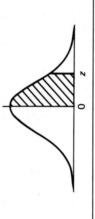

z	0	1	2	3	4	5	6	7	8	9
0·0	0·0000	0·0040	0·0080	0·0120	0·0160	0·0199	0·0239	0·0279	0·0319	0·0359
0·1	0·0398	0·0438	0·0478	0·0517	0·0557	0·0596	0·0636	0·0675	0·0714	0·0754
0·2	0·0793	0·0832	0·0871	0·0910	0·0948	0·0987	0·1026	0·1064	0·1103	0·1141
0·3	0·1179	0·1217	0·1255	0·1293	0·1331	0·1368	0·1406	0·1443	0·1480	0·1517
0·4	0·1554	0·1591	0·1628	0·1664	0·1700	0·1736	0·1772	0·1808	0·1844	0·1879
0·5	0·1915	0·1950	0·1985	0·2019	0·2054	0·2088	0·2123	0·2157	0·2190	0·2224
0·6	0·2258	0·2291	0·2324	0·2357	0·2389	0·2422	0·2454	0·2486	0·2518	0·2549
0·7	0·2580	0·2612	0·2642	0·2673	0·2704	0·2734	0·2764	0·2794	0·2823	0·2852
0·8	0·2881	0·2910	0·2939	0·2967	0·2996	0·3023	0·3051	0·3078	0·3106	0·3133
0·9	0·3159	0·3186	0·3212	0·3238	0·3264	0·3289	0·3315	0·3340	0·3365	0·3389
1·0	0·3413	0·3438	0·3461	0·3485	0·3508	0·3531	0·3554	0·3577	0·3599	0·3621
1·1	0·3643	0·3665	0·3686	0·3708	0·3729	0·3749	0·3770	0·3790	0·3810	0·3830
1·2	0·3849	0·3869	0·3888	0·3907	0·3925	0·3944	0·3962	0·3980	0·3997	0·4015

z	0·00	0·01	0·02	0·03	0·04	0·05	0·06	0·07	0·08	0·09
1·3	0·4032	0·4049	0·4066	0·4082	0·4099	0·4115	0·4131	0·4147	0·4162	0·4177
1·4	0·4192	0·4207	0·4222	0·4236	0·4251	0·4265	0·4279	0·4292	0·4306	0·4319
1·5	0·4332	0·4345	0·4357	0·4370	0·4382	0·4394	0·4406	0·4418	0·4429	0·4441
1·6	0·4452	0·4463	0·4474	0·4484	0·4495	0·4505	0·4515	0·4525	0·4535	0·4545
1·7	0·4554	0·4564	0·4573	0·4582	0·4591	0·4599	0·4608	0·4616	0·4625	0·4633
1·8	0·4641	0·4649	0·4656	0·4664	0·4671	0·4678	0·4686	0·4693	0·4699	0·4706
1·9	0·4713	0·4719	0·4726	0·4732	0·4738	0·4744	0·4750	0·4756	0·4761	0·4767
2·0	0·4772	0·4778	0·4783	0·4788	0·4793	0·4798	0·4803	0·4808	0·4812	0·4817
2·1	0·4821	0·4826	0·4830	0·4834	0·4838	0·4842	0·4846	0·4850	0·4854	0·4857
2·2	0·4861	0·4864	0·4868	0·4871	0·4875	0·4878	0·4881	0·4884	0·4887	0·4890
2·3	0·4893	0·4896	0·4898	0·4901	0·4904	0·4906	0·4909	0·4911	0·4913	0·4916
2·4	0·4918	0·4920	0·4922	0·4925	0·4927	0·4929	0·4931	0·4932	0·4934	0·4936
2·5	0·4938	0·4940	0·4941	0·4943	0·4945	0·4946	0·4948	0·4949	0·4951	0·4952
2·6	0·4953	0·4955	0·4956	0·4957	0·4959	0·4960	0·4961	0·4962	0·4963	0·4964
2·7	0·4965	0·4966	0·4967	0·4968	0·4969	0·4970	0·4971	0·4972	0·4973	0·4974
2·8	0·4974	0·4975	0·4976	0·4977	0·4977	0·4978	0·4979	0·4979	0·4980	0·4981
2·9	0·4981	0·4982	0·4982	0·4983	0·4984	0·4984	0·4985	0·4985	0·4986	0·4986
3·0	0·4987	0·4987	0·4987	0·4988	0·4988	0·4989	0·4989	0·4989	0·4990	0·4990
3·1	0·4990	0·4991	0·4991	0·4991	0·4992	0·4992	0·4992	0·4992	0·4993	0·4993
3·2	0·4993	0·4993	0·4994	0·4994	0·4994	0·4994	0·4994	0·4995	0·4995	0·4995
3·3	0·4995	0·4995	0·4995	0·4996	0·4996	0·4996	0·4996	0·4996	0·4997	0·4997
3·4	0·4997	0·4997	0·4997	0·4997	0·4997	0·4997	0·4997	0·4997	0·4997	0·4998
3·5	0·4998	0·4998	0·4998	0·4998	0·4998	0·4998	0·4998	0·4998	0·4998	0·4998
3·6	0·4998	0·4998	0·4999	0·4999	0·4999	0·4999	0·4999	0·4999	0·4999	0·4999
3·7	0·4999	0·4999	0·4999	0·4999	0·4999	0·4999	0·4999	0·4999	0·4999	0·4999
3·8	0·4999	0·4999	0·4999	0·4999	0·4999	0·4999	0·4999	0·4999	0·4999	0·4999
3·9	0·5000	0·5000	0·5000	0·5000	0·5000	0·5000	0·5000	0·5000	0·5000	0·5000

APPENDIX D

Percentage Points of the χ^2 Distribution

df \ α	0·50	0·25	0·10	0·05	0·025	0·01	0·005	0·001
1	0·4549	1·323	2·706	3·841	5·024	6·635	7·879	10·828
2	1·386	2·773	4·605	5·991	7·378	9·210	10·597	13·816
3	2·366	4·108	6·251	7·815	9·348	11·345	12·838	16·266
4	3·357	5·385	7·779	9·488	11·143	13·277	14·860	18·467
5	4·351	6·626	9·236	11·070	12·833	15·086	16·750	20·515
6	5·348	7·841	10·645	12·592	14·449	16·812	18·548	22·458
7	6·346	9·037	12·017	14·067	16·013	18·475	20·278	24·322
8	7·344	10·219	13·362	15·507	17·535	20·090	21·955	26·125
9	8·343	11·389	14·684	16·919	19·023	21·666	23·589	27·877
10	9·342	12·549	15·987	18·307	20·483	23·209	25·188	29·588
11	10·341	13·701	17·275	19·675	21·920	24·725	26·757	31·264

12	11·340	14·845	18·549	21·026	23·337	26·217	28·300	32·909
13	12·340	15·984	19·812	22·362	24·736	27·688	29·819	34·528
14	13·339	17·117	21·064	23·685	26·119	29·141	31·319	36·123
15	14·339	18·245	22·307	24·996	27·488	30·578	32·801	37·697
16	15·338	19·369	23·542	26·296	28·845	32·000	34·267	39·252
17	16·338	20·489	24·769	27·587	30·191	33·409	35·718	40·790
18	17·338	21·605	25·989	28·869	31·526	34·526	37·156	42·312
19	18·338	22·718	27·204	30·143	32·852	36·191	38·582	43·820
20	19·337	23·828	28·412	31·410	34·170	37·566	39·997	45·315
21	20·337	24·935	29·615	32·670	35·479	38·932	41·401	46·797
22	21·337	26·039	30·813	33·924	36·781	40·289	42·796	48·268
23	22·337	27·141	32·007	35·172	38·076	41·638	44·181	49·728
24	23·337	28·241	33·196	36·415	39·364	42·980	45·558	51·179
25	24·337	29·339	34·382	37·652	40·646	44·314	46·928	52·620
26	25·336	30·434	35·563	38·885	41·923	45·642	48·290	54·052
27	26·336	31·528	36·741	40·113	43·194	46·963	49·645	55·476
28	27·336	32·620	37·916	41·337	44·461	48·278	50·993	56·892
29	28·336	33·711	39·087	42·557	45·722	49·588	52·336	58·302
30	29·336	34·800	40·256	43·773	46·979	50·892	53·672	59·703
40	39·335	45·616	51·805	55·758	59·342	63·691	66·766	73·402
50	49·335	56·334	63·167	67·505	71·420	76·154	79·490	86·661
60	59·335	66·981	74·397	79·082	83·298	88·379	91·952	99·607
70	69·334	77·577	85·527	90·531	95·023	100·425	104·215	112·317
80	79·334	88·130	96·578	101·879	106·629	112·329	116·321	124·839
90	89·334	98·650	107·565	113·145	118·136	124·116	128·299	137·208
100	99·334	109·141	118·498	124·342	129·561	135·807	140·169	149·449

APPENDIX E

Percentage Points of the t-Distribution

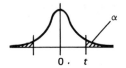

2α df	0·30	0·20	0·10	0·05	0·02	0·01	0·002	0·001
1	1·963	3·078	6·314	12·706	31·821	63·657	318·31	636·62
2	1·386	1·886	2·920	4·303	6·965	9·925	22·326	31·598
3	1·250	1·638	2·353	3·182	4·541	5·841	10·213	12·924
4	1·190	1·533	2·132	2·776	3·747	4·604	7·173	8·610
5	1·156	1·476	2·015	2·571	3·365	4·032	5·893	6·869
6	1·134	1·440	1·943	2·447	3·143	3·707	5·208	5·950
7	1·119	1·415	1·895	2·365	2·998	3·499	4·785	5·408
8	1·108	1·397	1·860	2·306	2·896	3·355	4·501	5·041
9	1·100	1·383	1·833	2·262	2·321	3·250	4·297	4·781
10	1·093	1·372	1·812	2·228	2·764	3·169	4·144	4·587
11	1·088	1·363	1·796	2·201	2·718	3·106	4·025	4·437
12	1·083	1·356	1·782	2·179	2·681	3·055	3·030	4·318
13	1·079	1·350	1·771	2·160	2·650	3·012	3·852	4·221
14	1·076	1·345	1·761	2·145	2·624	2·977	3·787	4·140
15	1·074	1·341	1·753	2·131	2·602	2·947	3·733	4·073
16	1·071	1·337	1·746	2·120	2·583	2·921	3·686	4·015
17	1·069	1·333	1·740	2·110	2·567	2·898	3·646	3·965
18	1·067	1·330	1·784	2·101	2·552	2·878	3·610	3·922
19	1·066	1·328	1·729	2·093	2·539	2·861	3·579	3·883
20	1·064	1·325	1·725	2·086	2·528	2·861	3·552	3·850
21	1·063	1·323	1·721	2·080	2·518	2·831	3·527	3·819
22	1·061	1·321	1·717	2·074	2·508	2·819	2·505	3·792
23	1·060	1·319	1·714	2·069	2·500	2·807	3·485	3·767
24	1·059	1·318	1·711	2·064	2·492	2·797	3·467	3·745
25	1·058	1·316	1·708	2·060	2·485	2·787	3·450	3·725
26	1·058	1·315	1·706	2·056	2·479	2·779	3·435	3·707
27	1·057	1·314	1·703	2·052	2·473	2·771	3·421	3·690
28	1·056	1·313	1·701	2·048	2·467	2·763	3·408	3·674
29	1·055	1·311	1·699	2·045	2·402	2·756	3·396	3·659
30	1·055	1·310	1·697	2·042	2·457	2·750	3·385	3·646
40	1·050	1·303	1·684	2·021	2·423	2·704	3·307	3·551
60	1·046	1·296	1·671	2·000	2·390	2·660	3·232	3·460
120	1·041	1·289	1·658	1·980	2·358	2·617	3·160	3·373
∞	1·0364	1·2815	1·6449	1·9600	2·3263	2·5758	3·0902	3·2905

Bibliography

The chapter in this book to which a reference applies is shown in square brackets

Abler R, Adams J S and Gould P R 1971 *Spatial Organization* (Englewood Cliffs, NJ: Prentice-Hall) [Ch. 5]

Alker H R 1969 'A typology of ecological fallacies' in *Quantitative Ecological Analysis in the Social Sciences* ed M Dogan and S Rokkan (Cambridge, Mass.: MIT Press) [Ch. 8]

Alonso W 1964 *Location and Land-Use* (Cambridge, Mass.: MIT Press) [Ch. 5]

Barr B and Smillie K 1972 'Some spatial interpretations of alternative optimal and sub-optimal solutions to the transportation problem' *Can. Geogr.* **16** 356–64 [Ch. 6]

Barry R G 1975 'Climate models in palaeoclimatic reconstruction' *Palaeogeography, Palaeoclimatology, Palaeoecology* **17** 123–37 [Ch. 4]

Baxter R S 1976 *Computer and Statistical Techniques for Planners* (London: Methuen) [Ch. 5]

Bennett R J 1975 'The representation and identification of spatio-temporal systems: an example of population diffusion in N.W. England'. *Trans. Inst. Br. Geogrs* **66** 73–92 [Ch. 8]

Bennett R J and Chorley R J 1978 *Environmental Systems: Philosophy, Analysis, and Control* (London: Methuen) [Chs 3 and 4]

Berry B J L and Marble D F 1968 *Spatial Analysis* (Englewood Cliffs, NJ: Prentice-Hall) [Ch. 8]

Besag J and McNeil D 1976 'On the use of exploratory data analysis in human geography' *Adv. Appl. Probability* **8** 652 [Ch. 8]

Bledsoe L J and Van Dyne G M 1971 'A compartment model simulation of secondary succession' in *Systems Analysis and Simulation in Ecology* vol 1, ed B C Patten (New York: Academic Press) pp 479–511 [Ch. 3]

Boast C W 1973 'Modelling the movement of chemicals in soils by water' *Soil Sci.* **115** 224–230 [Ch. 4]

Broadbent T A 1970 'Notes on the design of operational models' *Environment and Planning* **2** 469–76 [Ch. 5]

Briggs L I and Pollack H N 1967 'Digital model of evaporite sedimentation' *Science* **155** 453–6 [Ch. 4]

Burley D M 1974 *Studies in Optimization* (Leighton Buzzard: International Textbook Co.) [Ch. 5]

Burns I G 1974 'A model for predicting the redistribution of salts applied to fallow soils after excess rainfall or evaporation' *J. Soil Sci.* **25** 165–78 [Ch. 4]

Cassetti E 1965 'Optimal locations of steel mills serving the Quebec and Southern Ontario Steel Market' *Can. Geogr.* **10** 27–38 [Ch. 6]

Cesario F J 1974 'The interpretation and calculation of gravity model zone-to-zone adjustment factors' *Environment and Planning* **A6** 247–57 [Ch. 5]

Chernoff H and Moses L E 1959 *Elementary Decision Theory* (New York: Wiley) [Chs 7 and 9]

Chicago Area Transportation Study 1959 *Final Report* (Chicago, Illinois: CATS) [Ch. 5]

Chisholm M, Frey A E and Haggett P 1971 *Regional Forecasting* (London: Butterworths) [Ch. 8]

Chisholm M and O'Sullivan P 1973 *Freight Flows and Spatial Aspects of the British Economy* (London: Cambridge University Press) [Ch. 5]

Cliff A D 1968 'The neighbourhood effect in the diffusion of innovations' *Trans Inst. Br. Geogrs* **44** 75–84 [Ch. 8]

—— 1977 'Quantitative methods: time series methods for modelling and forecasting' *Progr. Human Geog.* **1** 492–502 [Ch. 8]

Cliff A D and Ord J K 1969 'The problem of spatial autocorrelation' in *London Papers in Regional Science* vol 1 ed A J Scott (London: Pion) pp25–55 [Ch. 8]

—— 1971 'Evaluating the percentage points of a spatial autocorrelation coefficient' *Geog. Anal.* **3** 51–62 [Ch. 8]

—— 1973 *Spatial Autocorrelation* (London: Pion) [Ch. 8]

Cochran W G 1973 *Sampling Techniques* (New York: Wiley) [Ch. 7]

Cox K R 1965 'The application of linear programming to geographic problems' *Tijdschr. Econ. Soc. Geogr.* **56** 228–36 [Ch. 6]

—— 1969 'The voting decision in a spatial context' *Progress in Geography* vol 1 ed C Board *et al* (London: Arnold) pp81–117 [Ch. 8]

Curry L 1966 'Seasonal programming and Bayesian assessment of atmospheric resources' *University of Chicago, Dept of Geography Research Papers No.* 105 pp127–40 [Ch. 9]

—— 1972 'A spatial analysis of gravity flows' *Regional Studies* **6** 131–47 [Ch. 5]

Dacey M F 1964 'Modified Poisson probability law for point patterns more regular than random' *Ann. Assoc. Am. Geogrs* **54** 559–65 [Ch. 8]

Dantzig G B 1963 *Linear Programming and Extensions* (Princeton, NJ: Princeton University Press) [Ch. 6]

Delft A Van and Nijkamp P 1977 *Multi-criteria Analysis and Regional Decision-making* (Leiden: Martinus Nijhoff) [Ch. 9]

Di Toro D M, O'Connor D J, Thomann R V and Mancini J L 1975 'Phytoplankton–zooplankton–nutrient integral model for western Lake Erie' in *Systems Analysis and Simulation in Ecology* vol. 3 ed B C Patten (New York: Academic Press) pp 423–74 [Ch. 3]

Dorfman R, Samuelson P A and Solow R M 1958 *Linear Programming and Economic Analysis* (New York: McGraw-Hill) [Ch. 6]

Evans A W 1971 'The calibration of trip distribution models with exponential, or similar cost functions' *Transportation Res.* **5** 15–38 [Ch. 6]

Evans S P 1973 'A relation between the gravity model and the transportation problem in linear programming' *Transportation Res.* 7 39–61 [Ch. 6]

Feller W 1957 *An Introduction to Probability Theory and its Application* Part I (New York: Wiley) [Ch. 7]

Fisher R A 1956 *Statistical Methods and Scientific Inference* (Edinburgh: Oliver and Boyd) [Ch. 7]

Forrester J W 1969 *Urban Dynamics* (Cambridge, Mass.: MIT Press) [Ch. 3]

— 1971 *World Dynamics* (Cambridge, Mass.: Wright-Allen Press) [Ch. 3]

Found W C 1971 *A Theoretical Approach to Rural Land-Use Patterns* (London: Arnold) [Ch. 9]

Garrison W L 1959, 1960 'Spatial Structure of the Economy: I, II, III' *Ann. Assoc. Am. Geogrs* **49** 232–9, 471–82; **50** 357–73 [Ch. 6]

Garrison W L and Marble D F 1958 'Analysis of highway networks: a linear programming formulation' *Proc. Highway Research Board* **37** 1–17 [Ch. 6]

Geary R C 1954 'The contiguity ratio and statistical mapping' *Incorporated Statistician* **5** 115–41 [Ch. 8]

Gersmehl P J 1976 'An alternative biogeography' *Ann. Assoc. Am. Geogrs* **66** 223–41 [Ch. 3]

Good I J 1962 'Subjective probability as a measure of a non-measurable set' in *Logic, Methodology, and the Philosophy of Science* ed E Nagel *et al* (Stanford, Calif.: Standford University Press) pp319–29 [Ch. 7]

Goodall D W 1970 'Simulating the grazing situation' in *Biomathematics* vol. 1 (*Concepts and Models of Biomathematics*) ed F Heinmets (New York: Dekker) pp211–36 [Ch. 3]

Gould P R 1963 'Man against his environment: A game theoretic framework' *Ann. Assoc. Am. Geogrs* **53** 290–7 [Ch. 9]

— 1965 'Wheat on Kilimanjaro: The perception of choice within game and learning model frameworks' *General Systems* **10** 157–66 [Ch. 9]

— 1972 'Pedagogic Review' *Ann. Assoc. Am. Geogrs* **62** 689–700 [Ch. 5]

Gould P R and Leinbach T R 1966 'An approach to the geographical assignment of hospital services' *Tijdschr. Econ. Soc. Geogr.* **57** 203–6 [Ch. 6]

Gray J R 1967 *Probability* (Edinburgh: Oliver and Boyd) [Ch. 7]

Gregory S 1978 *Statistical Methods and the Geographer* 4th edn (London: Longman) [Ch. 7]

Greig-Smith P 1952 'The use of random and contiguous quadrats in the study of the structure of plant communities' *Ann. Bot.* (new series) **16** 293–312 [Ch. 8]

— 1964 *Quantitative Plant Ecology* 2nd edn (London: Butterworth) [Ch. 8]

Haggett P 1976 'Hybridizing alternative models of an epidemic diffusion process' *Econ. Geog.* **52** 136–46 [Ch. 8]

Haggett P, Chorley R J and Stoddart D R 1965 'Scale standards in geographical research: a new measure of area magnitude' *Nature* **205** 844–7 [Ch. 2]

Haggett P, Cliff A D and Frey A 1977 *Locational Analysis in Human Geography* 2nd edn (London: Arnold) [Ch. 5 to 8]

Harbaugh J W and Bonham-Carter G 1970 *Computer Simulation in Geology* (New York: Wiley Interscience) [Ch. 4]

Harvey D W 1966 'Geographical processes and the analysis of point patterns' *Trans. Inst. Br. Geogrs* **40** 81–95 [Ch. 8]

— 1969 *Explanation in Geography* (London: Arnold) [Ch. 7]

Hay A 1977 *Linear Programming: Elementary Applications of the Transportation Problem* CATMOG 11 (Norwich: Geo Abstracts) [Ch. 6]

Henderson J M 1958 *The Efficiency of the Coal Industry: An Application of Linear Programming* (Cambridge, Mass.: Harvard University Press) [Ch. 6]

Hepple L W 1974 'The impact of stochastic process theory upon spatial analysis in human geography' in *Progress in Geography* vol. 6 ed C Board *et al* pp89–149 [Ch. 8]

Herbert D J and Stevens B H 1960 'A model for the distribution of residential activity in urban areas' *J. Regional Sci.* 2 21–36 [Ch. 6]

Hess S W, Weaver J B, Siegfelt H J, Whelan J N and Zitlau P A 1965 'Non-partisan re-districting by computer' *Operations Res.* 13 998–1006 [Ch. 6]

Hett J M and O'Neill R V 1974 'Systems analysis of the Aleut ecosystem' *Arctic Anthropology* 11 31–40 [Ch. 3]

Hirano M 1968 'A mathematical model of slope development – an approach to the analytical theory of erosional topography' *J. Geosci., Osaka City University* 2 13–52 [Ch. 4]

—— 1975 'Simulation of development process of interfluvial slopes with reference to graded form' *J. Geol.* 83 113–23 [Ch. 4]

Hitchcock F L 1941 'The distribution of a product from several sources to numerous localities' *J. Math. Phys.* 20 224–30 [Ch. 6]

Huff D L 1963 'A probabilistic analysis of shopping centre trade areas' *Land Econ.* 39 81–90 [Ch. 5]

Huggett R J 1975 'Soil landscape systems: a model of soil genesis' *Geoderma* 13 1–22 [Ch. 4]

—— 1980 *Systems Analysis in Geography* (London: Oxford University Press) [Chs 3 and 4]

Humphreys J 1973 'Curvilinear regression on glacial cross-valley profiles' *South Hampshire Geographer* 1973 21–8 [Ch. 2]

Hunt H W 1977 'A simulation model for decomposition in grasslands' *Ecology* 58 469–84 [Ch. 3]

Hyman G M 1969 'The calibration of trip distribution models' *Environment and Planning* 1 105–12 [Ch. 5]

Hyman G M and Gleave D 1978 'A reasonable theory of migration' *Trans. Inst. Br. Geogrs* 3 (new series) 179–201 [Ch. 5]

Jaynes E T 1957 'Information theory and statistical mechanics' *Phys. Rev.* 106 620–30 [Ch. 5]

Jones E and Eyles J 1977 *An Introduction to Social Geography* (London: Oxford University Press) [Ch. 8]

Jordan C F, Kline J R and Sasscer D S 1972 'A simple model of strontium and manganese dynamics in a tropical rain forest' *Health Phys.* 24 477–89 [Ch. 3]

King L J 1969 *Statistical Analysis in Geography* (Englewood Cliffs, NJ: Prentice-Hall) [Chs 7 and 8]

King R B 1975 'Geomorphic and soil correlation of land systems in Northern Luapula Provinces of Zambia' *Trans. Inst. Br. Geogrs* 64 67–76 [Ch. 8]

Kimball B A 1973 'Simulation of the energy balance of a greenhouse' *Agric. Meteorol.* 11 243–60 [Ch. 4]

Kirkby M J 1971 'Hillslope process–response models based on the continuity

equation' in *Slopes: Form and Process, Inst. Br. Geogrs Spec. Publ. No.* 3 ed
D Brunsden (London: Institute of British Geographers) pp15–30 [Ch. 4]

Kline J R 1973 'Mathematical simulation of soil–plant relationships and soil genesis'
Soil Sci. **115** 240–9 [Ch. 3]

Krumbein W C and Graybill F A 1965 *An Introduction to Statistical Models in
Geology* (New York: McGraw-Hill) [Ch. 2]

Lakshmanan T R and Hansen W G 1965 'A retail market potential model' *J. Am. Inst.
Planners* **31** 134–43 [Ch. 5]

Lee C 1973 *Models in Planning* (Oxford: Pergamon) [Ch. 5]

Leslie P H 1945 'The use of matrices in certain population mathematics' *Biometrika*
33 183–212 [Ch. 3]

—— 1948 'Some further notes on the use of matrices in population mathematics'
Biometrika **35** 213–45 [Ch. 3]

Lewis E G 1942 'On the generation and growth of a population' *Sankhya* **6** 93–6
[Ch. 3]

Liebetrau A M and Karr A F 1977 'The role of Maxwell–Boltzmann and Bose–
Einstein statistics in point pattern analysis' *Geog. Anal.* **9** 418–22 [Ch. 8]

Lindgren B W 1975 *Basic Ideas of Statistics* (New York: Macmillan) [Ch. 7]

Lindley D V 1965 *Introduction to Probability and Statstics* vols I and II (London:
Cambridge University Press) [Ch. 7]

—— 1971 *Making Decisions* (New York: Wiley) [Ch. 9]

Lloyd P E and Dicken P 1977 *Location in Space* (London: Harper and Row) [Ch. 9]

MacKinnon R D and Barber G M 1977 'Optimization models of transport network
improvement' *Progr. Human Geog.* **1** 387–412 [Ch. 6]

Macmillan W 1978 'Mathematical programming models and the introduction of time
into spatial economic theory' in *Time and Regional Dynamics* ed T Carlstein *et al*
(London: Arnold) pp51–65 [Ch. 6]

Manabe S, Smagorinsky J, Holloway J L and Stone H M 1970 'Simulated climatology
of a general circulation model with a hydrological cycle III: Effects of increased
horizontal resolution' *Mon. Weather Rev.* **98** 175–212 [Ch. 4]

Mansell R S, Selim H M and Fiskell J G A 1977 'Simulated transformations and
transport of phosphorus in soil' *Soil Sci.* **124** 102–9 [Ch. 5]

Martin R L and Oeppen J E 1975 'The identification of regional forecasting models
using space–time correlation functions' *Trans. Inst. Br. Geogrs* **66** 95–118 [Ch. 8]

Masser I 1972 *Analytical Models for Urban and Regional Planning* (London: David and
Charles) [Chs 3 and 5]

Masser I and Brown P J B 1975 'Hierarchical aggregation procedures for spatial
interaction data' *Environment and Planning* A7 509–23 [Ch. 6]

Mather P M 1976 *Computational Methods of Multivariate Analysis in Physical
Geography* (New York: Wiley) [Ch. 8]

Maxfield D W 1969 'An interpretation of the primal and dual solutions of linear
programming' *Prof. Geogr.* **21** 235–63 [Ch. 6]

—— 1972 'Spatial planning of school districts' *Ann. Assoc. Am. Geogrs* **62** 582–90
[Ch. 6]

McConnell H and Horn J M 1972 'Geographical processes and the analysis of Karst
depressions in limestone regions' in *Spatial Analysis in Geomorphology* ed R J
Chorley (London: Methuen) pp111–34 [Ch. 8]

Mead R 1974 'A test for spatial pattern at several scales using data from a grid of contiguous quadrats' *Biometrics* **30** 295–307 [Ch. 8]

Milner C 1972 'The use of computer simulation in conservation management' in *Mathematical Models in Ecology* ed J N R Jeffers (Oxford: Blackwell Scientific Publications) pp249–75 [Ch. 3]

Mintz Y 1968 'Very long term integration of the primitive equations of atmospheric motion: an experiment in climatic simulation' *Meteorological Monographs* **8** 20–36 [Ch. 4]

Moran P A P 1948 'The interpretation of statistical maps' *J. Roy. Statistical Soc., Ser. B* **10** 243–51 [Ch. 8]

—— 1950 'Notes on continuous stochastic phenomena' *Biometrika* **37** 17–23 [Ch. 8]

Morgan B S 1976 'The bases of family status segregation: a case study in Exeter' *Trans. Inst. Br. Geogrs* **1** (new series) 83–107 [Ch. 8]

Morrill R L and Garrison W L 1960 'Projections of inter-regional trade in wheat and flour' *Econ. Geog.* **36** 116–26 [Ch. 6]

Murray G D and Cliff A D 1977 'A stochastic model for measles epidemics in a multi-regional system' *Trans. Inst. Br. Geogrs* **2** (new series) 158–74 [Ch. 8]

Muth R F 1969 *Cities and Housing* (Chicago: University of Chicago Press) [Ch. 5]

Myrup L O 1969 'A numerical model of the urban heat island' *J. Appl. Meteorol.* **8** 908–18 [Ch. 4]

Neyman J 1947 'Raisonment inductif ou compontement inductif?' in *Proc. 25th Session Int. Statistical Inst.* vol 3 (Washington, DC) pp423–31 [Ch. 9]

Nijkamp P 1977 'Stochastic quantitative and qualitative multicriteria analysis of environmental design' *Pap. Reg. Sci. Assoc.* **39** 175–200 [Ch. 9]

Noble J V 1974 'Geographic and temporal development of plagues' *Nature* **250** 726–9 [Ch. 4]

Nunnally J C 1967 *Psychometric Theory* (New York: McGraw-Hill) [Ch. 2]

Olsson G 1965 *Distance and Human Interaction: A preview and and Bibliography* Regional Science Research Institute Bibliography Series No 2, Philadelphia, Pa. [Ch. 5]

O'Neill R V and Burke O W 1971 *A Simple Systems Model for DDT and DDE Movement in the Human Food-chain* Eastern Deciduous Biome, ORNL-IBP-71-9, Oak Ridge National Laboratory, Oak Ridge, Tenn. [Ch. 3]

Openshaw S 1976 'An empirical study of some spatial interaction models' *Environment and Planning* **A8** 23–41 [Ch. 5]

—— 1977 'A geographical solution to scale and aggregation problems in region-building, partitioning and spatial modelling' *Trans. Inst. Br. Geogrs* **2** (new series) 459–72 [Ch. 8]

Osayimwese I 1974 'An application of linear programming to the evacuation of groundnuts in Nigeria' *J. Transport Econ. and Policy* **8** 58–69 [Ch. 6]

Park R A *et al* 1974 'A generalized model for simulating lake ecosystems' *Simulation* August pp33–50 [Ch. 3]

Patten B C 1971 'A primer for ecological modeling and simulation with analog and digital computers' in *Systems Analysis and Simulation in Ecology* vol. 1 ed B C Patten (New York: Academic Press) pp3–121 [Ch. 3]

Pollard D 1978 'An investigation of the astronomical theory of the ice ages using a simple climate–ice sheet model' *Nature* **272** 233–5 [Ch. 4]

Popper K R 1959 *The Logic of Scientific Discovery* (London: Hutchinson) [Ch. 7]
Pressat R 1978 *Statistical Demography* (London: Methuen) [Ch. 3]

Quandt R E 1960 'Models of transportation and optimum network construction' *J. Regional Sci.* **2** 27–45 [Ch. 6]

Ravenstein E G 1885 'The laws of migration' *J. Roy. Statistical Soc.* **48** 167–235 [Ch. 5]
—— 1889 'The laws of migration' *J. Roy. Statistical Soc.* **52** 241–305 [Ch. 5]
Reichle D E, O'Neill R V, Kaye S B, Sollins P and Booth R S 1973 'Systems analysis as applied to modelling ecological processes' *Oikos* **24** 337–43 [Ch. 4]
Reilly W 1931 *The Law of Retail Gravitation* (New York: Pilsbury) [Ch. 6]
Reynolds D R 1974 'Spatial contagion in political influence processes' in *Locational Approaches to Power and Conflict* ed K R Cox *et al* (New York: Wiley) pp233–74 [Ch. 8]
Rhodes J and Kan A 1971 *Office Dispersal and Regional Policy* (London: Cambridge University Press) [Ch. 7]
Ridley T H 1969 'Reducing travel time in a transport network' in *Studies in Regional Science* ed A J Scott (London: Pion) pp73–87 [Ch. 6]
Robinson W S 1950 'Ecological correlation and the behaviour of individuals' *Am. Sociolog. Rev.* **15** 351–7 [Ch. 8]
Rogers A 1965 'A stochastic analysis of the spatial clustering of retail establishments' *J. Am. Statistical Assoc.* **60** 1094–103 [Ch. 8]
—— 1968 *Matrix Analysis of Interregional Population Growth and Distribution* (Berkeley: University of California Press) [Ch. 3]
—— 1972 *Matrix Methods in Urban and Regional Analysis* (San Francisco: Holden Day) [Ch. 3]
—— 1974 *The Statistical Analysis of Spatial Dispersion* (London: Pion) [Ch. 8]

Savage L J 1964 'The foundation of statistics reconsidered' in *Studies in Subjective Probability* ed H E Kyburg and H E Smokler (New York: Wiley) pp173–88 [Ch. 9]
Sayer R A 1976 'A critique of urban modelling' *Prog. Planning* **6** 187–254 [Ch. 5]
Schlesinger W H 1977 Carbon balance in terrestrial detritus' *Ann. Rev. Ecol. Systematics* **8** 51–81 [Ch. 3]
Scott A J 1969 'Combinational programming and the planning of urban and regional systems' *Environment and Planning* **1** 125–42 [Ch. 6]
—— 1971a 'An introduction to spatial allocation analysis' *Assoc. Am. Geogrs, Commission on College Geography, Resource Paper No. 9*, Washington DC [Ch. 6]
—— 1971b *Combinational Programming, Spatial Analysis and Planning* (London: Methuen) [Ch. 6]
Seinfeld J H and Kyan C P 1971 'Determination of optimal air pollution control strategies', *Socio-Economic Planning Sci.* **5** 173–90 [Ch. 4]
Sellers W D 1965 *Physical Climatology* (Chicago: University of Chicago Press) [Ch. 4]
—— 1969 'A global climatic model based on the energy balance of the earth–atmosphere system' *J. Appl. Meterol.* **8** 392–400 [Ch. 4]
—— 1973 'A new global climatic model' *J. Appl. Meteorol.* **12** 241–54 [Ch. 4]
Senior M L and Wilson A G 1974 'Some explorations and syntheses of linear programming and spatial interaction models of residential location' *Geog. Anal.* **6** 209–37 [Ch. 5]

Sibley D 1972 'Strategy and tactics in the selection of shop locations' *Area* **4** 151–6 [Ch. 7]

Siegel S 1956 *Non-parametric Statistics* (New York: McGraw-Hill) [Ch. 8]

Smith O L 1976 'Nitrogen, phosphorus, and potassium utilization in the plant–soil system: an analytical model' *Proc. Soil Sci. Soc. Am.* **40** 704–14 [Ch. 4]

Smith R H T, Taaffe E J and King L J 1968 *Readings in Economic Geography* (Chicago: Rand McNally) [Ch. 6]

Smith T R 1974 'A derivation of the hydraulic geometry of steady-state channels from conservation principles and sediment transport laws' *J. Geol.* **82** 98–104 [Ch. 4]

Smith T R and Bretherton F P 1972 'Stability and the conservation of mass in drainage basin evolution' *Water Resources Res.* **8** 1506–29 [Ch. 4]

Stevens B H 1961 'Linear programming and location rent' *J. Regional Sci.* **3** 15–25 [Ch. 6]

Sugden D and Hamilton P 1971 'Scale, systems and regional geography' *Area* **3** 139–44 [Ch. 2]

Sumner G N 1978 *Mathematics for Physical Geographers* (London: Arnold) [Ch. 2]

Swartzman G L and Singh J S 1974 'A dynamic programming approach to optimal grazing strategies using a succession model for a tropical grassland' *J. Appl. Ecol.* **1** 537–48 [Ch. 3]

Taylor P J 1973 'Some implications of the spatial organization of elections' *Trans. Inst. Br. Geogrs* **60** 121–36 [Ch. 6]

—— 1977 *Quantitative Methods in Geography* (Boston: Houghton Mifflin) [Ch. 8]

Terjung W H 1976 'Climatology for geographers' *Ann. Assoc. Am. Geogrs* **66** 199–222 [Ch. 5]

Theakstone W H and Harrison C 1970 *The Analysis of Geographical Data* (London: Heinemann) [Ch. 7]

Thomann R V 1973 'Effect of longitudinal dispersion on dynamic water quality response of streams and rivers' *Water Resources Res.* **9** 355–66 [Ch. 4]

Thomas R W 1975 'Some functional characteristics of British city central areas: an application of allometric principles' *Regional Studies* **9** 369–78 [Ch. 2]

—— 1977a *An Introduction to Quadrat Analysis* CATMOG 12 (Norwich: Geo Abstracts) [Ch. 8]

—— 1977b 'An interpretation of the journey-to-work on Merseyside using entropy-maximizing methods' *Environment and Planning* A9 817–34 [Ch. 5]

Thomas R W and Reeve D E 1976 'The role of Bose–Einstein statistics in point pattern analysis' *Geog. Anal.* **8** 113–36 [Ch. 8]

Thompson S P 1971 *Calculus Made Simple* (London: Macmillan) [Ch. 2]

Titus J *et al* 1975 'A production model for *Myriophyllum spicatum L*' *Ecology* **56** 1129–38 [Ch. 3]

Tukey J W 1977 *Exploratory Data Analysis* (New York: Addison–Wesley) [Ch. 8]

Vincent P, Sibley D, Ebdon D and Charlton B 1976 'Methodology by example: caution towards nearest neighbours' *Area* **8** 161–71 [Ch. 8]

Von Neumann J and Morgenstern O 1944 *Theory of Games and Economic Behaviour* (Princeton, NJ: Princeton University Press) [Ch. 9]

Wagner H M 1957 'On a class of capacitated transportation problems' *Management Sci.* **5** 304–18 [Ch. 6]

Warnes A M 1975 'Commuting towards city centres: a study of population and employment density gradients in Liverpool and Manchester' *Trans. Inst. Br. Geogrs* **64** 77–96 [Ch. 2]

Weaver W 1964 *Lady Luck* (London: Heinemann) [Ch. 7]

Webber M J 1976, 'Elementary entropy-maximizing probability distribution: analysis and interpretation' *Econ. Geog.* **52** 218–27 [Ch. 8]

—— 1977 'Pedagogy again: what is entropy?' *Ann. Assoc. Am. Geogrs* **67** 254–66 [Ch. 5]

Weigert R G 1975 'Simulation models of ecosystems' *Ann. Rev. Ecol. Systematics* **6** 311–88 [Ch. 3]

Williams J, Barry R G and Washington W W 1974 'Simulation of the atmospheric circulation using the NCAR global circulation model with Ice Age boundary conditions' *J. Appl. Meteorol.* **13** 305–17 [Ch. 4]

Wierenga P J and de Wit C T 1970 'Simulation of heat transfer in soils' *Proc. Soil Sci. Am.* **34** 845–8 [Ch. 4]

Wilson A G 1967 'A statistical theory of spatial distribution models' *Transportation Res.* **1** 253–69 [Ch. 5]

—— 1970 *Entropy in Urban and Regional Modelling* (London: Pion) [Ch. 5]

—— 1971 'A family of spatial interaction models and associated developments' *Environment and Planning* **3** 1–32 [Ch. 5]

—— 1974 *Urban and Regional Models in Geography and Planning* (New York: Wiley) [Chs 1 and 5]

Wilson A G and Kirkby M J 1975 *Mathematics for Geographers and Planners* (Oxford: Clarendon Press) [Ch. 7]

Wilson A G and Senior M L 1974 'Some relationships between entropy-maximizing models, linear programming models, and their duals.' *J. Regional Sci.* **14** 207–15 [Ch. 6]

White D J 1976 *Fundamentals of Decision Theory* (Amsterdam: North-Holland) [Ch. 9]

Whittle P 1976 *Probability* (New York: Wiley) [Ch. 7]

Yeates M H 1963 'Hinterland delimitation: A distance-minimizing approach' *Prof. Geogr* **15** 7–10 [Ch. 6]

—— 1974 *An Introduction to Quantitative Analysis in Human Geography* (New York: McGraw-Hill) [Ch. 7]

Zahl S 1974 'Applications of the S-method to the analysis of spatial pattern' *Biometrics* **30** 513–24 [Ch. 8]

Index

E3